Gulf Studies

Volume 24

This book series, indexed in Scopus, serves the growing scholarly interest in the dynamic, complex, and strategically important Arab Gulf region by offering an academic publication platform to scholars in the region and beyond. The series takes an interdisciplinary approach to documenting the changes occurring in Gulf societies. It aims to advance the understanding of Gulf societies and their interactions with the rest of the world from a non-Western perspective. The series explicitly gives scholars from the Arab Gulf region more voice and visibility in academic publications. It includes studies of the member states of the Gulf Cooperation Council (GCC), namely: Bahrain, Kuwait, Oman, Qatar, Saudi Arabia, and the United Arab Emirates, as well as Iran, Iraq, and Yemen. It accepts proposals for research monographs, edited volumes, and handbooks that explain how the Gulf societies' social, economic, and political landscapes have evolved over the past few decades. The series covers a wide range of topics within four broad themes, as follows:

(1) Social and Cultural Issues in the Gulf: Demography, migration, citizenship, gender, culture and identity, urbanization, education, new media, unemployment, youth, family, women empowerment, leadership, aging, human rights, sports, Islamic ethics, and Islam and society

(2) Politics and Security in the Gulf: Gulf defence and security, the GCC integration, Gulf politics and political systems, and the GCC's foreign policy

(3) Energy and Economics in the Gulf: The political economy of the Gulf, economic diversification, international investments, international trade, knowledge economy, energy security and geopolitics of energy, renewable energy development, environmental regulations, urban planning, and Islamic finance

(4) The GCC States and the World: The GCC and Iran, the GCC and Africa, the GCC and South Asia, the GCC and China, the GCC and East Asia, the GCC and Southeast Asia, and the GCC and Western powers.

Submission and Peer Review

All proposals will undergo peer review by the editorial board members. If accepted, the final manuscript will be peer reviewed internally by the editorial board as well as externally (single/double blind) by Springer ahead of acceptance and publication. Please submit your book proposal to the Series Editor, Md Mizanur Rahman: mizan@qu.edu.qa, and to Springer Publishing Editor, Alex Westcott Campbell: alexandra.campbell@springernature.com

Héla Miniaoui · Mondher Bellalah
Editors

Green Growth Opportunities and Sustainable Finance in the Gulf Cooperation Council Region

Strengthening the Response to Global Shocks

Springer

Editors
Héla Miniaoui
Lusail University
Doha, Qatar

Mondher Bellalah
CY Cergy Paris University and TheEMA
Paris, France

ISSN 2662-4494 ISSN 2662-4508 (electronic)
Gulf Studies
ISBN 978-981-96-8286-7 ISBN 978-981-96-8287-4 (eBook)
https://doi.org/10.1007/978-981-96-8287-4

This Springer imprint is published by the registered company Springer Nature Singapore Pte Ltd.
The registered company address is: 152 Beach Road, #21-01/04 Gateway East, Singapore 189721, Singapore

Introduction

Our world grapples with uncertainties, unprecedented environmental risks, fluctuating oil prices, and economic shifts, including energy transition. Moreover, the repercussions of the COVID-19 pandemic, coupled with climate risk as well as ongoing geopolitical tensions, have underscored the importance of the GCC region diversifying economically beyond oil and gas in order to enhance its resilience to global shocks.[1]

The nations of the Gulf Cooperation Council (GCC) stand at a crossroads. For decades, their economies have been inextricably linked to the hydrocarbon sector, with oil and gas exports fueling prosperity and rapid development. However, the twenty-first century presents a landscape fundamentally different from the one that propelled the GCC's rise.

The GCC countries are well-positioned to embrace green growth and sustainable finance. In fact, the region possesses unique advantages, mainly the wealth that allows for significant investments in green technologies and infrastructure. Moreover, its geographical position makes it ideal for large-scale solar and wind energy projects, with the potential to export clean energy to neighboring regions. Additionally, the existing infrastructure provides a strong foundation for integrating renewable energy sources and sustainable practices. The region has attracted significant foreign investment in technology and innovation, fostering a knowledge base that can be harnessed for green solutions.

Despite the advantages, transitioning to a green economy requires overcoming fundamental challenges, especially the regulatory framework, through establishing clear and consistent policies that incentivize green investments, enhancing partnerships with international technology leaders, and collaborating among GCC countries, which can enhance the efficiency and effectiveness of green initiatives. Nonetheless, these challenges also present a spectrum of opportunities that could be addressed to promote an enabling environment for green investments in the GCC region.

[1] Miniaoui, H., & Khayati, A. (2022). Socioeconomic effects of COVID-19 pandemic on the GCC countries: Threats and opportunities. In Sadiki, L. and Saleh, L. Eds. "Covid-19 and risk society across the MENA region: Assessing Governance, democracy and inequality (pp. 72–94). I.B. Tauris.

Regarding the journey toward green transformation, despite the many risks and uncertainties surrounding the future of hydrogen, Gulf countries see it as an opportunity to maintain their status as global energy leaders.[2] The GCC countries have already embarked on ambitious green growth initiatives. In fact, their national visions (2031 for the UAE, 2030 for KSA, Qatar, Kuwait, and Bahrain, and 2040 for Oman) prioritize renewable energy and sustainable development and acknowledge the need to reduce dependence on oil and gas while developing diversified and knowledge-based economies. Green growth presents a compelling alternative and outlines a development trajectory that drives economic growth alongside environmental protection and social well-being. However, the success of the energy transition in these countries will depend on the GCC governments' capacity to deliver economic diversification, investment in sustainable infrastructure, and the development of renewable energy, to enhance resilience and adaptation to climate-related risks.[3]

Given their geographical location and climate sensitivity, the GCC countries are especially vulnerable to the effects of global shocks. These events underscore the urgency of diversifying their economies, adopting sustainable development practices, and fostering regional collaboration in order to enhance resilience and mitigate potential risks.

The 28th UN Climate Change Conference (COP 28), hosted by the UAE in December 2023, issued a declaration on Climate, Relief, Recovery, and Peace[4] and concluded that: "*Having shown that progress was too slow across all areas of climate action—from reducing greenhouse gas emissions, to strengthening resilience to a changing climate, to getting the financial and technological support to vulnerable nations—countries responded with a decision on how to accelerate action across all areas by 2030. This includes a call on governments to speed up the transition away from fossil fuels to renewables such as wind and solar power in their next round of climate commitments.*[5]"

Countries continued their call to combat climate change during COP29, held from 11 to 22 November 2024 in Baku, Azerbaijan.

The GCC countries recognize and harness the potential of finance as a tool for combating climate change. The GCC region possesses a significant financial advantage. The region's sovereign wealth funds (SWFs) possess vast financial resources that can be strategically deployed to support sustainable development initiatives. In fact, SWFs can play a crucial role in the energy transition, as they already hold more than twice as many green assets[6] (investments in renewable energy and electric vehicles) as black assets (investments in carbon fossil fuels and mining). Nearly

---

[2] Al-Sarihi, A., Rasoulinezhad, E., and Sung, J. (2024) "The Gulf Touts Hydrogen, But is it Hype or Opportunity?", MEI Perspectives Series 35, June 20.

[3] Miniaoui, H. (2023) Challenges of GCC investment in the energy transition, The Economic Research Forum - ERF Policy Portal, December. Challenges of GCC investment in the energy transition - Economic Research Forum (ERF).

[4] https://www.cop28.com/en/cop28-declaration-on-climate-relief-recovery-and-peace.

[5] UN Climate Change Conference - United Arab Emirates | UNFCCC.

[6] Visualizing the Green Investments of Sovereign Wealth Funds (https://www.visualcapitalist.com).

half of these green assets are held by Gulf funds, which are channeling hydrocarbon revenue into sustainable green investments. However, the gap to reaching net zero carbon emissions by 2050 remains significant, and the United Nations and the International Finance Corporation, among others, are turning to SWFs in their efforts to achieve this goal and counter the negative effects of climate change.[7]

This timely volume brings together a diverse array of students, researchers, thought leaders, and practitioners at the forefront of discussing green growth opportunities and sustainable finance, with in-depth analyses and rich discussions of the effect of various global shocks, especially the COVID-19 pandemic and the climate change crisis.

This book is a collection of twelve chapters. Each contributes to analyzing green growth opportunities and providing a deep understanding of sustainable finance in the GCC region. These chapters suggest policy recommendations for decision-makers in the region to bring about improvement to perspectives on strengthening the response to global shocks.

Chapter 1 explores the critical importance of managing the water-energy-food nexus in the GCC region to enhance resources security and self-sufficiency, foster green economic growth, and ensure resilience against global shocks.

Chapter 2 examines the dynamic nexus among stock markets, environmental quality, and energy consumption across the GCC countries.

Chapter 3 discusses how coral reefs constitute a significant component of marine natural capital and play a vital role in the GCC countries and presents the substantial potential for a robust local blue economy.

Chapter 4 explains how financial technology (FinTech) is contributing to the advancement of financial inclusion and the establishment of a sustainable economy in the GCC countries, where the FinTech ecosystem offers both opportunities and challenges, providing a platform for growth and improvement.

Chapter 5 explores the concept and implementation of smart cities in the GCC countries, examining the unique challenges and opportunities presented by the region's economic, social, and technological landscape. It analyzes the motivations behind smart city initiatives in the GCC, including urbanization, economic diversification, and sustainability goals.

Chapter 6 discusses the transparency performance of central banks and how this enhances financial resilience in the Gulf region. It underscores that Gulf central banks can bolster transparency by adopting credible reform agendas that will be a powerful tool in building more sustainable financial systems, that can weather global shocks more effectively and pave the path to sustainable finance.

Chapter 7 reveals that by focusing on compressing artificial intelligence (AI) models and making them both energy-efficient and cost-effective, the GCC region can take meaningful steps toward reducing the carbon footprint of its rapidly expanding AI ecosystem.

---

[7] Miniaoui, H. and Bechri, M. (2024). Gulf sovereign wealth fund investment in the energy transition – The Arab Gulf States Institute in Washington, August.

Chapter 8 explores the potential for green growth opportunities and the implementation of a sustainable finance framework within the healthcare sector across the GCC region. It highlights the pivotal role of sustainable finance in catalyzing the adoption of green initiatives within the healthcare industry, ultimately leading to improved health outcomes and a reduced environmental footprint.

Chapter 9 examines the motivations behind the development of sustainable finance in the GCC region and studies the risks and the returns of the environmental, social, and governance (ESG) investments by considering the performance of equity, debt, and mean-variance optimized portfolios.

Chapter 10 explores the adoption of AI for sustainable development goals in higher education within the Gulf region and contributes to the ongoing discourse on sustainable education by elucidating the interplay between sustainable development principles and AI adoption.

Chapter 11 highlights key policy initiatives promoting renewable energy, energy efficiency, and sustainable financial systems as vital to mitigating climate change and ensuring long-term environmental sustainability.

Chapter 12 examines the GCC countries' economic diversification and resilience to the COVID-19 pandemic and discusses their institutional governance, with the aim of achieving green growth and prosperity.

Doha and Paris

Héla Miniaoui
Mondher Bellalah

Contents

Chapter 1
The Water, Energy, Food Nexus in the GCC: Overcoming Challenges to Unlock Opportunities

Rumaitha Al Busaidi

Abstract With the Arabian Peninsula notably experiencing the effects of climate change in the form of desertification, biodiversity loss, frequent cyclones, and water scarcity, all six members of the GCC are among the top 10 countries predicted by the World Resources Institute (WRI) to suffer a severe water crisis by 2040. As a result, we observe these countries struggling to find answers to their water insecurity issues as the need for water increases due to socioeconomic development and population growth.

The challenges associated with chronic water problems are unique. Understanding water insecurity involves looking beyond supply shortages to the political, social, and economic landscape. Water is vital to health and livelihood, from sanitation services to economic activities like agriculture, energy production, and transportation. Lack of economic stability and growth will result in societies being unable to provide for themselves. This in turn will lead to urbanization, increased negative impacts of climate change, and a rise in protest that may result in political tensions.

It is also evident that there is a need to change the global energy mix, especially considering the Russia-Ukraine war, which has triggered an energy crisis worldwide and may have caused us to revert back a few steps toward decarbonization. An upside to the current energy crisis, however, is the growing impetus for alternative energy storage in the form of low-carbon hydrogen fuel cells, which have been shown to be an effective substitute. The GCC countries are capitalizing on their superior renewable energy resources as a competitive advantage to position themselves as early movers in the low-carbon hydrogen area with the goal of becoming global superpowers in the export of green energy. All of this ensures that their significant energy expertise is utilized, and presents a fantastic opportunity to improve water accessibility and security, which will not only enable further growth and innovation but also provide food security to a region that is currently experiencing food insecurity.

R. Al Busaidi (✉)
Environment Society of Oman, Muscat, Oman
e-mail: rumaitha.albusaidi@eso.org.om

H. Miniaoui and M. Bellalah (eds.), *Green Growth Opportunities and Sustainable Finance in the Gulf Cooperation Council Region*, Gulf Studies 24,
https://doi.org/10.1007/978-981-96-8287-4_1

Keywords Gulf Cooperation Council · GCC · Water scarcity · Climate change · Renewable energy · Low-carbon hydrogen · Food security · Water-Energy-Food nexus · Socioeconomic development · Decarbonization

1.1 Introduction

The Arabian Peninsula, home to the six Gulf Cooperation Council (GCC) countries—Bahrain, Kuwait, Oman, Qatar, Saudi Arabia, and the United Arab Emirates—has in recent decades become an economic powerhouse driven by vast oil and gas resources. This economic boom has led to unprecedented growth, significantly improving infrastructure, services, and quality of life. However, this growth comes with unique and urgent environmental challenges, including harsh geographical and climatic conditions. The GCC is characterized by extreme aridity, limited freshwater resources, rapidly growing urban populations, and increasing environmental stressors such as desertification and biodiversity loss. According to the World Resources Institute (WRI), all GCC nations are among the top 10 countries most likely to experience a severe water crisis by 2040. Climate change exacerbates these risks, leading to increased desertification and more frequent extreme weather events like cyclones and droughts. The pressing global demands for environmental sustainability and economic diversification further compound these issues. Despite these challenges, the rapid socioeconomic development in GCC nations continues to put immense pressure on their scarce water resources.

This water crisis is deeply intertwined with energy and food security, forming a complex and interdependent Water-Energy-Food (WEF) nexus. Water scarcity directly affects agriculture and food production, while energy is essential for desalination and water pumping. Conversely, energy production is water-intensive and heavily dependent on fossil fuels, particularly in the power and manufacturing sectors, while food imports rely on stable energy infrastructure. The GCC's dependence on desalination to meet water needs further complicates the situation, as desalination is energy-intensive and often depends on fossil fuels.

The WEF nexus framework is essential for understanding and addressing these interconnected challenges, offering pathways toward sustainability and resilience in the GCC region. The nexus approach recognizes that water security, energy efficiency, and food sovereignty are profoundly interlinked and that actions in one area invariably influence outcomes in others (Lazaro et al., 2022). For the GCC, a region characterized by scarce water resources, high energy consumption, and substantial food import dependency, the WEF nexus offers a pathway toward enhancing sustainability and resilience (Herrera-Franco et al., 2024; Siderius et al., 2020).

Addressing the WEF nexus is imperative to ensure sustainable development and resilience in this region. Recent global events, such as the COVID-19 pandemic and the Russia-Ukraine conflict, have disrupted global supply chains and highlighted the vulnerabilities of food and energy security in the GCC, emphasizing the urgent need for integrated solutions. This unique confluence of factors has made the WEF nexus an essential paradigm for understanding and addressing the region's long-term sustainability and green growth (Abbas et al., 2023; Al-Saidi & Saliba, 2019; Jägerskog & Barghouti, 2022). Green growth strategies aim to foster economic growth and development while ensuring that natural assets continue to provide the resources and environmental services on which the population's well-being depends. The GCC's pursuit of green growth is evident in its aggressive renewable energy targets and commitments to sustainable urban development, which are vital to reducing carbon footprints and enhancing energy security (IRENA, 2024).

1.2 Geographic and Climatic Challenges

The GCC stands at a crossroads of significant environmental and economic challenges. Expansive deserts, rugged mountains, and coastal plains define the region's geography. The region's climate features exceptionally high summer temperatures, low precipitation rates, and frequent droughts, resulting in severe water scarcity and making the GCC one of the most water-stressed regions in the world, as shown in Fig. 1.1. These conditions present both significant economic challenges and opportunities. In this predominantly arid region, facing extreme temperatures and scarce water resources, water security is a critical concern. Economically, GCC countries have historically depended heavily on oil and gas revenues, fueling rapid urbanization and infrastructural development. However, this economic model is under increasing pressure due to volatile oil prices and global shifts to renewable energy sources, making economic diversification a strategic imperative for these nations. The push toward economic diversification has become a strategic imperative for these nations, as evidenced by initiatives like Saudi Arabia's Vision 2030, Oman's Vision 2040, and the UAE's post-oil economic strategies.

The reliance on energy-intensive desalination plants further complicates sustainable water management, highlighting the intricate water-energy nexus. According to a report by the World Resources Institute, the GCC region's water crisis is exacerbated by climate change, leading to higher temperatures and increased water evaporation rates, which further accelerates the need for urgent action (WRI, 2024). Furthermore, the GCC countries depend on groundwater, which is being depleted at an alarming rate due to unsustainable extraction practices. In some countries, groundwater levels have dropped significantly, and the remaining reserves are increasingly saline, negatively impacting agriculture and local food production (Allouche, 2024; Al-Saidi & Saliba, 2019; Saif et al., 2014). Desalination has emerged as the primary solution to the region's freshwater needs. However, it is an energy-intensive process that directly links water and energy consumption.

On the energy front, the GCC countries are embarking on ambitious renewable energy projects. Saudi Arabia, for instance, aims to achieve 50% of its energy from renewable sources by 2030 as part of its Vision 2030, aligning with global trends toward reducing carbon footprints and enhancing energy security (IRENA, 2024). Similarly, the UAE has invested significantly in solar power, exemplified by the

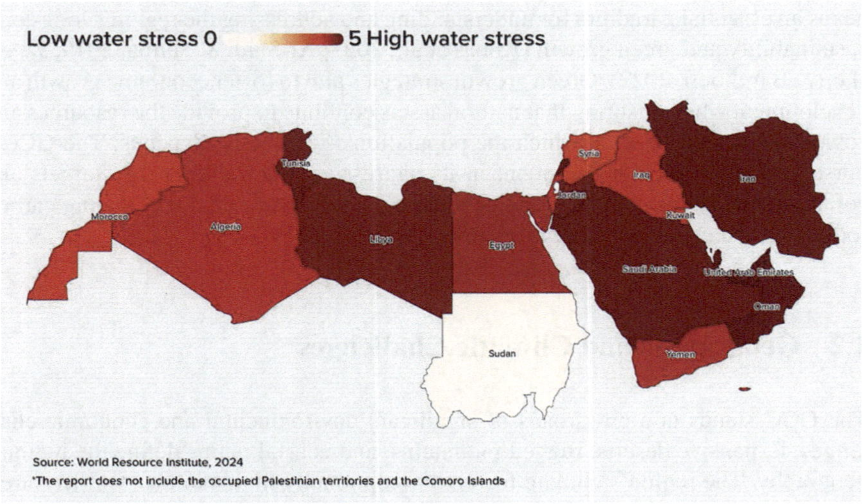

Fig. 1.1 Water stress levels in the Middle East and North Africa by 2030. The GCC countries are all shown to expect severe water stress by 2030 (WRI, 2024)

Mohammed bin Rashid Al Maktoum Solar Park, one of the largest solar parks in the world.

Food security is another critical aspect of the GCC's environmental-economic nexus. With limited arable land and freshwater resources, the region mostly depends on food imports, which makes it vulnerable to global supply chain disruptions. Initiatives to enhance food security through technology-driven agricultural practices like vertical farming and hydroponics are being adopted to reduce this dependency (Ahmed et al., 2024; Ajjur & Al-Ghamdi, 2022; Elzaki, 2024).

The GCC is also a region expected to be hit with more frequent climate-change-related events like Cyclone Shaheen, which impacted Oman in October 2021 and caused economic losses of around 520 million USD (Times News Service, 2021), the extreme drought conditions impacting Saudi Arabia most recently in 2017 (Syed et al., 2022), and the floods of May 2024 that affected most GCC countries, with the hardest hit being Bahrain, Oman, and the UAE, where the country witnessed rainfall of 250 mm in 1 day, an event not seen since meteorological records began in the region in 1949 (Johnson, 2024). The frequency of such previously unprecedented events is likely to increase, with Oman reporting an expectation of a major cyclone disrupting the country once every three to 4 years (Ibrahim et al., 2022). These changes and impacts bring into the fold further nuanced approaches to addressing the WEF nexus and its importance moving forward.

1.3 Evolution of the WEF Nexus Framework

The WEF nexus concept has evolved significantly over several decades, reflecting a growing awareness of the interconnectedness of these critical resources and the need for integrated management strategies. Early recognition of the interdependencies between water, energy, and food resources dates back to the 1960s, with the Harvard Water Program promoting interdisciplinary approaches to water research, laying the groundwork for understanding the complex interactions between these sectors (Cairns & Krzywoszynska, 2016). The 1970s marked a significant era in the evolution of integrated resource management, with influential events such as the United Nations Conference on the Human Environment in Stockholm (1972) and the UN Conference on Water in Mar del Plata (1977) bringing international attention to the need for comprehensive approaches to managing environmental and natural resource issues (Wichelns, 2017).

The formal concept of the WEF nexus began gaining wider recognition in the aftermath of the 2007–2008 financial crisis. This era marked the formalization of the WEF nexus as a tool for addressing global challenges such as climate change, resource scarcity, and sustainable development (Bazilian et al., 2011). The nexus approach fostered cooperation among different sectors, enhancing resource security and promoting sustainable development.

As the concept gained traction, it became increasingly linked to the United Nations Sustainable Development Goals (SDGs). The Food and Agriculture Organization (FAO) played a crucial role in this process, highlighting the nexus approach as essential for food security and sustainable agriculture. In recent years, the nexus approach has become critical for enhancing resilience and adaptive capacity in the face of climate variability and extreme events. Integrated climate policies that address the interconnected risks to water, energy, and food security have become a focal point. This shift reflects a broader recognition of the need for integrated solutions to complex, interrelated challenges. A comprehensive review by Lazaro et al. (2022) of 10 years of WEF nexus research highlights the evolution of topics and the increasing complexity of the interrelationships being studied. The findings underscore the importance of continued innovation and collaboration in advancing the WEF nexus framework.

In the GCC context, adopting a WEF nexus approach offers a pathway to address resource security challenges and achieve sustainable development goals. The region's unique environmental conditions make the nexus approach particularly relevant. Implementing integrated strategies can enhance resource efficiency, reduce vulnerabilities, and promote sustainable growth in the GCC (Fig. 1.2).

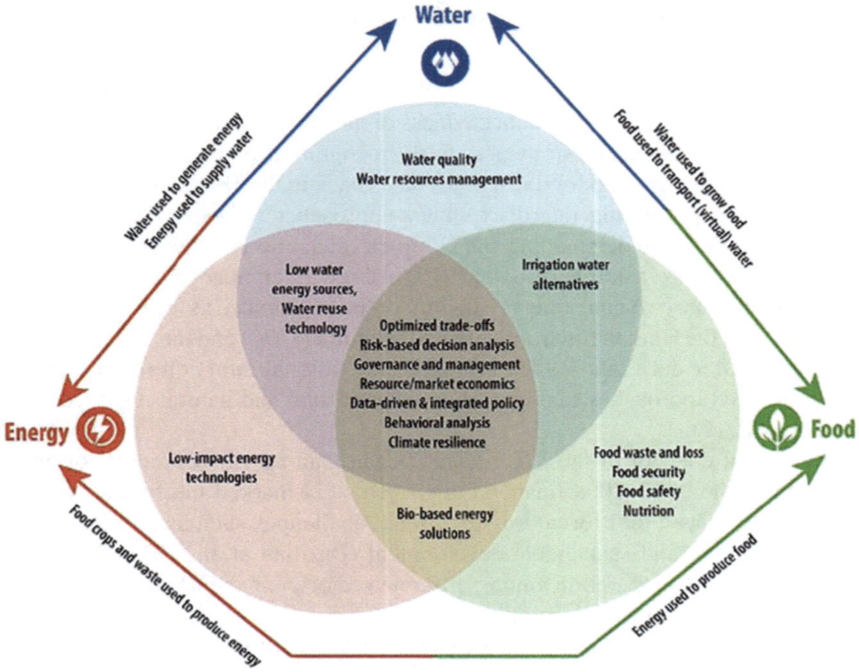

Fig. 1.2 A schematic layout of the Water-Energy-Food (WEF) nexus as depicted by Baker et al. (2021). The linkages between the three resource sectors include water for food, food for energy, energy for food, energy for water, and water for energy. These interdependencies demonstrate that if the security of one sector is compromised, the stability of the other two sectors will also be at risk

1.4 Current State of the Water-Energy-Food Nexus in the GCC

1.4.1 Water Scarcity: A Closer Look at the GCC Context

Water scarcity in the GCC is a critical issue deeply intertwined with the region's energy and food sectors. The unique geographical, climatic, and socioeconomic characteristics of the GCC exacerbate the challenges associated with water availability and management.

1.4.2 Natural Scarcity and Climatic Conditions

As mentioned earlier, the GCC region is characterized by an arid to hyper-arid climate, with deficient annual precipitation levels and high evaporation rates. The average annual rainfall in the region is less than 100 mm, significantly lower than the global average, and some areas receive even less, while evaporation rates are extremely high due to the high temperatures, further reducing the availability of surface water (Jägerskog & Barghouti, 2022). These climatic conditions severely limit the availability of renewable water resources, making groundwater the primary source of fresh water. Groundwater reserves are critical, but they are being depleted at an unsustainable rate due to over-extraction for agricultural, industrial, and domestic uses (Salberg, 2023). This over-extraction is driven by the need to support growing populations and economic activities.

The slow natural recharge processes of groundwater further exacerbate the scarcity issue. Recharge rates are insufficient to replenish the extracted volumes, leading to a continuous decline in groundwater levels. Additionally, the quality of the remaining groundwater is deteriorating due to increased salinity and contamination, often from agricultural runoff and industrial activities. These factors collectively contribute to the severe water scarcity situation in the GCC, highlighting the need for careful management and conservation of these precious resources (Al Rashed et al., 2023; Saif et al., 2014).

1.4.3 Rapid Population Growth and Urbanization

The GCC countries have experienced rapid population growth and urbanization over the past few decades, driven by both natural population increase and significant migration due to economic opportunities. This rapid growth has substantially increased water demand for domestic, industrial, and agricultural purposes. Urban areas, in particular, have seen a dramatic rise in water consumption, necessitating the expansion of water supply infrastructure to meet the needs of burgeoning cities.

The increased water demand has placed immense pressure on the already scarce water resources. The existing water infrastructure is often stretched to its limits, and in many cases, it is inadequate to meet the growing needs. This situation has led to over-exploitation of available water sources, further depleting groundwater reserves and increasing reliance on desalination. The rapid pace of urbanization also brings challenges related to water distribution and management, including reducing water losses through improved infrastructure and technology (Hameed et al., 2019; Sherif et al., 2023).

1.4.4 Desalination Dependency

Desalination has become a cornerstone of water supply in the GCC, with the region hosting some of the largest desalination plants in the world. Saudi Arabia and the United Arab Emirates have developed extensive desalination infrastructure, with Saudi Arabia leading the world in desalinated water production (Moossa et al., 2022). These plants are critical for providing potable water to the region's growing populations and industries. Desalination processes remove salt and other impurities from seawater, making it suitable for human consumption and other uses. Due to their lack of natural freshwater resources, the GCC countries have invested heavily in desalination technology, which now accounts for a significant portion of their water supply.

However, desalination is an energy-intensive process, contributing significantly to the high energy consumption patterns observed in the region. The energy required for desalination is typically sourced from fossil fuels, leading to high greenhouse gas emissions and other environmental impacts. Additionally, the desalination process generates brine, a highly concentrated salt solution, which poses disposal challenges and can negatively impact marine ecosystems if not properly managed (Lahlou et al., 2023; Salberg, 2023). The operational costs of desalination are also considered high, often necessitating government subsidies to keep water prices affordable for consumers, imposing a long-term fiscal burden on national budgets (Al-Saidi & Saliba, 2019; Borgomeo et al., 2018). Despite these challenges, desalination remains a crucial component of the GCC's water strategy due to the severe natural scarcity of freshwater resources (Ajjur & Al-Ghamdi, 2022; Saif et al., 2014).

1.4.5 Agricultural Water Use

Historically, groundwater has been the primary source of water, particularly for agriculture. However, unsustainable extraction practices have led to widespread depletion of groundwater and deterioration in its quality due to salinity. The agricultural sector accounts for about 80% of fresh water consumption in the GCC, although contributing minimally to the GDP (Ben Hassen & El Bilali, 2019). Agriculture relies heavily on this diminishing resource and traditional irrigation practices, which are often inefficient and wasteful, exacerbate the problem. Efforts to increase agricultural productivity and achieve food self-sufficiency have intensified water usage, putting additional strain on the limited water resources.

The result has been a significant decline in groundwater tables across the region, particularly in Oman, Bahrain, and Saudi Arabia, where many wells now run dry or contain water unfit for use without expensive treatment (Saif et al., 2014). Despite this situation, the agricultural water demand in the GCC is expected to continue to rise, further straining the already limited water resources. This is particularly concerning given the region's climatic conditions and unsustainable water usage practices. There is a critical need for more efficient water management practices within the agricultural sector to ensure the sustainability of water resources. This includes adopting modern irrigation techniques and improving water use efficiency to reduce the overall water footprint of agriculture in the region (Ben Hassen & El Bilali, 2019; Sherif et al., 2023).

1.4.6 Energy Landscape in the GCC: A Focus on Current Status

The region's vast fossil fuel resources shape the GCC's energy landscape. The GCC countries collectively possess around 30% of the world's oil reserves and over 20% of global natural gas reserves (Al-Sarihi & Mansouri, 2022; IEA, 2023b). As a result, fossil fuels remain the predominant energy source for domestic consumption and export, leading to several defining characteristics of the GCC energy landscape.

1.4.7 High Energy Consumption Rates

Hydrocarbons are the primary energy sources, accounting for the vast majority of energy production and consumption in the region, and have historically driven its economic growth. For instance, Saudi Arabia and Kuwait are among the top oil producers in the world, with oil and gas revenues constituting a significant portion of their GDP. The extensive fossil fuel extraction, processing, and export infrastructure underscores the region's dependency on these resources (Amer et al., 2017).

Energy consumption in the GCC is characterized by high per capita rates, primarily driven by industrial activities, residential cooling demands due to the hot climate, and water desalination processes. The availability of cheap energy due to subsidies has further exacerbated consumption levels. In Qatar, for example, the per capita energy consumption is one of the highest in the world, reflecting both the wealth generated from fossil fuels and the intensive energy needs of the country's industries and residential sectors (Al-Saidi & Saliba, 2019).

Generous government subsidies on electricity and fuel have kept energy prices low, fostering an environment of high per capita energy consumption. This subsidy-driven model makes the GCC countries some of the world's most energy-intensive economies, particularly in residential electricity use and transportation (IEA, 2023b).

The industrial sector is a significant energy consumer in the region and contributes to its large carbon footprint. Industries such as petrochemicals, aluminum, and cement require vast amounts of energy for their operations. For example, in Kuwait, energy consumption is driven by the needs of the petrochemical industry, leading to a substantial energy demand (Sherif et al., 2023). Additionally, the residential sector contributes significantly to the high energy demand due to the widespread use of air-conditioning systems to combat the intense heat, which can reach over 50 degrees Celsius in summer months. The transportation sector also adds to the overall energy consumption, primarily using gasoline and diesel for vehicular travel.

The region relies heavily on natural gas for electricity generation, with oil playing a secondary but significant role. In 2021, natural gas accounted for approximately 73% of the GCC's electricity generation, while oil comprised 26% (Al-Sarihi & Mansouri, 2022). This heavy reliance on fossil fuels for electricity generation is a significant factor in the region's high carbon emissions.

1.4.8 High Carbon Emissions and Renewable Energy Potential

The reliance on fossil fuels and the associated subsidies have resulted in high carbon dioxide emissions per capita. Qatar, Kuwait, and the United Arab Emirates rank among the world's highest per capita emitters (Jägerskog & Barghouti, 2022). The high carbon emissions are a direct result of the energy-intensive lifestyles and industrial activities prevalent in these countries. The environmental impact of these emissions is significant, contributing to global climate change and local air pollution.

Despite the dominance of fossil fuels, there is a growing recognition of the need to diversify the energy mix in the GCC. The region has substantial potential for renewable energy, particularly solar and wind. Most of the GCC countries have also announced ambitious net-zero targets to address climate change and reduce carbon emissions. For example, Bahrain, Kuwait, and Saudi Arabia aim to achieve net-zero emissions by 2060, and Oman and the UAE have set this target for 2050. Qatar is left as the only GCC country with no announced carbon neutrality targets. Despite these commitments, the adoption of renewable energy has been relatively slow, and the current energy mix remains heavily dominated by fossil fuels. Renewable energy projects are underway, but their contribution to the total energy supply is minimal compared to traditional energy sources (Derouez & Ifa, 2024). The primary challenges include high initial investment costs, integrating renewable sources into existing power grids, and technological advancements (Al-Saidi & Saliba, 2019).

1.5 Food Security in the GCC: A Focus on Current Challenges

The GCC countries face significant challenges in ensuring food security. The region's arid climate, limited arable land, and dependence on food imports shape the current status of food security, which is closely intertwined with water and energy resources.

1.5.1 High Dependence on Food Imports

The GCC region itself has limited agricultural capacity. The harsh climatic conditions, including high temperatures and low rainfall, restrict the ability to produce sufficient food domestically. As a result, the GCC is dependent on food imports, with some estimates suggesting that up to 85–90% of the region's food needs are met through imports (FAO, 2022a). This dependence on external sources makes the region vulnerable to global market fluctuations and supply chain disruptions. For instance, the 2007–2008 global food crisis highlighted the GCC's vulnerability, exposing the risks associated with heavy reliance on food imports (Ben Hassen & El Bilali, 2019).

The GCC states have historically leveraged their substantial oil revenues to secure food imports, ensuring food availability despite domestic production limitations. This financial strength has allowed them to buffer against price volatility and secure a stable food supply. However, the increasing global demand for food and potential geopolitical tensions pose a continuous threat to the stability of food imports. The need to diversify food supply sources and improve local production capabilities in order to mitigate these risks is becoming more apparent.

1.5.2 Limited Arable Land and Agricultural Production

Several natural factors, including scarce water resources and poor soil quality, constrain agricultural production in the GCC. Less than 2% of the GCC's total land area is arable due to desertification and salinity issues, limiting the variety and volume of crops that can be cultivated domestically (WRI, 2024). The region's agriculture relies heavily on irrigation, which puts additional pressure on already scarce water resources. Groundwater, a primary source of irrigation, is being depleted at unsustainable rates, further threatening agricultural viability. In the UAE, for example, groundwater resources are projected to be insufficient to meet future agrarian needs by 2030, highlighting the critical situation (Shahin & Salem, 2015).

Crops traditionally cultivated in the region, such as dates and fodder, are water-intensive, further straining the already limited groundwater resources. Efforts to boost local agricultural production face significant challenges. Introducing modern farming techniques such as hydroponics and aquaponics offers some potential to increase productivity. However, these methods require substantial investment and technological expertise, which is not yet widespread in the region. Additionally, the high energy demands of these technologies link agricultural sustainability directly to the energy sector, emphasizing once more the interconnectedness of the WEF nexus (Ajjur & Al-Ghamdi, 2022).

Despite the constraints, governments have initiated programs to boost local food production. The UAE's hydroponic farming projects and Qatar's investment in controlled environment agriculture (CEA) showcase efforts to bolster food security through technology-driven agriculture (Shahin & Salem, 2015; Shubbar et al., 2024).

1.5.3 Fisheries and Aquaculture

Fisheries and aquaculture have historically been essential to the region's food supply. The GCC's geographic location, bordering the Arabian Gulf and the Arabian Sea, provides significant potential for fish production. Traditional fishing remains vital for coastal communities in the region, especially in Oman, a net exporter of fish, with the majority of its fish exports ending up in the neighboring GCC countries. However, overfishing and environmental degradation have led to the depletion of fish stocks. In the Arabian Gulf, fish populations have declined by over 50% in the past 30 years due to overfishing and habitat destruction (FAO, 2022b).

Aquaculture is growing in importance in addressing declining wild fish stocks and meeting rising demand. In Saudi Arabia, for instance, Vision 2030 aims to expand aquaculture production to 600,000 tons by 2030. Oman, too, has prioritized aquaculture by establishing the Aquaculture Development Plan and an entire investment arm focusing on the sector's development to promote sustainable fish farming. Despite the potential, aquaculture faces environmental challenges. Fish farming can lead to habitat destruction and pollution. As a result, the region's emerging aquaculture sector must adopt sustainable practices to ensure long-term viability.

1.5.4 Vulnerability to Climate Change

Climate change exacerbates the challenges of food security in the GCC. Rising temperatures and changing precipitation patterns affect crop yields and livestock productivity, further straining the limited agricultural output. Additionally, extreme weather events, such as droughts and heat waves, can disrupt food production and supply chains, increasing the region's food insecurity risk (Borgomeo et al., 2018).

The GCC countries have recognized the need to address the impacts of climate change on food security. National strategies, such as Qatar's National Food Security Program, aim to enhance agricultural resilience and reduce dependence on food imports. However, progress is slow, and significant investments are required to implement effective adaptation measures. Therefore, the region's efforts to achieve food security must integrate climate resilience strategies if they are to ensure long-term sustainability (Al-Saidi & Saliba, 2019).

1.6 Water-Energy-Food Nexus Interdependencies

The interdependencies between water, energy, and food in the GCC are profound and complex. The WEF nexus framework underscores the need for integrated resource management approaches to address these interdependencies effectively. Understanding these linkages is crucial for developing sustainable policies and strategies.

1.6.1 Water-Energy Linkages

In the GCC, the energy sector's reliance on water is substantial. For instance, the energy sector relies heavily on water for cooling processes in power plants, while the water sector depends on significant energy, primarily from fossil fuels, for desalination processes, water pumping, and treatment, linking water production directly to energy consumption. This creates a cycle where energy production depends on water and water production relies on energy. The extensive use of desalination drives up energy consumption and increases greenhouse gas emissions, contributing to climate change (Ajjur & Al-Ghamdi, 2022; Derouez & Ifa, 2024).

Additionally, the environmental consequences of desalination, such as brine discharge into marine environments, exacerbate the problem by increasing the salinity of water bodies and affecting marine life (Lahlou et al., 2023). The rising salinity levels and the contamination of groundwater resources further intensify water scarcity issues. Addressing these water-energy linkages requires adopting energy-efficient desalination technologies and sustainable water management practices to mitigate the environmental impacts and ensure resource sustainability.

1.6.2 Energy-Food Linkages

Food production in the GCC is heavily reliant on energy, from the mechanization of farming practices to the transportation of agricultural products. Although increasing local food production, modern agricultural techniques require substantial energy inputs. This dependency on energy for food production highlights the close connection between these two sectors.

Moreover, the significant energy requirements for irrigation and agricultural processes contribute to the region's high energy consumption. This linkage emphasizes the need to balance energy use with food production efficiency. Integrating renewable energy sources into agricultural operations could help reduce the environmental footprint and ensure a more sustainable food production system. However, this requires significant investment and technological advancement to be effective.

1.6.3 Food-Water Linkages

Though a minor contributor to GDP, agriculture in the GCC consumes a large proportion of the region's limited water resources. The sector's heavy dependence on groundwater for irrigation leads to substantial depletion of these reserves. Traditional irrigation methods often result in inefficient water use, further exacerbating water scarcity; for instance, cultivating water-intensive crops like dates and fodder places additional stress on groundwater supplies (Ben Hassen & El Bilali, 2019; Shubbar et al., 2024).

The salinization of groundwater due to over-extraction and contamination poses further challenges to agricultural sustainability. Technological solutions like hydroponics and aquaponics offer potential alternatives but require substantial energy and investment. Managing these food-water linkages effectively demands a shift toward more sustainable agricultural practices that conserve water while maintaining productivity. This approach would help alleviate the pressure on water resources and support long-term food security in the GCC.

The considerable energy requirements for water and food production further complicate this nexus, emphasizing the importance of understanding and managing the trade-offs between these sectors. The WEF nexus framework provides a holistic approach to managing these resources, recognizing that actions in one sector can significantly impact the others. This interconnectedness necessitates integrated policies and strategies that consider the mutual dependencies and potential trade-offs between water, energy, and food (Abulibdeh et al., 2019; Hameed et al., 2019).

1.6.4 Strategies for Integrated Resource Management within the WEF Nexus

The GCC countries proactively address the interconnected challenges of water scarcity, energy sustainability, and food security through a strategic blend of technological innovations and policy reforms. Each GCC nation—Bahrain, Kuwait, Oman, Qatar, Saudi Arabia, and the United Arab Emirates—has developed comprehensive vision documents that articulate long-term objectives to transform their natural resource management and ensure sustainable development. These documents reflect the commitment to meeting the UN Sustainable Development Goals and highlight the unique approaches tailored to each country's specific ecological and socioeconomic context, as explained in Table 1.1.

Each GCC country's vision serves as a roadmap for implementing the substantial reforms required in these sectors. They emphasize the importance of integrated solutions and are crucial for guiding policy decisions, informing infrastructure investments, and fostering international collaborations to leverage technological innovations to tackle the pressing issues of water and food security while transitioning to sustainable energy solutions. The subsequent sectoral implementation details

Table 1.1 Summary of each GCC nation's vision and focus area

Country	Vision document	Overview	Key focus areas
Saudi Arabia	Vision 2030	Aims to reduce oil dependency, promote renewable energy, and ensure sustainable water management. Built around three main themes: a vibrant society, a thriving economy, and an ambitious nation	Significant investments in renewable energy, water conservation projects, sustainable agricultural practices, economic diversification, improving government efficiency, and enhancing quality of life
United Arab Emirates	We the UAE 2031	A national plan for the next decade focusing on social, economic, investment, and development aspects	Doubling GDP, increasing non-oil exports, enhancing tourism, positioning Emirati cities among the best globally in quality of life. Four pillars: Forward society, forward economy, forward diplomacy, and forward ecosystem
Oman	Vision 2040	Focuses on creating a society of creative individuals, a competitive economy, responsible state agencies, and an environment with sustainable components	Economic diversification, sustainability, efficient resource management, water demand management, energy efficiency, smart, sustainable cities
Qatar	National Vision 2030	Aims to transform Qatar into an advanced society capable of sustaining its development and providing a high standard of living	Addresses modernization and preservation of traditions, needs of current and future generations, managed growth, size and quality of the expatriate labor force, economic growth, social development, and environmental management. Four interconnected pillars: human development, economic development, environmental management, and social development

(continued)

Table 1.1 (continued)

Country	Vision document	Overview	Key focus areas
Kuwait	Vision 2035	Aims to transform Kuwait into a financial and trade hub, making it more attractive to investors with a private-sector-led economy	Enhancing infrastructure, fostering a competitive private sector, aligning with UN sustainable development goals, improving living standards, developing non-oil sectors, supporting human and social development
Bahrain	Vision 2030	Focuses on sustainable development and economic diversification. Launched in 2008 with the principles of sustainability, fairness, and competitiveness	Private-sector-driven economy, human capital development, environmental sustainability, improved living standards, minimal government intervention, adhering to sustainable development goals

contained within these strategies highlight specific projects and initiatives already underway, demonstrating the practical steps being taken toward these ambitious goals.

By integrating these strategic frameworks, the GCC countries are positioning themselves to tackle the interconnected challenges of the WEF nexus through coordinated and forward-thinking policies and initiatives. This section will explore the policy responses and governance strategies, technological innovations, regional cooperation, and indigenous knowledge that can drive sustainable solutions in the GCC region.

1.7 Policy and Technological Responses in the GCC

1.7.1 Integrated Resource Management Policies

Integrated resource management policies are crucial for optimizing the use of the GCC region's water, energy, and food resources. Implementing a WEF nexus framework promotes coordinated planning and decision-making across sectors, ensuring that the interdependencies of the WEF nexus are considered.

For instance, Bahrain, a small island nation with significant water scarcity challenges, has developed comprehensive water management policies integrating desalination with wastewater treatment and reuse. Bahrain's Vision 2030 emphasizes the need for a sustainable water management system that combines advanced desalination technologies with efficient wastewater treatment to ensure a reliable water supply for domestic and industrial use. This integrated approach helps balance the limited freshwater resources with the growing demand for water in various sectors.

Similarly, Qatar has adopted integrated resource management strategies to enhance food production while conserving water and energy. Qatar's National Vision 2030 focuses on balancing economic growth with environmental protection, including integrating advanced agricultural technologies such as hydroponics and aquaponics. These technologies allow for high-efficiency food production with minimal water use, simultaneously addressing food security and water scarcity challenges (Ajjur & Al-Ghamdi, 2022).

Kuwait has also taken significant steps toward integrated resource management. The country's focus has been on evaluating the influence of socioeconomic development and climate change on water, energy, and food resources through scenario-based assessments. Kuwait's integrated policies are directed toward demand-side management and efficiency approaches to mitigate the trade-offs within the WEF nexus (Al-Adwani et al., 2024).

1.7.2 Regulatory Reforms to Improve Efficiency

Regulatory reforms are essential for improving the efficiency of water and energy use in the GCC. These reforms often involve updating existing regulations and introducing new ones to promote conservation, enhance efficiency, and support the adoption of sustainable practices. The push toward net-zero emissions has significantly influenced the regulatory landscape in the GCC, emphasizing the interconnectedness of the WEF nexus in achieving sustainability goals.

Five Gulf Arab states—Bahrain, Kuwait, Oman, Saudi Arabia, and the UAE—have pledged to reach net-zero emissions by or around the mid-twenty-first century. These announcements are underpinned by robust techno-economic modeling but are brought to life through comprehensive policy frameworks and a blend of sector-specific policies, such as investment in renewables and incentives for greener energy infrastructure, alongside cross-sectoral policies like carbon pricing and regulation to support electrification. A set of examples is provided below to show a range of such efforts currently underway in the GCC.

Water and Energy Efficiency Reforms Kuwait is implementing progressive water tariffs to manage demand more effectively. By aligning the price of water with its production cost, these reforms incentivize both industries and households to use water more judiciously. Similarly, Saudi Arabia has begun phasing out water subsidies so that prices will reflect the actual cost of water production and distribution, encouraging conservation and reducing waste.

Despite the reduction or removal of subsidies being critical to encouraging better practices, this is often a sensitive political issue that requires careful management, especially in light of an economic situation that is plagued with higher interest rates that impact the everyday life of individuals. An example would be Oman's announcement during the COVID-19 pandemic that it intended to move ahead with its plan to remove electricity subsidies by 2025, followed a year later by a reconsideration of the policy due to an overwhelmingly negative reception of the announcement by the public (see Fact Box 1.1).

Renewable Energy Targets and Incentives Energy diversification is a cornerstone of the GCC's strategy to enhance energy security and reduce carbon footprints. Saudi Arabia's Vision 2030 aims to generate 50% of the country's energy from renewable sources by 2030, significantly reducing reliance on fossil fuels. The UAE's Energy Strategy 2050 plans to cut carbon dioxide emissions by 70% through an energy mix of 44% clean energy. Table 1.2 below showcases the main efforts in pushing toward more renewable energy in all the GCC countries. These targets are supported by regulatory frameworks that provide incentives for renewable energy projects, such as subsidies, tax breaks, and grants.

Sustainable Agriculture and Food Security Oman has banned the export of certain fish species to ensure that local fish production meets domestic demand and supports the sustainability of marine resources through the introduction of closed seasons for fishing. This policy aims to balance the nexus by protecting local fish stocks, ensuring food security, and maintaining the health of aquatic ecosystems. Additionally, recognizing the pressure on freshwater and arable land, some GCC countries are turning to sustainable aquaculture as a strategic solution for increasing protein production, minimizing environmental impacts, and making efficient use of resources. Oman aims to increase its seafood farming production through aquaculture and has announced production targets of 200,000 tons annually by 2040. Similarly, Saudi Arabia is focusing on increasing its annual per capita fish consumption from 9 kg to 12 kg by 2030 as a means of contributing to economic growth and green development (FAO, 2022b).

The UAE has also implemented significant reforms in sustainable agriculture. Through initiatives like the Emirates Food Security Strategy 2051, the UAE aims to promote modern farming techniques to increase agricultural productivity while conserving water and energy resources. The strategy includes financial incentives and subsidies for farmers adopting these innovative practices, aiming to reduce the nation's dependence on food imports and enhance local food production sustainability.

Table 1.2 Summary of all 6 GCC countries and their renewable energy targets and net-zero ambitions

Country	Net-zero target	Renewable energy targets	Emissions reduction target	National climate strategy	Other climate initiatives/governance entities
Bahrain	Yes, by 2060	5% by 2025, 10% by 2035	N/A	No	Joint National Committee on Climate Change (2007)
Kuwait	Yes, by 2060	15% by 2030	N/A	No	Kuwait National Committee on climate change
Oman	Yes, by 2050	10% by 2025, 30% by 2030	Reduce GHG emissions by 7% relative to BAU by 2030	National Strategy for Adaptation and Mitigation to Climate Change, 2020–2040; National Carbon Neutral Strategy	Regulations for the management of climate affairs (2016); National Climate Strategy; Oman sustainability Center
Qatar	No	20% by 2030	Reduce 25% of GHG emissions by 2030	Yes	National Climate Change Committee (chaired by the Ministry of Environment)
Saudi Arabia	Yes, by 2060	50% by 2030	Reduce, avoid, and remove 278 million tons of CO_2 annually by 2030	Yes (National Circular Carbon Economy Program)	National Committee for the Clean Development Mechanism/ Designated National Authority (2009); Saudi Green Building Forum (2010); Saudi Energy Efficiency Center (2012); PIF Regional Voluntary Carbon Market Company
UAE	Yes, by 2050	Clean energy 50% (44% RE, 6% nuclear) by 2050	31% reduction compared to BAU by 2030	Yes (2017: Green Growth Strategy/UAE Green Agenda 2015–2030)	Dubai Integrated Energy Strategy 2030; Abu Dhabi carbon trading exchange and carbon clearing

Fact Box 1.1 Oman's Energy Subsidy Reform

The gradual reduction or removal of energy subsidies is a strategic move designed to decrease domestic energy demand and foster the adoption of renewable energy sources. However, this policy shift often involves delicate political and economic considerations due to its direct impact on consumer costs and lifestyle.

In Oman, the government initially set out a plan to phase out electricity subsidies by 2025, intending to streamline energy consumption and enhance the financial sustainability of the energy sector. This reform was part of a broader strategy aimed at promoting energy efficiency and supporting the transition to renewable energy sources. Utility tariffs were to be progressively increased until the subsidies were completely removed, a change expected to affect all consumer categories, including residential, industrial, and government sectors.

However, the implementation of these reforms encountered significant public resistance. By August 2021, following the recalibration of electricity bills which resulted in substantial increases, the Oman News Agency reported that the Authority of Public Services Regulation had received over 5000 complaints, highlighting the public's sensitivity to such financial adjustments. The government responded by reconsidering its approach to subsidy removal. It planned amendments that included recalculating bills for recent months and adjusting tariffs to allow more access to supply at lower rates, particularly benefiting consumers who were previously on lower tariffs.

Simultaneously, Oman introduced a National Support System to cushion the economic impact on lower-income households. Based on data from the 2019 census, this support system considers factors such as household income and size to adjust electricity tariffs more equitably. This targeted approach aims to ensure that while subsidies are phased out, vulnerable populations do not bear the brunt of the changes.

This case illustrates the complexities involved in reforming energy subsidies. While such measures are essential for promoting sustainable energy use and fiscal health, they require careful management to balance economic objectives with social equity. The Omani government's iterative approach to policy implementation and its adjustments in response to public feedback reflect the nuanced governance needed to navigate the socioeconomic landscapes affected by these reforms.

1.7.3 Technological Responses

Water Management The GCC countries are adopting advanced desalination technologies to address water scarcity more sustainably. Energy-efficient desalination techniques, such as reverse osmosis, combined with the use of solar energy, are crucial for reducing desalination's environmental impact and energy consumption. These innovations significantly enhance the region's water production sustainability.

In addition to desalination, wastewater treatment and reuse is gaining prominence as a viable solution to water scarcity. Treated wastewater can be reused for irrigation and industrial purposes, reducing the demand for fresh water. Saudi Arabia, UAE, and Bahrain, have successfully implemented such systems, contributing to water sustainability by providing alternative water sources and reducing the burden on freshwater resources (Al Rashed et al., 2023).

Rainwater harvesting projects are being developed to capture and utilize rainwater, particularly in urban areas. These advanced rainwater harvesting systems help mitigate water scarcity by providing additional water resources. By integrating these systems into urban planning, GCC countries can enhance their water resilience and reduce reliance on traditional water sources. Cloud seeding is another innovative technique to enhance precipitation and address water scarcity issues. The UAE has been actively using cloud seeding, and other GCC countries are exploring its potential benefits and challenges. This technique involves introducing substances into the atmosphere to encourage cloud formation and precipitation, thereby increasing rainfall and supplementing water supplies.

1.7.4 Energy Solutions

Renewable energy integration is a cornerstone of the GCC's strategy to reduce dependency on fossil fuels and mitigate climate change impacts. Large-scale solar and wind projects, such as the Mohammed bin Rashid Al Maktoum Solar Park in the UAE and the NEOM project in Saudi Arabia, demonstrate the region's commitment to renewable energy. These projects provide clean energy and reduce the water consumption associated with conventional power generation methods.

Developing a hydrogen economy is also gaining traction in the GCC. Green hydrogen, produced using renewable energy, offers a clean energy source for various sectors, including transportation and industry. Oman's hydrogen ambitions and Saudi Arabia's plans to scale up hydrogen production highlight the region's potential in this area (see Fact Box 1.2). The production and use of green hydrogen can significantly reduce carbon emissions and contribute to the global transition to a low-carbon economy (IEA, 2023b).

Nuclear energy is another component of the GCC's energy strategy. The UAE's Barakah Nuclear Power Plant reduces reliance on natural gas and enhances energy security. By diversifying its energy mix to include nuclear power, the UAE can ensure a stable and reliable energy supply while reducing greenhouse gas emissions. Nuclear energy provides a low-carbon alternative to fossil fuels, supporting the region's efforts to combat climate change.

Improving energy efficiency in desalination plants and agricultural operations reduces overall energy consumption and promotes sustainability. Advanced technologies and practices can enhance the efficiency of these processes, reducing their environmental impact. For instance, energy-efficient desalination technologies, such as forward osmosis and hybrid systems, are being explored to minimize energy use in water production (Jägerskog & Barghouti, 2022).

1.7.5 Food Security

Sustainable agriculture is critical for enhancing food security in the GCC. Modern agricultural techniques, such as hydroponics, aquaponics, and controlled environment agriculture (CEA), can increase food production with minimal water use. These methods are particularly well suited to the arid conditions of the GCC region. Hydroponics involves growing plants in nutrient-rich water solutions without soil, significantly reducing water requirements. Aquaponics combines hydroponics with fish farming, creating a closed-loop system where fish waste provides nutrients for plants, and the plants help filter the water for the fish. CEA involves growing crops in controlled environments, such as greenhouses, where conditions can be optimized for plant growth. These techniques are being implemented in Qatar and the UAE, demonstrating their potential to enhance food security while conserving water (Shahin & Salem, 2015; Shubbar et al., 2024).

Expanding sustainable aquaculture practices in countries like Oman and Saudi Arabia can enhance food security and reduce the environmental impact of traditional fishing methods. Sustainable aquaculture involves farming fish and other aquatic organisms in ways that minimize environmental harm and promote resource efficiency. For example, integrated multi-trophic aquaculture (IMTA) systems combine different species, such as fish, shellfish, and seaweed, to create a balanced ecosystem that improves water quality and resource use (FAO, 2022a). These practices can provide a reliable source of protein while reducing overfishing and habitat destruction in coastal areas.

Fact Box 1.2 GCC's Leadership in the Low-Carbon Hydrogen Economy

Low-carbon hydrogen, particularly green hydrogen produced using renewable energy, is vital for reducing global carbon emissions and meeting climate goals. It offers a sustainable alternative to fossil fuels and is essential for the energy transition. The IEA's annual report titled "Global Hydrogen Review" emphasizes hydrogen's potential to decarbonize sectors such as transportation, industry, and power generation.

The GCC countries are leveraging their abundant solar and wind resources to become leaders in the green hydrogen economy. Key initiatives include:

- Oman: Oman is slated as the most ambitious of them all, being the first country to launch a hydrogen strategy in October 2022. It endeavors to produce up to 1.5 million tons of green hydrogen annually by 2030, increasing to 8.5 million tons by 2050. Oman is also collaborating internationally to enhance its production and export capabilities and setting itself apart as a government by investing alongside the developers by providing what they call a Common-Use Infrastructure.
- Saudi Arabia: Saudi Arabia is investing heavily in green hydrogen, with the $5 billion NEOM project set to be the largest hydrogen plant globally, producing 650 tons/day. As the first mega green hydrogen project to receive Final Investment Decision in the world, construction is currently underway, and all eyes will be on NEOM to follow its progress. The country aims to produce 2.9 million tons/year by 2030 and four million tons/year by 2035.
- United Arab Emirates (UAE): The UAE launched the Dubai Hydrogen Alliance and aims to become a major hydrogen hub after announcing its strategy 2 days before COP 28. Significant projects, such as the Masdar green hydrogen production facilities, highlight the UAE's commitment to green hydrogen.
- Qatar: Qatar focuses on using its natural gas and carbon capture to power electrolyzers while exploring partnerships and investments to support hydrogen production technologies.

COP28 Initiatives:

At COP28, GCC countries signed a comprehensive Declaration of Intent on Mutual Recognition of Certification Schemes for Hydrogen and Hydrogen Derivatives. This initiative aims to create a unified market and streamline hydrogen trade globally. The declaration, backed by more than 30 countries, is set to unlock the socioeconomic benefits of hydrogen and its derivatives, covering over 80% of the future global market.

By focusing on these strategic initiatives, the GCC countries are well on their way to becoming global leaders in the hydrogen economy.

1.7.6 Global Shocks and the WEF Nexus: Integrating Macro and Micro Perspectives

The WEF nexus in the GCC operates within a broader global context, where various global shocks and trends significantly influence local and regional strategies. This section explores how global economic crises, climate change, geopolitical tensions, and pandemics impact the WEF nexus in the GCC while also connecting these global dynamics to the overarching theme of green growth.

1.7.7 Economic Crises and the WEF Nexus

Global economic crises, such as the 2008 financial crisis, the oil price crash of 2016, and the economic downturns triggered by the COVID-19 pandemic, profoundly affect the WEF nexus. These crises disrupt supply chains, affect resource availability, and alter consumption patterns. For instance, the 2008 financial crisis reduced investments in infrastructure and technology, negatively impacting the efficiency and sustainability of water, energy, and food systems. Similarly, the COVID-19 pandemic exposed vulnerabilities in global supply chains, emphasizing the need for resilient and self-sufficient WEF systems (Al-Saidi & Hussein, 2021).

In the GCC, economic dependency on oil revenues exacerbates these challenges. A clear example was the oil price crash of 2016, from which some GCC countries like Oman and Bahrain are still recovering on the credit rating scale. Fluctuations in global oil prices can destabilize the economies of GCC countries, highlighting the importance of diversifying energy sources and investing in sustainable technologies. Policies promoting economic resilience, such as renewable energy adoption and sustainable agricultural practices, are essential for stabilizing the WEF nexus amid economic uncertainties (Al-Saidi & Saliba, 2019).

1.7.8 Climate Change and Environmental Shocks

Climate change significantly threatens the WEF nexus by exacerbating water scarcity, increasing energy demand, and affecting agricultural productivity. Rising temperatures, changing precipitation patterns, and the increasing frequency of extreme weather events disrupt the availability and quality of water resources, impact crop yields, and place additional stress on energy systems. For the GCC, where water scarcity is already a critical issue, climate change intensifies the need for innovative water management solutions (Hindiyeh et al., 2023). As explained in earlier sections, integrated approaches that consider the interdependencies within the WEF nexus are crucial for developing effective climate adaptation strategies.

1.7.9 Geopolitical Tensions and Supply Chain Disruptions

Geopolitical tensions and conflicts can lead to significant disruptions in the WEF nexus by affecting the availability and distribution of resources. The GCC region, situated in a geopolitically sensitive area, is particularly vulnerable to such disruptions. Conflicts in neighboring regions can impede the flow of goods and resources, leading to shortages and increased prices (Ben Hassen & El Bilali, 2019).

The Russia-Ukraine war and the Suez Canal blockage are recent geopolitical events causing significant disruption. These events have strained global supply chains, resulting in delayed shipments and increased costs for essential goods. The GCC countries, heavily dependent on wheat imports from Russia and Ukraine, experienced severe disruptions in their food supply chains. This disruption not only affected access to raw materials but also had a cascading effect on animal production due to soaring prices of animal feed, exacerbating the food security crisis (Al-Saidi, 2023). Additionally, the blockade of Qatar by neighboring GCC countries highlighted similar vulnerabilities within the WEF nexus, leading to an increased reliance on domestic food production and the development of alternative supply routes. This compounded impact underscores the importance of enhancing local food production resilience and diversifying import sources to mitigate future risks.

Developing strategic reserves and improving infrastructure resilience can help mitigate the impact of geopolitical shocks on the WEF nexus (Derouez & Ifa, 2024; Jägerskog & Barghouti, 2022). In the energy sector, diversifying energy sources and investing in renewable energy can enhance energy security and align with global sustainability goals.

1.7.10 Pandemics and the WEF Nexus

The COVID-19 pandemic has significantly impacted the WEF nexus, highlighting the interconnected nature of these sectors and the vulnerabilities within them. The pandemic increased water usage for hygiene purposes, increasing the energy demand required for water treatment and distribution. Additionally, the surge in medical waste has implications for water pollution and management (Al-Saidi & Hussein, 2021).

The pandemic also disrupted food supply chains, prompting a shift toward local food production and self-sufficiency. This shift necessitates water and energy usage changes, particularly in agricultural practices. The COVID-19 pandemic underscores the need for resilient WEF systems that can adapt to such global health crises and provide essential resources sustainably (Al-Saidi & Hussein, 2021).

1.7.11 Integrating Global Perspectives into Regional Strategies

To effectively address the challenges posed by global shocks, GCC countries must integrate global perspectives into their regional strategies for the WEF nexus. This involves adopting best practices worldwide, leveraging international collaborations, and participating in global initiatives to enhance sustainability and resilience. Aligning regional policies with global frameworks such as the United Nations Sustainable Development Goals (SDGs) and the Paris Agreement on climate change ensures that efforts to address the WEF nexus align with international standards and contribute to global sustainability targets.

Investments in research and development, as well as knowledge exchange with global experts, can drive the development of new technologies and solutions tailored to the unique challenges of the GCC region. International partnerships and collaborative projects can enhance the capacity of GCC countries to respond to global shocks and build resilient WEF systems.

1.7.12 Enhancing the Green Economy and Resilience to Global Shocks

To drive the adoption of sustainable practices within the WEF nexus, GCC countries have implemented a range of policies that provide incentives for innovation. These incentives are designed to encourage the development and deployment of technologies and practices that enhance resource efficiency, promote sustainability, and reduce environmental impacts.

One key area of focus is renewable energy. Financial incentives such as subsidies, tax breaks, and grants are offered to support the development of solar, wind, and other renewable energy projects. For example, the UAE's Energy Strategy 2050 includes significant investments in renewable energy infrastructure and provides financial incentives for adopting clean energy technologies. These incentives aim to reduce reliance on fossil fuels and lower carbon emissions, thereby contributing to the broader goals of energy diversification and sustainability (IRENA, 2024).

In the water sector, incentives are provided for adopting advanced water treatment and conservation technologies. This includes subsidies for wastewater treatment and reuse systems, which help reduce the demand for freshwater resources. Additionally, investments are made in research and development to promote innovative solutions, such as desalination technologies that are more energy-efficient and environmentally friendly. These incentives support the sustainable management of water resources, which is critical in the arid climates of the GCC (Jägerskog & Barghouti, 2022; Siderius et al., 2020).

The food sector also benefits from incentives aimed at promoting sustainable agricultural practices. Financial support is offered for implementing modern agricultural techniques such as hydroponics and aquaponics, which use significantly less water than traditional farming methods. Subsidies and grants are provided to encourage the development of sustainable aquaculture practices, helping to increase local food production and reduce the environmental impact of fishing. By supporting these innovative practices, GCC countries aim to enhance food security while minimizing resource use and environmental degradation (FAO, 2022b).

1.7.13 Linking Nexus Management to Financial Stability

Effective management of the WEF nexus can lead to excellent financial stability, attract sustainable investments, and foster economic growth. According to industry analysts, the potential market for green hydrogen in the GCC could be worth up to $200 billion annually by 2050 (IEA, 2023a). This highlights the significant economic potential of investing in sustainable energy solutions. Investments in renewable energy projects are projected to generate an estimated economic return of $10 trillion globally by 2050, with the GCC aiming to secure a substantial share of this market (IEA, 2023b; IRENA, 2024).

By prioritizing sustainable practices and leveraging economic incentives, GCC countries can enhance their financial stability. For instance, water pricing reforms can create economic incentives for reducing water wastage and promoting efficient water use. Similarly, subsidies for renewable energy can lower the cost of clean energy production, making it more competitive with fossil fuels. Incentives for water-efficient technologies can drive innovation and reduce the overall cost of water management.

Sustainable finance frameworks play a crucial role in supporting these investments. By integrating environmental, social, and governance (ESG) criteria into financial decision-making, GCC countries can attract local and international investors and increasingly prioritize sustainability. This alignment with global sustainability trends can enhance the attractiveness of GCC markets and drive economic growth.

1.7.14 Financing Mechanisms for Resilience and Growth

Investments in sustainable water, energy, and food systems enhance resilience to global shocks and contribute to green growth. Financing mechanisms that support these investments include green bonds, sustainability-linked loans, and climate finance. These instruments can provide the capital for projects that enhance resource efficiency and sustainability within the WEF nexus.

For instance, green bonds can be used to finance renewable energy projects, water treatment facilities, and sustainable agricultural practices. One successful example is the UAE's issuance in 2023 of green bonds and sukuk totaling $7.95 billion (WAM, 2024). These bonds financed projects like the Mohammed bin Rashid Al Maktoum

Solar Park and advanced water treatment facilities, showcasing the potential of green bonds to drive sustainability in the energy and water sectors. This initiative not only supported the UAE's transition to a green economy but also set a benchmark for the region, demonstrating how financial instruments can be leveraged to achieve significant environmental and economic benefits.

Integrating sustainable finance frameworks into national development plans can attract local and international investors. By aligning financial incentives with sustainability goals, GCC countries can foster economic growth while enhancing their resilience to external shocks. This approach not only supports the green economy but also positions the GCC as a leader in sustainable development.

1.7.15 Global Role

The GCC countries have increasingly positioned themselves actively in global sustainable finance initiatives. By aligning their national strategies with international sustainability goals, GCC nations contribute to global efforts to address climate change and promote green growth. Participation in forums such as the United Nations Climate Change Conferences (COP) and collaboration with international organizations underscore the region's commitment to sustainable development.

For instance, the UAE's hosting of the World Expo 2020, focusing on sustainability, and COP 28 highlighted the country's role in promoting green technologies and sustainable practices globally. Similarly, Saudi Arabia's involvement in the G20 Sustainable Finance Working Group demonstrates its leadership in advancing sustainable finance agendas internationally. These initiatives reflect the GCC's proactive stance in fostering global cooperation on sustainability issues and showcase the region's potential as a hub for sustainable investments.

Engaging with international climate funds and development banks also provides additional financing opportunities for WEF nexus projects. By accessing global funding sources, GCC countries can support research, development, and deployment of innovative technologies that enhance resource efficiency and resilience. This contributes to the region's sustainable development and aligns with global efforts to build a resilient and sustainable future.

In conclusion, aligning the WEF nexus with sustainable finance and green growth initiatives is pivotal for the GCC's economic stability and resilience. By leveraging innovative financing mechanisms, fostering public-private partnerships, and actively participating in global sustainability initiatives, the GCC can enhance its resource management practices, attract sustainable investments, and drive long-term economic growth. This comprehensive approach ensures that the region is well-equipped to address future challenges and capitalize on emerging opportunities in the green economy.

1.8 Conclusion and Recommendations

Throughout this chapter, we have explored the critical importance of managing the WEF nexus in the GCC region to enhance resources security and self-sufficiency, foster economic growth, and ensure resilience against global shocks. The integration of sustainable practices within the WEF nexus, such as investments in renewable energy, advanced water treatment technologies, and sustainable agricultural practices, plays a vital role in promoting economic diversification and environmental sustainability. The potential market for alternative renewable energy highlights the significant economic opportunities associated with sustainable energy investments. By prioritizing renewable energy projects, the GCC can leverage its abundant natural resources to reduce carbon emissions, attract substantial foreign investment, and create new economic opportunities. Similarly, adopting modern agricultural techniques and water-efficient technologies can enhance food security while minimizing environmental impacts and securing the scarcest resource in the region: water.

Effective WEF nexus management also supports sustainable finance and green growth goals. The adoption of ESG criteria in investment practices, the issuance of green bonds, and the establishment of public-private partnerships are essential mechanisms for financing sustainable projects. These initiatives not only attract local and international investors but also align the GCC's economic strategies with global sustainability trends. The chapter also addressed the impacts of global shocks such as the COVID-19 pandemic, the Russia-Ukraine war, and climate change on the WEF nexus. These events underscore the vulnerability of the GCC to external disruptions and highlight the need for resilient resource management strategies. By investing in local food production systems, diversifying energy sources, and implementing adaptive water management practices, the GCC can enhance its resilience and ensure long-term sustainability.

1.9 Future Outlook on Sustainable Development

The future outlook for sustainable development in the GCC hinges on the continued implementation of integrated resource management policies and the adoption of innovative technologies. The region's commitment to ambitious renewable energy targets and the development of sustainable agricultural practices will be crucial in enhancing resilience against global shocks. The ongoing shift toward local food production and the diversification of energy sources are positive trends that need to be accelerated. Additionally, the integration of climate policies with WEF nexus strategies will play a vital role in achieving long-term sustainability and mitigating the impacts of climate change.

1.10 Call to Action for Policymakers and Stakeholders

To ensure the GCC region's sustainable future, policymakers and stakeholders must take decisive action in several key areas:

1. Enhance integrated resource management: Develop and implement policies that promote the coordinated management of water, energy, and food resources. This includes investing in advanced technologies for desalination, wastewater treatment, and sustainable agriculture.
2. Promote renewable energy: Accelerate the transition to renewable energy sources by providing financial incentives and regulatory support for solar, wind, and other clean energy projects. Enhance energy efficiency across all sectors to reduce reliance on fossil fuels.
3. Strengthen food security: Increase local food production through innovative agricultural practices such as hydroponics, aquaponics, and sustainable aquaculture. Develop strategic food reserves and diversify import sources to mitigate the risks of global supply chain disruptions.
4. Foster regional collaboration: Encourage regional cooperation on food security, water management, and renewable energy projects. Share best practices and leverage collective resources to enhance resilience and sustainability across the GCC.
5. Invest in research and development: Support research and development initiatives that focus on sustainable technologies and practices. Collaborate with international experts and institutions to drive innovation and address the unique challenges of the GCC region.
6. Implement climate adaptation strategies: Integrate climate adaptation measures into national development plans, focusing on reducing vulnerability to extreme weather events and other climate-related risks. Align these strategies with global sustainability frameworks such as the United Nations Sustainable Development Goals (SDGs) and the Paris Agreement.

References

Abbas, F., Al-Naemi, S., Farooque, A. A., & Phillips, M. (2023). A review on the water dimensions, security, and governance for two distinct regions. *Water, 15*(1), 208. https://doi.org/10.3390/w15010208

Abulibdeh, A., Zaidan, E., & Al-Saidi, M. (2019). Development drivers of the water-energy-food nexus in the Gulf cooperation council region. *Development in Practice, 29*(5), 582–593. https://doi.org/10.1080/09614524.2019.1602109

Ahmed, A. E., Al-Khayri, J. M., & Elbushra, A. A. (2024). *Food and nutrition security in the Kingdom of Saudi Arabia, volume 2: Macroeconomic policy and its implication on food and nutrition security*. Springer International Publishing. https://doi.org/10.1007/978-3-031-46704-2

Ajjur, S. B., & Al-Ghamdi, S. G. (2022). Towards sustainable energy, water and food security in Qatar under climate change and anthropogenic stresses. *Energy Reports, 8*, 514–518. https://doi.org/10.1016/j.egyr.2022.02.099

Al Rashed, M., Sefelnasr, A., Sherif, M., Murad, A., Alshamsi, D., Aliewi, A., & Ebraheem, A. A. (2023). Novel concept for water scarcity quantification considering nonconventional and virtual water resources in arid countries: Application in gulf cooperation council countries. *Science of the Total Environment, 882*, 163473. https://doi.org/10.1016/j.scitotenv.2023.163473

Al-Adwani, A., Karnib, A., Elsadek, A., & Al-Zubari, W. (2024). Scenario-based assessment of the water-energy-food nexus in Kuwait: Insights for effective resource management. *Computational Water, Energy, and Environmental Engineering, 13*(1), 38–57. https://doi.org/10.4236/cweee.2024.131003

Allouche, J. (2024). Nexus framing of sustainability issues: Feasibility, synergies, and trade-offs in terms of water-energy-food. *Annual Review of Environment and Resources*. https://doi.org/10.1146/annurev-environ-112321-112445

Al-Saidi, M. (2023). Caught off guard and beaten: The Ukraine war and food security in the Middle East. *Frontiers in Nutrition, 10*, 983346. https://doi.org/10.3389/fnut.2023.983346

Al-Saidi, M., & Hussein, H. (2021). The water-energy-food nexus and COVID-19: Towards a systematization of impacts and responses. *Science of the Total Environment, 779*, 146529. https://doi.org/10.1016/j.scitotenv.2021.146529

Al-Saidi, M., & Saliba, S. (2019). Water, energy and food supply security in the gulf cooperation council (GCC) countries—A risk perspective. *Water, 11*(3), 455. https://doi.org/10.3390/w11030455

Al-Sarihi, A., & Mansouri, N. (2022). Renewable energy development in the Gulf cooperation council countries: Status, barriers, and policy options. *Energies, 15*, 1923. https://doi.org/10.3390/en15051923

Amer, K., Adeel, Z., Böer, B., & Saleh, W. (2017). *The water, energy, and food security nexus in the Arab region*. Springer International Publishing. https://doi.org/10.1007/978-3-319-48408-2

Baker, J., Van Houtven, G., Cai, Y., Moreda, F., Wade, C., Henry, C., Redmon, J., & Kondash, A. J. (2021). *A hydro-economic methodology for the food-energy-water nexus: Valuation and optimization of water resources*. RTI International. https://doi.org/10.3768/rtipress.2021.mr.0044.2105

Bazilian, M., Rogner, H., Howells, M., Hermann, S., Arent, D., Gielen, D., Steduto, P., Mueller, A., Komor, P., Tol, R. S. J., & Yumkella, K. K. (2011). Considering the energy, water and food nexus: Towards an integrated modelling approach. *Energy Policy, 39*(12), 7896–7906. https://doi.org/10.1016/j.enpol.2011.09.039

Ben Hassen, T., & El Bilali, H. (2019). Food security in the gulf cooperation council countries: Challenges and prospects. *Journal of Food Security, 7*(5), 159–169. https://doi.org/10.12691/jfs-7-5-2

Borgomeo, E., Jagerskog, A., Talbi, A., Wijnen, M., Hejazi, M., & Miralles-Wilhelm, F. (2018). *The water-energy-food nexus in the middle east and north Africa*. World Bank.

Cairns, R., & Krzywoszynska, A. (2016). Anatomy of a buzzword: The emergence of 'the water-energy-food nexus' in UK natural resource debates. *Environmental Science & Policy, 64*, 164–170. https://doi.org/10.1016/j.envsci.2016.07.007

Derouez, F., & Ifa, A. (2024). Sustainable food security: Balancing desalination, climate change, and population growth in five Arab countries using ARDL and VECM. *Sustainability, 16*(6), 2302. https://doi.org/10.3390/su16062302

Elzaki, R. M. (2024). Does fish production influence the GDP and food security in gulf cooperation council countries? Evidence from the dynamic panel data analysis. *Aquaculture, 578*, 740058. https://doi.org/10.1016/j.aquaculture.2023.740058

FAO. (2022a). *The state of food security and nutrition in the world 2022*. FAO. https://doi.org/10.4060/cc0639en

FAO. (2022b). *The state of world fisheries and aquaculture 2022*. Food Agriculture Organization. https://doi.org/10.4060/cc0461en

Hameed, M., Moradkhani, H., Ahmadalipour, A., Moftakhari, H., Abbaszadeh, P., & Alipour, A. (2019). A review of the 21st century challenges in the food-energy-water security in the middle east. *Water, 11*(4), 682. https://doi.org/10.3390/w11040682

Herrera-Franco, G., Bravo-Montero, L., Caicedo-Potosí, J., & Carrión-Mero, P. (2024). A sustainability approach between the water–energy–food nexus and clean energy. *Water, 16*(7), 1017. https://doi.org/10.3390/w16071017

Hindiyeh, M., Albatayneh, A., & AlAmawi, R. (2023). Water energy food nexus to tackle future Arab countries water scarcity. *Air, Soil and Water Research, 16*, 117862212311609. https://doi.org/10.1177/11786221231160906

Ibrahim, O. R., Al-Amir, M., & Al-Maghawry, S. (2022). Tracking the damages of the Shaheen cyclone in the Sultanate of Oman. *Water Practice and Technology, 17*(12), 2548–2553. https://doi.org/10.2166/wpt.2022.138

IEA. (2023a). *Renewable hydrogen from Oman: A producer economy in transition*. OECD. https://doi.org/10.1787/d6bc1fc9-en

IEA. (2023b). *World energy outlook 2023—Analysis*. International Energy Agency (IEA). Retrieved from https://www.iea.org/reports/world-energy-outlook-2023

IRENA. (2024). *The Global Atlas for Renewable Energy: A decade in the making*. Retrieved from https://www.irena.org/Publications/2024/Apr/The-Global-Atlas-for-Renewable-Energy-A-decade-in-the-making

Jägerskog, A., & Barghouti, S. (2022). *Advancing knowledge of the water-energy nexus in the GCC countries*. World Bank. https://doi.org/10.1596/38296

Johnson, A. (2024). *Heavy rainfall causes extreme flooding In UAE, Oman, Bahrain. Forbes*. Retrieved from https://www.forbes.com/sites/ariannajohnson/2024/04/17/historic-floods-in-uae-dubai-oman-soaked-by-severe-rains-photos/

Lahlou, F.-Z., AlNouss, A., Govindan, R., Hazrat, B., Mackey, H. R., & Al-Ansari, T. (2023). Water and sludge resource planning for sustainable agriculture: An energy-water-food-waste nexus approach. *Sustainable Production and Consumption, 38*, 130–148. https://doi.org/10.1016/j.spc.2023.03.027

Lazaro, L. L. B., Bellezoni, R. A., Puppim De Oliveira, J. A., Jacobi, P. R., & Giatti, L. L. (2022). Ten years of research on the water-energy-food nexus: An analysis of topics evolution. *Frontiers in Water, 4*, 859891. https://doi.org/10.3389/frwa.2022.859891

Moossa, B., Trivedi, P., Saleem, H., & Zaidi, J. (2022). Desalination in the GCC countries- A review. *Journal of Cleaner Production, 357*, 131717. https://doi.org/10.1016/j.jclepro.2022.131717

Saif, O., Mezher, T., & Arafat, H. A. (2014). Water security in the GCC countries: Challenges and opportunities. *Journal of Environmental Studies and Sciences, 4*(4), 329–346. https://doi.org/10.1007/s13412-014-0178-8

Salberg, F. (2023). *Turning oil into water: Water security in the Middle East and North Africa*. Uppsala University.

Shahin, S. M., & Salem, M. A. (2015). The challenges of water scarcity and the future of food security in The United Arab Emirates (UAE). *Natural Resources and Conservation, 3*(1), 1–6. https://doi.org/10.13189/nrc.2015.030101

Sherif, M., Liaqat, M. U., Baig, F., & Al-Rashed, M. (2023). Water resources availability, sustainability and challenges in the GCC countries: An overview. *Heliyon, 9*(10), e20543. https://doi.org/10.1016/j.heliyon.2023.e20543

Shubbar, H. T., Tahir, F., & Al-Ansari, T. (2024). Bridging Qatar's food demand and self-sufficiency: A system dynamics simulation of the energy–water–food nexus. *Sustainable Production and Consumption, 46*, 382–399. https://doi.org/10.1016/j.spc.2024.02.017

Siderius, C., Conway, D., Yassine, M., Murken, L., Lostis, P.-L., & Dalin, C. (2020). Multi-scale analysis of the water-energy-food nexus in the Gulf region. *Environmental Research Letters, 15*(9), 094024. https://doi.org/10.1088/1748-9326/ab8a86

Syed, F. S., Adnan, S., Zamreeq, A., & Ghulam, A. (2022). Identification of droughts over Saudi Arabia and global teleconnections. *Natural Hazards: Journal of the International Society for the Prevention and Mitigation of Natural Hazards, 112*(3), 2717–2737.

Times News Service. (2021). *Cyclone Shaheen damages estimated at OMR 200 million | times of Oman—Times of Oman.* Times of Oman. https://timesofoman.com/article/110437-cyclone-shaheen-damages-estimated-at-omr-200-million

WAM. (2024). *UAE tops regional green bond league tables with $10.7 billion.* Emirates News Agency. https://wam.ae/en/article/b1bvln9-uae-tops-regional-green-bond-league-tables-with

Wichelns, D. (2017). The water-energy-food nexus: Is the increasing attention warranted, from either a research or policy perspective? *Environmental Science & Policy, 69*, 113–123. https://doi.org/10.1016/j.envsci.2016.12.018

WRI. (2024). *Aqueduct water risk atlas [map].* Retrieved from https://www.wri.org/applications/aqueduct/water-risk-atlas/#/?advanced=false&basemap=hydro&indicator=w_awr_def_tot_cat&lat=-14.445396942837744&lng=-142.85354599620152&mapMode=view&month=1&opacity=0.5&ponderation=DEF&predefined=false&projection=absolute&scenario=optimistic&scope=baseline&timeScale=annual&year=baseline&zoom=2.

Chapter 2
The Role of the Red Sea Coral Reefs in the Development of a Sustainable and Nature-Positive Economy in the GCC Region

Nathalie Hilmi, Aurélien Calas, and Alain Safa

Abstract Globally, coral reefs harbor an impressive one-quarter of marine biodiversity despite occupying merely 1% of the ocean. However, climate-change-induced impacts such as sea surface temperature anomalies and ocean acidification have brought reef-building corals close to or beyond their bleaching threshold worldwide. Remarkably, due to the Red Sea's unique oceanographical features and historical background, its coral reefs emerge as exceptionally resilient. In a landscape where 70–90% of global reefs face critical damage by 2050, even with adherence to the 1.5 °C goals of the Paris Agreement, the Red Sea's highly tolerant holobionts stand out as crucial actors for the survival of marine life, demanding urgent protection. Nonetheless, the burgeoning interest in "blue" economic sectors along the Red Sea's periphery has led to an increase in coastal development and economic activities, notably coastal tourism, the fishing industry, and shipping. These endeavors, often mismanaged, pose local threats to coral reefs as associated stressors such as overfishing, pollution, and mining bring these organisms closer to their bleaching threshold, accelerating and exacerbating the impact of climate change.

Coral reefs deliver primordial ecosystem services to the Red Sea region and its societies, with particular emphasis on fisheries support, coastal and maritime tourism, coastal protection, and emerging biotechnology, all of which are economically and socially vital for the region, but also iconic values underscoring their cultural importance. With its distinctive reef assemblages, the Red Sea stands as one of the richest aquatic environments in the world. Its diverse reef morphologies host a range of

N. Hilmi (✉) · A. Calas
Department of Environmental Economics, Centre Scientifique de Monaco, Monaco, Monaco
e-mail: hilmi@centrescientifique.mc

A. Calas
e-mail: aurelien.calas@imbrsea.eu

A. Safa
GRM, IAE, Université Côte d'Azur, Nice, France
e-mail: alain.safa@univ-cotedazur.fr

H. Miniaoui and M. Bellalah (eds.), *Green Growth Opportunities and Sustainable Finance in the Gulf Cooperation Council Region*, Gulf Studies 24,
https://doi.org/10.1007/978-981-96-8287-4_2

commercially significant species, supporting a substantial portion of the commercial fishing industry. Additionally, the region is a premier global tourist destination, with tourism emerging as a primary source of income for numerous countries in the vicinity. The Red Sea's extensive coral reef systems have been luring tourists for decades, while its shoreline, characterized by fringing and barrier reefs, furnishes natural coastal protection and stabilizes sediment structures, mitigating erosion and safeguarding human natural assets. In addition to their ecological and economic importance, Red Sea coral reefs also hold significant cultural value, especially by playing a pivotal role in efforts to prevent the global collapse of coral reefs, thanks to their remarkable tolerance. This iconic value highlights their intangible benefits, emphasizing their role as a major cultural asset.

This chapter will commence by elucidating the ecological status of Red Sea coral reefs. Subsequently, it will underscore the urgent need for their protection through an ecosystem services approach, highlighting how coral reefs constitute a significant component of marine natural capital and play a vital role in the Gulf Cooperation Council (GCC) economy.

Nonetheless, these ecosystems face multifaceted threats from climate change, overfishing, and pollution, necessitating concerted efforts to safeguard their ecological and economic benefits. Finally, the chapter will run through tools and policy recommendations aimed at coral reef protection, underscoring their contribution to realizing the objectives of the Paris Agreement and the United Nations Sustainable Development Goals, thereby fostering sustainable and nature-positive development for the Red Sea shore of Saudi Arabia.

2.1 Introduction

Globally, coral reefs harbor an impressive one-quarter of marine biodiversity despite occupying merely 1% of the ocean. Climate-change-induced impacts (sea surface temperature anomalies, ocean acidification…) have pushed reef-building corals worldwide close to or beyond their bleaching threshold (Fine et al., 2019; Hughes et al., 2007; Savary et al., 2021). Due to the Red Sea's particular oceanographical features and historical background, Red Sea coral reefs appear to be the most resilient in a global context where 70–90% of reefs are expected to be critically damaged by 2050 even if the 1.5 °C goal of the Paris agreement is achieved (Kleinhaus et al., 2020). After the last Ice Age, recolonization of the Red Sea from the south required crossing maritime zones with particularly high temperatures. This natural selection process favored heat-resistant coral larvae. This historical adaptation is believed to be the key to the remarkable thermal tolerance of today's Red Sea corals, which have a significantly higher bleaching threshold than the norm (Kleinhaus et al., 2020).

However, the growing interest in "blue" economic sectors on the coast, especially coastal tourism, fishing, and shipping, still threatens coral reefs locally as the associated stressors (overfishing, pollution, mining…) bring the organisms closer to their bleaching threshold (Kleinhaus et al., 2020). The first modern human occupation of the shores of the Red Sea dates back to the Middle Stone Age when our ancestors began to exploit marine food resources (Walter et al., 2000). Around the Red Sea basin, coastal populations were initially scarce due to local drought and harsh climatic conditions (Fine et al., 2019). By offering natural resources and essential trade routes, the Red Sea has shaped human dispersal in the African continent, enabled cultural interaction, and contributed to societies' progress (Cziesielski et al., 2021). The Red Sea has been attracting countless merchants, tourists, and investors for thousands of years, thanks to its crucial location, magnificent landscapes, and rich marine resources. The region's development and demographic landscape shifted due to the expansion of maritime transportation, notably after the inauguration of the Suez Canal in 1869 (Fine et al., 2019).

The Red Sea is one of the richest aquatic environments thanks to its unique reef assemblages. Its diversified reef morphologies host a range of species of commercial interest and support a significant part of the commercial fishing industry (Bruckner et al., 2009; Kleinhaus et al., 2020; Marshall et al., 2010). The Red Sea is also one of the major tourist destinations worldwide, and the tourism industry is one of the main sources of income for several countries in the region (Carvalho et al., 2019). The Red Sea is known to contain one of the most extensive coral reef systems in the world, containing not less than c.a. 3.8% of the world's coral reefs, which have attracted tourists to the area for decades (Carvalho et al., 2019; Fine et al., 2019; Rasul et al., 2015). The Red Sea shoreline is characterized by fringing and barrier reefs (Langodan et al., 2020), providing natural coastal protection and stabilizing sediment structures (Hilmi et al., 2018). These reef-building structures prevent erosion and protect human natural assets (Frihy et al., 2004; Hilmi et al., 2018; Pascal et al., 2016).

Ecosystem services are the essential benefits natural ecosystems provide to human societies, supporting survival, well-being, and economic activities. These services encompass four main categories: provisioning services like food, water and bioactive compounds; regulating services such as climate regulation, shoreline protection and disease control; cultural services including recreational, aesthetic, and spiritual benefits; and supporting services like nutrient cycling, habitat provision and soil formation. By maintaining biodiversity and ecosystem health, these services ensure the resilience and sustainability of both natural environments and human communities. The coral reefs of the Red Sea are therefore fundamental ecosystems that provide numerous ecosystem services. The objective is to highlight the critical need to assess the services they offer to ensure their preservation in the face of increasing threats and to maintain these vital functions. Economic assessments value the global benefits of coral reefs at approximately US$30 billion annually, underlining the urgency of conservation efforts against the backdrop of climate change and unsustainable practices, and pointing to the necessity for prompt action.

Human-induced direct disturbances to coral reefs jeopardize these significant economic benefits and threaten the livelihoods of 28 million people throughout the region. The implementation of existing solutions is hindered by a glaring lack of funding, due to the gap between available public funds and the investment required for their preservation. Protecting the reefs is essential to mitigate the effects of human pressures, notably through the development of nature-based solutions (NBSs), the creation of marine protected areas (MPAs), and the establishment of coral conservatories. To rehabilitate deteriorated coral reefs, it is vital to adopt strategies such as assisted evolution, aquaculture of resistant coral varieties found in the Red Sea, and transplantation—crucial approaches for the long-term survival of these precious ecosystems. These strategies must encompass reef health monitoring, awareness-raising, the use of socio-economic tools, and the promotion of sustainable practices within a viable blue economy. Such initiatives require robust transnational and multi-disciplinary cooperation, concerted action against global warming, and, above all, additional policy support and financial input. The valuation of ecosystem services is a pivotal step in quantifying the economic benefits provided by coral reefs. This is crucial for influencing policymakers and for the potential export of these benefits to financial markets, such as through biodiversity credits. Such initiatives aim to attract private investment by showcasing the tangible economic value of preserving biodiversity. This is essential in order to implement existing solutions and ensure the continued provision of vital ecosystem services for present and future generations.

The GCC's Red Sea shore has the particularity to exhibit a relatively important proportion of pristine and healthy coral reefs (Chalastani et al., 2020; Hilmi et al., 2020), representing an abundant "blue natural capital" resource with strong potential to support a local blue economy. Its coastline is the longest of the basin and hosts spectacular marine life. "Blue natural capital" refers to the valuable marine natural resources that contribute significantly to the economic and environmental well-being of the region. The potential for sustainable ocean-based economic development, also called blue growth, is very promising on the Kingdom of Saudi Arabia's western coastline, considering these major assets. In line with Saudi Vision 2030, the Kingdom needs to diversify its economy toward a post-oil era, and developing a sustainable blue economy represents a major pillar of this strategy (Hilmi et al., 2020). Saudi Vision 2030 is a strategic framework aimed at diversifying Saudi Arabia's economy and reducing its dependence on oil. It focuses on three pillars: a vibrant society, a thriving economy, and an ambitious nation. Central to the vision is the protection of the environment and the sustainable use of natural resources. Initiatives include promoting renewable energy, implementing sustainable agricultural practices, and enhancing conservation efforts. The framework emphasizes environmental stewardship to ensure long-term sustainability and improved quality of life for future generations. The GCC countries have the potential to become the pioneering nations in the Middle East and North Africa region to showcase this novel economic paradigm. The GCC also has the opportunity to be a catalyst for a transformative era to show other nations the advantages and benefits of a nature-positive economy.

2.2 Sustaining Red Sea Coral Reefs: Guardians of Ecosystem Services in a Changing World

2.2.1 From Nature to Human Benefits

In this section, the aim is to identify the ecosystem goods and services provided by the Red Sea coral reefs and the importance of maintaining their delivery to humans. Productive and healthy ecosystems generate a flow of services, resulting in goods that are either for consumption or crucial for human well-being or survival (Turner, 2015). Ecosystem services benefits are therefore the "export" from the ecosystem sector to the human economic and social sectors (Boyd & Banzhaf, 2005; Turner, 2015). Coral reefs are considered a form of national capital and their total economic value stems from all the goods and services delivered by this ecosystem (Hilmi et al., 2018). Ecosystem services identification and economic valuation are becoming crucial to properly managing and protecting coral reefs (Pascal et al., 2016). Coral reefs play pivotal roles as biodiversity havens, natural shields against hydrodynamic forces, vital economic assets through tourism and fisheries, a promising element for the future of biotechnology, and a crucial component of human cultural values—among numerous other functions. However, rapid coral decline imperils these services, impacting over half a billion people reliant on reefs for sustenance, income, and coastal protection (Kleinhaus et al., 2020). Goods and services provided by global coral reefs, including coastal protection, building materials, fisheries, and tourism industries, among others, have been evaluated at c.a. US$30 billion yearly (Chen et al., 2015; Moberg & Folke, 1999). Protection and conservation of key ecosystems such as coral reefs are therefore imperative in the context of climate change and unsustainable practices to maintain the delivery of such benefits and avoid economic damage to crucial industries (Burke & Spalding, 2004).

2.2.2 Fisheries

The fishing industry provides massive economic backing globally and supports approximately 260 million jobs around the globe (Wahlig & Sommer, 2021). However, the peak of 86.4 million tons of marine catch reached in 1996 is unlikely to happen again, as 34.2% of global fish stocks are overfished (Jolliffe et al., 2021).

The Red Sea stands out as one of the most diverse aquatic environments, characterized by its distinctive reef formations and varied reef structures that support various commercially valuable species (Bruckner et al., 2009; Kleinhaus et al., 2020; Marshall et al., 2010). The coral reefs of the Red Sea play a crucial role in the region's commercial fishing sector. More importantly, they have supported artisanal and subsistence fishing activities sustaining the basic needs of numerous local people for thousands of years (Marshall et al., 2010). However, rising demand due to the rapidly growing population and higher trade in coral reef species has led to overexploitation and the consequent depletion of many targeted valuable species (Bruckner et al., 2009). In other words, coral reefs provide habitat to many high-valued species and overfishing is damaging them and bringing them closer to their bleaching threshold, threatening the fisheries industry itself (Fine et al., 2019; Hughes et al., 2007; Savary et al., 2021). The fisheries industry is considered one of the most important economic activities relying on the coral reefs in the Red Sea (Fine et al., 2019). Thanks to these precious ecosystems, fisheries landed value reaches c.a. US$230 million yearly in the region (Kleinhaus et al., 2020).

Saudi Arabia, as the largest country bordering the Red Sea, encompasses the majority of the eastern Red Sea basin. While large-scale commercial fishing is limited in this region, the overfishing crisis still affects Saudi Arabia. Non-compliance with regulations is a widespread issue seen in many tropical and reef-associated fisheries. In Saudi Arabia, fisheries support from local coral reefs represents the main ecosystem service delivered by these ecosystems, as reef fisheries account for about three-quarters of total fishery landings in the country (Shellem, 2020). Commercial exploitation of coral reefs used to be limited to small artisanal fisheries because of the Red Sea's topography but the development of new destructive gear types fosters much larger-scale commercialization (Bruckner et al., 2009). The destruction of such habitats and the extensive targeting of species of interest have catastrophic effects on trophic structure and ecological function (Bruckner et al., 2009), putting whole coral reef ecosystems in danger as well as people depending on reef fisheries as an ecosystem service.

2.2.3 Coastal and Maritime Tourism

Coastal and maritime tourism is the fastest-growing sector within the global tourism industry, the latter being the world's largest industry. Globally, ocean revenues include some US $161 billion annually from coastal and marine tourism (Ababouch & Fipi, 2015). This sector has evolved significantly and has brought new opportunities since the beginning of the century, especially with the implementation of marine strategy and the increasing popularity of blue ecotourism (Guo et al., 2020).

The Red Sea ranks as one of the leading global tourist destinations, with tourism serving as a primary source of revenue for several nations within the region (Carvalho et al., 2019). Annually, tourism in the Red Sea basin generates approximately US$12 billion (Kleinhaus et al., 2020). These reefs have been a major attraction for tourists for many years (Carvalho et al., 2019; Fine et al., 2019; Rasul et al., 2015). For the Red Sea, tourism represents the main generator of ecosystem services flow with an 85% contribution, followed by fisheries with 11% (Hilmi et al., 2018). The region is gifted with a pleasant climate, rich and colorful marine life, unrivaled coral reef diversity, and calm and transparent waters with important light penetration (Romaniv & Yarmolyuk, 2021). All these favorable conditions make the Red Sea one of the best diving spots in the world (Hilmi et al., 2018). Compared to other coral reefs experiencing massive bleaching events around the world, the current exceptional condition of Red Sea reefs, especially in the northern part, remains a weighty argument for tourism attractiveness (Amer et al., 2021). Reef-based tourism will therefore continue to rise and local coral reefs will certainly continue to grow as a major economic asset (Kleinhaus et al., 2020). However, the entire reef-tourism ecosystem service is at risk, as the carrying capacity of the marine system is believed to have been surpassed due to pressures from coastal development, including construction, various forms of pollution, and population density (Carvalho et al., 2019; Hasler & Ott, 2008; Zakai & Chadwick-Furman, 2002). Significant stressors occur very locally, such as boat anchoring, toxic sunscreen penetrating coral tissues, and direct interactions with divers, who sometimes break off pieces of coral to keep as a souvenir (Carvalho et al., 2019), increasing the reefs' sensitivity to bleaching. When corals bleach, a whole ecosystem is in critical danger. Bleached corals are not attractive anymore, neither to marine organisms nor to humans, and the whole ecosystem, as well as the associated industry, becomes severely threatened.

While tourism development in GCC countries has been relatively limited, the region's growing appeal has triggered extensive coastal development, featuring major tourist facilities, particularly in cities such as Jeddah, Saudi Arabia. Before 2019, international tourists faced significant barriers to visiting Saudi Arabia, as visas were usually only granted for employment, education, religious pilgrimage, and business (Robitzch et al., 2023). Red Sea marine tourism in Saudi Arabia was mainly concentrated in the coastal city of Jeddah. The objective of the Saudi Vision 2030 plan is not only to foster a vibrant society but also to enhance the Kingdom's tourism sector as an aspect of diversifying its economy. Currently contributing only 3% to the overall economy, tourism development is viewed as a substantial opportunity to generate employment, boost economic activity, and decrease Saudi Arabia's reliance on revenues from hydrocarbons (Chalastani et al., 2020). In pursuit of this vision, numerous giga-projects, including three along the Red Sea (The Red Sea Project, AMAALA, NEOM), are underway that will revolutionize the national coastal tourism sector. These initiatives aspire to create cutting-edge, ultra-luxurious tourist destinations and economic centers while fostering environmental protection and good practices toward net-zero carbon emissions. The Red Sea Project, for example, anticipates the generation of 70,000 new jobs, making a yearly contribution of over 22 billion SAR to the national economy (Robitzch et al., 2023). The project aims not only to achieve a net positive impact but also to enhance the conservation value of the site by 30% (Chalastani et al., 2020). Conversely, extensive tourism projects along the coast entail the landscape modifications and alterations associated with significant coastal development, posing potential threats to marine life. Reports have already highlighted the impact on the Saudi Arabian coast (Aljahdali & Alhassan, 2020). Coastal and maritime tourism is nonetheless considered a very promising industry in the area given the length of the coastline (the longest on the Red Sea) and the spectacular diversity and attractiveness of its coral reefs (Fine et al., 2019). Valuation of coastal tourism as a coral reef ecosystem service would represent a crucial step toward the development of a sustainable local tourism economy.

2.2.4 Coastal Protection

Coastal areas are among the most dynamic natural environments. Seaboards are subject to major wind, hydrodynamic, and stochastic climate events. The increase in frequency and intensification of these dynamic processes, combined with sea level rise, can lead to coastal erosion and accretion occurring more and more worldwide because of climate change (Gracia et al., 2018). Coastal development increases the risk of erosion and can reduce habitats for native ecosystems that play a vital role in the prevention of sediment structure weakening. As a response to such issues, coastal management plans often include costly and heavily engineered structures such as seawalls and breakwaters which have already raised environmental concerns linked to their negative impacts on the marine environment (Gracia et al., 2018).

The Red Sea shoreline is characterized by fringing and barrier reefs (Langodan et al., 2020) which provide natural coastal protection and stabilization of sediment structures (Hilmi et al., 2018). Their disappearance may therefore induce net coastal erosion (Hilmi et al., 2018). Protection of human assets against beach erosion is a major ecosystem service provided by coral reefs (Pascal et al., 2016). Red Sea reefs, thanks to their physical features, also protect the associated recreational activities behind them (Frihy et al., 2004). Although shoreline erosion is minor in the Red Sea because of few coarse sands and relatively weak currents, coral reefs act as submerged breakwaters (Frihy et al., 2004). Local reefs dissipate 40–50% of wave energy, preventing any coastal erosion (Frihy et al., 2004). Wave energy is one of the main stressors to shoreline structure, as it increases sediment transport to the deep and fosters erosion. Maintaining Saudi Arabia's coral reefs in good health and considering them as a nature-based management solution should prevent local coastal hazards and avoid the substantial expenditure and environmental damage implied by beach replenishment and the installation of engineered structures (Menn et al., 2003).

2.2.5 Blue Biotechnology

Coral reefs are very complex supraorganisms given their indispensable symbiotic relationship with *Symbiodiniceae* microalgae and the important density and diversity of the bacterial fraction of their microbiota (van de Water et al., 2022). This complexity enhances their potential as ecosystem service providers in blue biotechnology, a field that explores marine organisms for industrial and commercial uses. Corals and their associated microbiomes are a source of crucial pharmaceutical compounds, fostering nature-based drug discovery (van de Water et al., 2022). Additionally, their intricate 3D skeleton structures, which optimize light absorption and prevent shading, inspire human applications such as bioproducts, algae reactors, biofuels, and wastewater treatment (through bionic 3D-printed corals) (Hossain et al., 2019).

The Red Sea is characterized by historical physical-biochemical changes (i.e., periods of drought, hypersalinity, and isolation) (DiBattista et al., 2016) and the consequent unique assemblages of its coral reefs. The Red Sea is an oligotrophic environment with exceptionally high salinity, UV radiation, and temperatures. These remarkable conditions have shaped the evolution of microbial communities of local corals and are believed to have fostered the acquisition of specific environmental adaptations (Behzad et al., 2016; El Samak et al., 2018). Red Sea coral reef systems enjoy exceptional microbial diversity with unique primary and secondary metabolites (El Samak et al., 2018). The special interest of the Red Sea's coral holobionts for blue biotechnologies comes from the unique molecules and metabolites linked to their high endemism and adaptation capacity to environmental changes (Kleinhaus et al., 2020). Such compounds and marine bioactive products could be crucial for finding new medicines to cure cancer, viral infections, and many further diseases (El Samak et al., 2018; Kleinhaus et al., 2020).

2.2.6 *Iconic Value*

Beyond their ecological and economic roles, coral reefs also hold iconic value as symbols of natural heritage and cultural identity (O'Mahoney et al., 2017). This intangible dimension complements their ecosystem services by reinforcing societal support for reef conservation through non-use values such as existence, bequest, and moral responsibility which denote the non-monetary advantages bestowed by an ecosystem. The iconic value underscores the cultural significance of Saudi Red Sea coral reefs on the national and global scale. The Red Sea holds significance as a climate change refuge for coral reefs, exhibiting remarkable resilience and vitality compared to coral reefs worldwide that are experiencing severe bleaching. Consequently, the Red Sea plays a pivotal role in efforts to prevent a global collapse of coral reefs, contributing iconic value to the Red Sea coral reef ecosystem.

2.2.7 *Crucial Ecosystem Services to Safeguard*

The Red Sea has therefore unique coastal biodiversity, responsible for the delivery of ecosystem services crucial to economic growth and which support human well-being through diverse applications (Hilmi et al., 2018). To safeguard these ecosystem services of major importance and ensure their sustainability, it is now a priority to protect them effectively. Evaluating the economic benefits of these services is a crucial first step in providing evidence of their importance to policymakers, promoting conservation efforts, and considering their potential for integration into financial markets. While this chapter highlights some key ecosystem services provided by coral reefs, it is important to recognize that these represent only a subset of the many ecological, economic, and sociocultural functions coral reefs support.

2.2.8 Ecosystem Services Valuation

The evaluation of ecosystem services provided by the Red Sea coral reefs in the Kingdom of Saudi Arabia is a complex yet critical endeavor necessary for ensuring sustainable management, conservation, and progress toward a nature-positive economy. In this section, we present examples of monetary valuation methods that could be applied to assess ecosystem services in the Red Sea context. Among the many services coral reefs provide, we illustrate potential primary approaches using four ecosystem services relevant to this case: fisheries, coastal and maritime tourism, coastal protection, and iconic value.

For evaluating fisheries support, the direct market pricing method is relevant (Brander et al., 2024), focusing on key market sales data related to reef-associated species and operational costs, including salaries, taxes, and fees. This approach requires data on fishing statistics, fishing effort, and the fish residence index. This provides a comprehensive view of the economic contribution of reef fisheries to the local economy. It is necessary to address the sustainability of fishing practices by distinguishing between sustainable catch volumes and overfishing impacts. The valuation of tourism and recreation services could rely on both revealed preference methods (Brander, 2023; Ncaves & Maia, 2022) and stated preference methods (Mukanjari et al., 2022). Surrogate market methods, such as the travel cost method (which estimates the recreational value of a site based on visitors' travel expenses) and hedonic pricing (which infers environmental value from variations in property prices or hotel rates linked to proximity to coral reefs in this case), utilize existing market data to estimate value indirectly. Hypothetical market approaches, including contingent valuation (which elicits individuals' willingness to pay for specific environmental attributes or preservation scenarios) and choice modeling (which assesses preferences across different hypothetical options), construct scenarios to gauge perceived value in the absence of market prices. Some of these approaches require detailed investment data on current and future tourism and proprieties projects and their reef-associated components. Key variables such as room rates, occupancy rates, and other service-related costs (reef recreation and MPAs data) are decisive to help understand how the proximity of coral reefs influences tourism economics, offering insights into the willingness to pay for reef-related experiences. To estimate the value of coral reefs in natural coastal protection, methods like replacement cost and avoided damage cost can be applied. The avoided damage method estimates the value of ecosystem services by assessing the cost of damages that would occur in the absence of the coral reefs (e.g., flood, beach erosion) by isolating the relative amount of wave energy dissipation by coral reefs. The replacement cost method, on the other hand, approximates value based on the cost of man-made alternatives that would provide similar protection (e.g., artificial dikes). Both approaches are commonly used to value coastal protection services provided by coral reefs although replacement cost method often yields relatively high estimates and should therefore be applied with caution (Pascal et al., 2016). These methods necessitate gathering extensive data on reefs, hydrodynamics, historical damage, coastline vulnerability, erosion rate, engineered protection structures, and assets potentially at risk. The iconic value of coral reefs is best assessed through the contingent valuation method or the choice modeling method (stated preference) (O'Mahoney et al., 2017), which involve designing and administering local surveys to determine public willingness to pay for reef conservation. This highlights the cultural and non-economic benefits of coral reefs that resonate with both locals and tourists.

Managing uncertainties is an integral part of ecosystem service valuation. Challenges such as incomplete data, changing environmental conditions, and economic fluctuations necessitate robust gap analysis. This process is crucial as it identifies data deficiencies and outlines strategies to mitigate these through continuous data improvement. When primary data is unavailable and the planned method is not applicable, an ecosystem service valuation could integrate benefit transfer methods, ensuring first that valuations are based on the best possible data. Overall, this structured evaluation process underscores the multifaceted benefits of coral reefs and supports the formulation of targeted conservation strategies. By systematically addressing each service's economic and ecological contributions and acknowledging inherent uncertainties, this approach facilitates informed decision-making for the sustainable use and protection of the Red Sea's coral reef ecosystems and the potential export of this natural capital to financial markets.

2.3 Potential Solutions to Protect and Restore Red Sea Coral Reefs

The valuation of ecosystem services is crucial for demonstrating the economic benefits of biodiversity conservation to policymakers and attracting private investment. It supports the implementation of existing solutions to conserve biodiversity and ensures the sustained provision of vital ecosystem services for current and future generations. This section outlines various strategies to protect and restore coral reefs in the GCC. Given that Saudi Arabia has the longest coastline and is the only GCC country bordering the Red Sea, we emphasize these tools specifically for Saudi Arabia.

2.3.1 Local Context

Considering the variation in water temperature across the latitudinal gradient of the Red Sea is crucial due to its impact on coral resilience. This is particularly important given the extensive length of the Saudi Arabian coastline, which at 1760 km comprises nearly 80% of the eastern shoreline. Management and solutions must be tailored accordingly. While the southernmost reefs face ongoing threats from bleaching, climate change is not expected to cause significant coral mortality in the northern Red Sea for the coming decades, according to IPCC scenarios, especially in the northernmost regions (Fine et al., 2019; Osman et al., 2018). However, local anthropogenic pressure can bring coral closer to their thermal bleaching threshold (Kleinhaus et al., 2020; Osman et al., 2018). Acting on the local scale is therefore an urgent imperative to maintain the healthy status of these crucial ecosystems. The nature and sources of local pressures vary across the region in terms of intensity (Amer et al., 2021). The major causes of coral coverage decline in the northern part are coastal development, tourism, and the multiple associated pollutions, whereas, in the central part, overfishing, land runoff, metals, and anti-fouling paint pollution are considered the main local stressors on coral reefs (Amer et al., 2021; Fine et al., 2019; Spaet & Berumen, 2015). Except for climate-change-induced bleaching occurring in the southernmost section, the major anthropogenic pressures are overfishing, sewage treatment, eutrophication, anti-fouling paint, and poor management in general (Amer et al., 2021; Fine et al., 2019). Tourism is a very profitable activity in the region but it is also very harmful to the marine environment without proper management (Fine et al., 2019). Coastal tourism in Saudi Arabia, while still in its nascent stages, presents a unique opportunity to embed sustainable practices right from the beginning. The country has all the necessary resources, tools, and capabilities to fulfill its Vision 2030 for the coastal tourism industry. However, this endeavor requires careful and effective management and regulation to ensure it proceeds without causing significant environmental or social impacts. Achieving this vision calls for a meticulous and sensitive approach to balancing development with conservation needs. The increasing human settlement and tourist immigration in many places around the basin have fostered massive coastal development and anthropic activity intensification. The resultant runoff, waste and sewage issues, sedimentation perturbation, other kinds of pollution, and many industrial activities such as mining and overfishing, severely disturb coral reefs and other marine ecosystems (Bruckner et al., 2009; Fine et al., 2019). The consequent degradation of the natural landscape will also lead to lower attractiveness for tourists, thus impacting the tourism industry itself (Fine et al., 2019). The increase in fishing activities to support the needs of the growing population has led to overexploitation and the depletion of many valuable reef species, disturbing entire coral reef ecosystems (Bruckner et al., 2009). Therefore, an efficient management strategy is needed to encompass innovative and relevant solutions, to avoid a conflict between economic prosperity and environmental sustainability (Fine et al., 2019). Red Sea reefs' survival implies mitigating climate change, protecting them from direct local pressure, repairing them, and adapting through a coherent strategy including ecological solutions backed by policies and socio-economic tools and instruments. In this section, we will present a set of potential solutions which could be adopted by the region, to include in an integrated strategy for the survival of these priceless ecosystems and to maintain the delivery of their crucial services.

2.3.2 From Sciences to Solutions

Climate change and anthropogenic local impacts on coral reefs highlight the urgency to act for the survival of these crucial ecosystems. To get there, relevant and coherent solutions should be included in an integrated strategy. Actions to increase coral resiliency on the local scale must accompany global climate change mitigation. The implementation of these actions is only possible once local stressors have been well managed (Allemand et al., 2015; Hilmi et al., 2020).

Mitigation seeks to stabilize and reduce greenhouse gas (GHG) emissions, through CO2 sequestration for example. This strategy might appear to be the most efficient if efforts are made on a global scale. Solutions to mitigation include replanting and protecting seagrass meadows and mangroves, which are particularly efficient carbon sinks and can further mitigate ocean acidification by buffering the pH drop locally. Marine geoengineering is another approach to mitigating climate change's impact on coral reefs locally. The addition of alkaline material to seawater has been proposed as a way of acting on ocean acidification (Allemand et al., 2015), and the propagation of degradable polymers at the surface has been suggested to mitigate solar irradiation and reduce heat stress, but this technique still presents risks and uncertainties.

Measures aimed at protection are indispensable to reduce the impact of anthropogenic stressors and, thus, to safeguard the existing reefs. The identification of key areas such as refuges (e.g., the Gulf of Aqaba), mesophotic corals, and tolerant species, in general, is crucial to expanding and developing further MPAs. MPAs and networks of MPAs have been proven to be efficient management tools for protecting coral reefs by increasing resilience and reducing stress (Abelson et al., 2016; Allemand et al., 2015; White et al., 2014). The use of ecological mooring buoys has also been suggested to protect reefs of particular interest. Moreover, the creation of a coral repository and coral conservatory (similar to the Monaco World Coral Conservatory) for scientific and mediation purposes should play a crucial role in coral resilience and assisted evolution research, as well as in raising awareness.

To repair degraded coral ecosystems, potential restoration measures include assisted coral evolution and breeding based on resistant coral from refuges and coral aquaculture for rehabilitation. Restoration measures should include resistant coral breeding to increase resilience, and assisted evolution and spread of heat tolerant corals through natural selection in areas where human local stressors are already well managed. Enhancing coral tolerance to warm temperatures and low pH through assisted evolution should improve reef restoration significantly on both local and global scales (Krueger et al., 2017). Moreover, assisted fertilization and recruitment is a part of the strategy, as well as fostering species diversity, genetic variation enhancement among populations, and management to attain the optimal best-adapted species assemblages. However, coral restoration technologies may adapt to different reef systems (fringing, barrier, patchy) and different objectives (coral aquaculture, marine life nursery, reproduction hubs, coral connectivity) (Fine et al., 2019). The efficiency of such restoration action relies importantly on the connectivity of coral populations, as the aim is to accelerate heat tolerant coral propagation and focus on where dispersal and larvae recruitment can trigger year-to-year and scaled-up propagation. Indeed, coral reef ecosystems are intrinsically fragmented. Connectivity is of crucial importance as the population mix of reef-associated species promotes persistence, resilience, and recovery. Connectivity is recognized as a critical parameter in conservation and management decisions, especially for coral reef ecosystems.

Adaptation solutions should include coral reef health monitoring, promoting awareness through educational programs for local people and tourists, and promoting sustainable blue economy activities such as fishing, coastal tourism, and waste management. Divers' access to certain reefs (especially the less studied) must be regulated to avoid direct damage. The adaptation will also occur through policy changes to support these actions.

The implementation of these potential solutions will require a solid cross-border and multidisciplinary collaboration. Moreover, these actions cannot be effective in the whole Red Sea without global warming mitigation and strong backing from public interest, investment, and especially socio-economic tools and policy adaptation to enforce regulations and improve management.

2.3.3 Socio-Economic Tools and Governance

This section reviews policies and socio-economic instruments for decision-making on different scales to address coral reef issues and their impact on society. As highlighted previously, the need for reef socio-economic valuation through the obtention of data in statistical systems (economic activities relying on these ecosystems and damage/loss measures) is fundamental. It provides evidence to decision-makers on coral reefs' economic provisioning services. Coral reef protection is also crucial for human and marine environmental welfare, which should be prioritized in policymaking.

Rather than focusing on the development of new agreements and protocols, there are a colossal number of existing ones that could synergize coherently and be adapted, if possible, for regional implementation around the Red Sea through cross-border collaboration. To do so, the public and decision-makers must be involved in the current and projected impacts on coral reefs and their associated ecosystem services delivery. Regional compliance with global agreements such as the Paris Agreement is crucial to reaching the objectives and strengthening regional collaboration. Another important approach consists of gathering existing norms dealing with pollution and adapting them to Red Sea coral reefs' major threats, such as land runoff, sewage discharges, anti-fouling paint, and plastic (Fine et al., 2019), to mitigate local stressors. Payment for ecosystem services is an additional existing tool involving the household's willingness to protect ecosystem services or structures proactive in coral stress mitigation such as MPAs. It would allow the provision of incentives to business owners, fostering good practices, and enhancing environmental management to generate ecosystem services. Such an approach could come with certification for best practices, encouraging other business owners to do likewise. Coupled with the proper enforcement of regulatory measures, this tool could also be a potential solution to improving fisheries management and promoting sustainability (Bladon et al., 2016; Chen et al., 2018).

There are also measures applicable by governments at the national level to better adapt mitigation to the major local stressor, which appears to be variable from one country to another around the Red Sea. Addressing coastal development pressures, including urbanization, population density, pollution, and physical disturbance, represents a substantial challenge, especially for the northern section of the Red Sea. Governments could use zoning regulations to limit the development of activities along the shoreline and attach premiums to licenses that develop residential and commercial facilities. The resulting receipts could be used for conservation purposes. Taxing commercial activities of major importance within the blue economy such as high-traffic shipping could also provide financial support to help in the protection and restoration of coral reef ecosystems. Coastal and maritime tourism is the source of many stressors that severely impact coral reefs. The whole industry should be rethought and should be more and more considered as a tool to be the vector of social change, sustainability, and restoration financing. National governments need to promote ecotourism, regenerative tourism, and school programs adapted to each country's capacities, but following the same aspirations of raising awareness and protecting coral reefs. Social initiatives strengthening regulations, citizen scientists, and education could represent stepping stones toward sustainable coastal tourism. Certain tools can be implemented to foster the mitigation of coastal tourism's adverse impacts, such as certification for tourism service providers and employees on their knowledge and understanding of ecotourism principles. Developing educational programs and allocating a fund to promote better environmental practices awareness among public sector employees and fishermen should mitigate damage to coral reefs. At school, ocean literacy programs could be implemented to raise awareness and motivate further innovative conservation and progress toward a sustainable blue economy. New pricing signals and incentives could also be implemented to promote sustainable consumption and production in alignment with SDG 12 to reduce pressure on coral reefs (pesticides on land, overfishing at sea).

The links between local stressors and overall coral reef vulnerability shine a light on the urgency of reducing unsustainable practices such as overfishing and pollution through targeted local management actions. Such efforts should seriously consider land-based management, considering the important adverse impact of agricultural pollutants, nutrients, and sediment runoff on coral reefs (Kroon et al., 2016). Land-based management measures that include long-term monitoring throughout the watershed can significantly reduce runoff pressures. Issues like destructive fishing, often due to lax management and poor support and awareness, predominantly affect less developed countries in the southern part of the basin, such as Sudan and Yemen. Conversely, long-term stressors that impact water quality and biodiversity, such as coastal development and pollution, are more common in industrialized countries like Egypt, Israel, and Jordan. Saudi Arabia must learn from the experiences of these neighboring countries to enhance its coastal management strategies and align them with its Vision 2030 goals.

Fishery management should include a special focus on key species of importance for coral reef resilience, such as grazers (e.g., rabbitfish) which control seaweed growth and thus prevent corals from being outcompeted (Hughes et al., 2007). Such measures could include the closing of sensitive areas, restrictions on size and number of fish caught, and sales bans on such species, including incentivization of a shift in targeted species. Allowing greater flexibility to fishermen is another crucial aspect of fishery management improvement. "Equitable" fishery subsidies could allow switching to more sustainable gear and could provide advanced technologies to help fish further out and change fishing locations and targeted species, preventing overexploitation. Such subsidies could also cover high-skill learning programs for fishermen about ecology, fishing locations, gear types, and other potential targeted species in the area. Social-ecological interdependency should be taken into account in improving fisheries management, especially with the Red Sea coast presenting significant differences in terms of development from one country to another. The suggested potential solutions are not standardized and depend on local adaptive capacities, especially for stakeholders relying on marine resources (for livelihood in the case of subsistence fishing). To reduce the sector's vulnerability, it must lessen its impact on coral reefs by switching fishing gear and areas. Simultaneously, the ecosystem's productivity should be enhanced through effective management measures like reducing fishing efforts. The primary goal of local adaptation strategies is to prevent the cumulative effect of direct stressors that push corals closer to their bleaching threshold. To reduce fishing efforts, it is key to implement well-enforced measures. Potential solutions include licensing systems to control access to water, seasonal closures, and MPA implementations. Environmental economists also suggest output control methods such as ITQs (Individual Transferable Quotas), where quotas are allocated to harvest a fixed amount of fish in specific periods and can be traded with other fishermen. The aim is to slow down the race-to-fish and this method has been shown to be efficient in U.S fisheries. To be effective, all of these potential strategies require cross-border cooperation between institutes and stakeholders in the fishery sector.

It is crucial to encourage stakeholders, whose livelihoods depend on fishing, to actively participate in strategies that minimize their impact on coral reefs. This involves integrating them into the policy implementation and management processes. Such strategies could be supported by facilitating access to alternative employment opportunities in small businesses related to tourism and cultural activities, particularly around community-managed protected areas such as LMMAs, Locally Managed Marine Areas. This approach not only helps protect the coral reefs but also fosters economic diversity within the community. Local stakeholders should also have easy access to public services such as healthcare and education on scientific knowledge. Programs like the Vamizi Island Conservation Project in East Africa and the Reef Guardian program in Australia exemplify successful strategies. These programs elevate the social status of fishermen and promote sustainable practices, thereby protecting coral reefs and inspiring broader community involvement, which in turn positively impacts both the environment and the industry.

MPAs provide a regulatory framework to control and reduce the impact on the marine environment. This is an efficient tool to protect coral reefs from local stressors and to improve their resilience to climate change. It is also a trigger for the political imperative to implement the relevant measures and address local impact issues such as land-based pollution. It is a scientific instrument of crucial importance to monitor diversity, migration, and marine communities overall. To not be considered "paper parks" (MPAs that exist only on paper and are inefficient due to weak or no management and the absence of restriction enforcement and law compliance), MPAs need sustainable financing and effective implementation. Innovative financing mechanisms such as Private-Public Partnerships (PPP) and entrepreneurial MPAs (EMPAs) offer strong management to these structures and attract private investors, mainly through impact investment, as such structures are identified as a potential asset class for investors. It is a management area that is primarily funded by a profit-bearing business model, typically associated with nature tourism. This kind of emerging solution allows sustainable financing of the area and a guaranteed positive impact.

Given the current scarcity of financial resources in marine conservation and the urgency of the situation, it is advisable to establish a framework for prioritizing mitigation efforts. Conducting cost-benefit analyses with numerical estimates at international, national, and local levels is essential for decision-makers to determine priorities effectively. None of the aforementioned tools are silver bullets, rather; effectively addressing the challenges facing coral reefs will require a combination of complementary mechanisms—including innovative tools and policies, adaptation and improvement to existing frameworks and instruments, technological advancement and sustainable investments. Scientific knowledge and multidisciplinary research are imperative to understand the threats to and impacts on coral reefs and to trigger the appropriate governance responses. Transformation of economies is also crucial to allow sustainable consumption and protection of coral reefs, especially in climate change refuges such as the Red Sea. Development policies should adapt to guarantee coral reefs protection in line with the sustainable development framework, instead of merely focusing on economic growth. Now, more than ever, coral reef conservation is a global challenge where regional actions in the Red Sea are fundamental.

2.4 Conclusion: Developing a GCC Economy that Is both Sustainable and Nature-Positive

2.4.1 Toward a Regenerative Blue Economy

Since the late twentieth century, substantial hopes have been placed on the concept of "blue" growth, both economically and socially, to stimulate a "blue" economy, ultimately aimed at developing what is termed a "blue" GDP. The concept of the blue economy, which is continuously evolving, refers to the economic use of the ocean, also known as the ocean economy (le Gouvello & Simard, 2024). This approach has incorporated traditional maritime activities from an anthropocentric perspective. However, since 2012, the concept has shifted toward a "sustainable blue economy" that integrates environmental, social, and governance concerns with a more sustainable and inclusive vision. Efforts to safeguard, repair, and restore marine and coastal ecosystems and their services are incorporated into this approach. As climate change and biodiversity loss have become more pressing, the concept has further evolved since 2020 into a "regenerative blue economy," also referred to as an "ocean positive economy." This newer model fosters climate change resilience and biodiversity recovery and promotes best practices. It therefore excludes activities deemed incompatible with the carbon reduction goals of the 2015 Paris Agreement or those too threatening to marine ecosystems, such as the oil and gas sector. Under this regenerative framework, sectors like fishing, aquaculture, and tourism are encouraged to adapt their practices to align with broad ecosystem-based principles. Economically, a regenerative blue economy aims for robust sustainability, prioritizing the preservation of blue natural capital, and aligning with the SDGs to ensure healthy oceans and the sustainable use of marine resources. The Red Sea coastline of the GCC is characterized by a notably diverse marine life, which contributes to an abundant "blue natural capital" and presents substantial potential for a robust local blue economy.

2.4.2 GCC Countries Have the Potential to Become the Pioneers in the MENA Region, Showcasing a Novel Economic Paradigm

Utilizing natural asset valuation services can provide valuable support to the Saudi Green Initiative (SGI), Saudi Vision 2030, and more generally to the GCC countries and the SDGs, enabling the transformation and diversification of the economy to reduce the dependency on the oil industry and foster sustainability. This transformation aims to incorporate wealth creation and employment opportunities that are intricately linked to nature conservation and restoration, aligning with global endeavors to combat climate change and restore biodiversity.

This new economic paradigm envisions conservation as a viable source of capital for fostering economic development. This approach can be successfully replicated across various natural assets found within the GCC, such as forests, mangroves, seagrass, and more. Moreover, it has the potential to be adopted by other countries within the region and globally. GCC countries have the opportunity to act as a catalyst for a transformative era in which nations recognize the advantages of a flourishing and sustainable environment, promoting a regenerative way of life. Such a demonstration would compellingly highlight the societal benefits derived from investing in nature.

Moreover, for a viable opportunity to remain below the 1.5 °C warming threshold, it is imperative to invest in climate mitigation and prioritize nature- and people-positive activities. This can be achieved through the implementation of nature-based solutions. By directing resources toward these initiatives, we can effectively address the challenges posed by climate change while ensuring the well-being of both ecosystems and communities.

As a result of the Kunming-Montreal Global Biodiversity Framework and the emergence of biodiversity credit markets, we can anticipate the arrival of fresh commitments and disclosures concerning the impact on nature. The demand for carbon and biodiversity credits is steadily increasing, consequently driving the need for skilled professionals in various fields to develop and implement these projects effectively. It is crucial to protect the reefs thanks to innovations to increase coral resilience. Effort and funding are needed to save coral reefs (Kleypas et al., 2021).

The valuation of "blue natural capital" in GCC countries can therefore catalyze significant multifaceted benefits. It enhances conservation efforts, ensuring the preservation of diverse flora, fauna, and ecosystems for future generations. Economically, this valuation introduces a new, consistent revenue stream by monetizing ecosystem services, which helps decrease the costs related to conservation projects and improves overall fiscal health. Additionally, engaging local communities in these efforts not only fosters a deeper connection with the environment but also stimulates economic growth by creating new job opportunities in conservation and restoration. This inclusion enhances community health and well-being by promoting a cleaner environment. On a broader scale, incorporating environmental, social, and governance (ESG) criteria into financial services enriches the investment landscape, driving financial innovation and economic stability, and reducing reliance on fluctuating income sources. Such diversification strengthens economic resilience and supports long-term viability, demonstrating the comprehensive benefits of ecosystem service valuation in fostering sustainable development in the GCC region.

References

Ababouch, L., & Fipi, F. (2015). Fisheries and aquaculture in the context of blue economy. *Feeding Africa, 2*(21–23), 13.

Abelson, A., Nelson, P. A., Edgar, G. J., Shashar, N., Reed, D. C., Belmaker, J., Krause, G., Beck, M. W., Brokovich, E., France, R., & Gaines, S. D. (2016). Expanding marine protected areas to include degraded coral reefs. *Conservation Biology, 30*(6), 1182–1191. https://doi.org/10.1111/cobi.12722

Aljahdali, M. O., & Alhassan, A. B. (2020). Ecological risk assessment of heavy metal contamination in mangrove habitats, using biochemical markers and pollution indices: A case study of Avicennia marina L. in the Rabigh lagoon, Red Sea. *Saudi Journal of Biological Sciences, 27*(4), 1174–1184. https://doi.org/10.1016/j.sjbs.2020.02.004

Allemand, D., Hilmi, N., Kavanagh, C., Laffoley, D., Metian, M., Osborn, D., & Reynaud, S. (2015). *Bridging the gap between ocean acidification impacts and economic valuation: Regional impacts of ocean acidification on fisheries and aquaculture.* Retrieved from https://policycommons.net/artifacts/1374000/bridging-the-gap-between-ocean-acidification-impacts-and-economic-valuation/1988239/

Amer, M., Arrigoni, R., Attalla, T., Helmi, A., Abu Awali, A., Al-Tawaha, M., Abaker, K., Berumen, M., Bouwmeester, J., Coker, D., Elsawy, N., Freiwald, J., Harvey, C., Johnson, A., & Terraneo, T. (2021). *Status of coral reefs of the world 2020* (p. Chapter 3. Status and trends of coral reefs of the Red Sea and Gulf of Aden).

Behzad, H., Ibarra, M. A., Mineta, K., & Gojobori, T. (2016). Metagenomic studies of the Red Sea. *Gene, 576*(2), 717–723. https://doi.org/10.1016/j.gene.2015.10.034

Bladon, A. J., Short, K. M., Mohammed, E. Y., & Milner-Gulland, E. J. (2016). Payments for ecosystem services in developing world fisheries. *Fish and Fisheries, 17*(3), 839–859. https://doi.org/10.1111/faf.12095

Boyd, J., & Banzhaf, H. (2005). The architecture and measurement of an ecosystem services index. *Sustainability, 4*, 430. https://doi.org/10.3390/su4040430

Brander, L. M. (2023). *Guidance on the economic valuation of ecosystem services and natural capital of the area (part II of the report).* International Seabed Authority. Retrieved from https://www.isa.org.jm/wp-content/uploads/2023/06/Report-on-Valuation-of-ecosystem-services.pdf

Brander, L. M., de Groot, R., Schägner, J. P., Guisado-Goñi, V., van Hoff, V., Solomonides, S., McVittie, A., Eppink, F., Sposato, M., Do, L., Ghermandi, A., Sinclair, M., & Thomas, R. (2024). Economic values for ecosystem services: A global synthesis and way forward. *Ecosystem Services, 66*, 101606. https://doi.org/10.1016/j.ecoser.2024.101606

Bruckner, A. W., Alnazry, H. H., & Faisal, M. (2009). A paradigm shift for fisheries management to enhance recovery, resilience, and sustainability of coral reef ecosystems in the Red Sea. In *Sustainable fisheries: Multi-level approaches to a global problem.* American Fisheries Society.

Burke, L., & Spalding, J. M. (2004). *Reefs at risk in the Caribbean.* Retrieved from https://www.wri.org/research/reefs-risk-caribbean.

Carvalho, S., Kürten, B., Krokos, G., Hoteit, I., & Ellis, J. (2019). Chapter 3—The red sea. In C. Sheppard (Ed.), *World seas: An environmental evaluation* (2nd ed., pp. 49–74). Academic Press. https://doi.org/10.1016/B978-0-08-100853-9.00004-X

Chalastani, V. I., Manetos, P., Al-Suwailem, A. M., Hale, J. A., Vijayan, A. P., Pagano, J., Williamson, I., Henshaw, S. D., Albaseet, R., Butt, F., Brainard, R. E., Coccossis, H., Tsoukala, V. K., & Duarte, C. M. (2020). Reconciling tourism development and conservation outcomes through marine spatial planning for a Saudi Giga-project in the red sea (the red sea project, vision 2030). *Frontiers in Marine Science, 7*, 490664. https://doi.org/10.3389/fmars.2020.00168

Chen, J.-L., Lin, Y.-S., & Chuang, C.-T. (2018). Improving the management of Taiwanese fishery resource conservation zones based on public perceptions and willingness to pay for ecosystem services. *Journal of Coastal Conservation, 22*(2), 385–398. https://doi.org/10.1007/s11852-017-0586-5

Chen, P.-Y., Chen, C.-C., Chu, L., & McCarl, B. (2015). Evaluating the economic damage of climate change on global coral reefs. *Global Environmental Change, 30*, 12–20. https://doi.org/10.1016/j.gloenvcha.2014.10.011

Cziesielski, M., Duarte, C., Aalismail, N., Al-Hafedh, Y., Anton, A., Baalkhuyur, F., Baker, A., Balke, T., Baums, I., Berumen, M., Chalastani, V., Cornwell, B., Daffonchio, D., Diele, K., Ehtsaam, F., Gattuso, J.-P., He, S., Lovelock, C., Mcleod, E., & Aranda Lastra, M. (2021).

Investing in blue natural capital to secure a future for the red sea ecosystems. *Frontiers in Marine Science*. https://doi.org/10.3389/fmars.2020.603722

DiBattista, J. D., Roberts, M. B., Bouwmeester, J., Bowen, B. W., Coker, D. J., Lozano-Cortés, D. F., Howard Choat, J., Gaither, M. R., Hobbs, J.-P. A., Khalil, M. T., Kochzius, M., Myers, R. F., Paulay, G., Robitzch, V. S. N., Saenz-Agudelo, P., Salas, E., Sinclair-Taylor, T. H., Toonen, R. J., Westneat, M. W., & Berumen, M. L. (2016). A review of contemporary patterns of endemism for shallow water reef fauna in the Red Sea. *Journal of Biogeography, 43*(3), 423–439. https://doi.org/10.1111/jbi.12649

El Samak, M., Solyman, S. M., & Hanora, A. (2018). Antimicrobial activity of bacteria isolated from Red Sea marine invertebrates. *Biotechnology Reports, 19*, e00275. https://doi.org/10.1016/j.btre.2018.e00275

Fine, M., Cinar, M., Voolstra, C. R., Safa, A., Rinkevich, B., Laffoley, D., Hilmi, N., & Allemand, D. (2019). Coral reefs of the red sea—Challenges and potential solutions. *Regional Studies in Marine Science, 25*, 100498. https://doi.org/10.1016/j.rsma.2018.100498

Frihy, O. E., El Ganaini, M. A., El Sayed, W. R., & Iskander, M. M. (2004). The role of fringing coral reef in beach protection of Hurghada, Gulf of Suez, Red Sea of Egypt. *Ecological Engineering, 22*(1), 17–25. https://doi.org/10.1016/j.ecoleng.2003.11.004

Gracia, A., Rangel-Buitrago, N., Oakley, J. A., & Williams, A. T. (2018). Use of ecosystems in coastal erosion management. *Ocean & Coastal Management, 156*, 277–289. https://doi.org/10.1016/j.ocecoaman.2017.07.009

Guo, W., Zhang, X., & Xie, J. (2020). The efficiency and evolution mechanism of tourism based on the coastal cities. *Journal of Coastal Research, 112*, 93–96. https://doi.org/10.2112/JCR-SI112-027.1

Hasler, H., & Ott, J. A. (2008). Diving down the reefs? Intensive diving tourism threatens the reefs of the northern Red Sea. *Marine Pollution Bulletin, 56*(10), 1788–1794. https://doi.org/10.1016/j.marpolbul.2008.06.002

Hilmi, N., Farahmand, S., & Bélaïd, F. (2020). *Why should Saudi Arabia diversify its economy?* (pp. 89–109). Bentley University. https://doi.org/10.1007/978-981-15-6058-3_5

Hilmi, N., Safa, A., Reynaud, S., & Allemand, D. (2018). Coral-based tourism in Egypt's red sea. In *Coral reefs: Tourism, conservation and management*. Routledge.

Hossain, N., Mahlia, T. M. I., & Saidur, R. (2019). Latest development in microalgae-biofuel production with nano-additives. *Biotechnology for Biofuels, 12*(1), 125. https://doi.org/10.1186/s13068-019-1465-0

Hughes, T. P., Rodrigues, M. J., Bellwood, D. R., Ceccarelli, D., Hoegh-Guldberg, O., McCook, L., Moltschaniwskyj, N., Pratchett, M. S., Steneck, R. S., & Willis, B. (2007). Phase shifts, herbivory, and the resilience of coral reefs to climate change. *Current Biology, 17*(4), 360–365. https://doi.org/10.1016/j.cub.2006.12.049

Jolliffe, J., Jolly, C., & Stevens, B. (2021). *Blueprint for improved measurement of the international ocean economy: An exploration of satellite accounting for ocean economic activity*. OECD. https://doi.org/10.1787/aff5375b-en

Kleinhaus, K., Al-Sawalmih, A., Barshis, D., Genin, A., Grace, L., Hoegh-Guldberg, O., Loya, Y., Meibom, A., Osman, E., Ruch, J.-D., Shaked, Y., Voolstra, C., Zvuloni, A., & Fine, M. (2020). Science, diplomacy, and the red sea's unique coral reef: It's time for action. *Frontiers in Marine Science, 7*, 90. https://doi.org/10.3389/fmars.2020.00090

Kleypas, J., Allemand, D., Anthony, K., Baker, A., Beck, M., Hale, L., Hilmi, N., Hoegh-Guldberg, O., Hughes, T., Kaufman, L., Kayanne, H., Magnan, A., Mcleod, E., Mumby, P., Palumbi, S., Richmond, R., Steneck, R., Voolstra, C., & Gattuso, J.-P. (2021). Designing a blueprint for coral reef survival. *Biological Conservation, 257*, 109107. https://doi.org/10.1016/j.biocon.2021.109107

Kroon, F. J., Thorburn, P., Schaffelke, B., & Whitten, S. (2016). Towards protecting the great barrier reef from land-based pollution. *Global Change Biology, 22*(6), 1985–2002. https://doi.org/10.1111/gcb.13262

Krueger, T., Horwitz, N., Bodin, J., Giovani, M.-E., Escrig, S., Meibom, A., & Fine, M. (2017). Common reef-building coral in the northern Red Sea resistant to elevated temperature and acidification. *Royal Society Open Science, 4*, 170038. https://doi.org/10.1098/rsos.170038

Langodan, S., Antony, C., Pr, S., Dasari, H. P., Abualnaja, Y., Knio, O., & Hoteit, I. (2020). Wave modeling of a reef-sheltered coastal zone in the Red Sea. *Ocean Engineering, 207*, 107378. https://doi.org/10.1016/j.oceaneng.2020.107378

Le Gouvello, R., & Simard, F. (2024). *Towards a regenerative blue economy*. IUCN. Retrieved from https://portals.iucn.org/library/node/51442

Marshall, N. A., Marshall, P. A., Abdulla, A., & Rouphael, T. (2010). The links between resource dependency and attitude of commercial fishers to coral reef conservation in the red sea. *Ambio, 39*(4), 305–313. https://doi.org/10.1007/s13280-010-0065-9

Menn, I., Junghans, C., & Reise, K. (2003). Buried alive: Effects of beach nourishment on the infauna of an erosive shore in the North Sea. *Marine Biodiversity, 32*, 125–145. https://doi.org/10.1007/BF03043089

Moberg, F., & Folke, C. (1999). Ecological goods and services of coral reef ecosystems. *Ecological Economics, 29*(2), 215–233. https://doi.org/10.1016/S0921-8009(99)00009-9

Mukanjari, S., Ntuli, H., & Muchapondwa, E. (2022). Valuation of nature-based tourism using contingent valuation survey: Evidence from South Africa. *Journal of Environmental Economics and Policy, 11*(3), 331–349. https://doi.org/10.1080/21606544.2021.2010604

Ncaves, & Maia. (2022). *Monetary valuation of ecosystem services and ecosystem assets for ecosystem accounting: Interim version* (1st ed.). United Nations Department of Economic and Social Affairs, Statistics Division.

O'Mahoney, J., Simes, R., Redhill, D., Heaton, K., Atkinson, C., Hayward, E., & Nguyen, M. (2017). *At what price? The economic, social and icon value of the great barrier reef (Deloitte access economics)*. Great Barrier Reef Marine Park Authority. Retrieved from https://elibrary.gbrmpa.gov.au/jspui/handle/11017/3205

Osman, E. O., Smith, D. J., Ziegler, M., Kürten, B., Conrad, C., El-Haddad, K. M., Voolstra, C. R., & Suggett, D. J. (2018). Thermal refugia against coral bleaching throughout the northern Red Sea. *Global Change Biology, 24*(2), e474–e484. https://doi.org/10.1111/gcb.13895

Pascal, N., Allenbach, M., Brathwaite, A., Burke, L., Port, G., & Clua, E. (2016). Economic valuation of coral reef ecosystem service of coastal protection: A pragmatic approach. *Ecosystem Services, 21*, 72–80. https://doi.org/10.1016/j.ecoser.2016.07.005

Rasul, N. M. A., Stewart, I. C. F., & Nawab, Z. A. (2015). Introduction to the red sea: Its origin, structure, and environment. In N. M. A. Rasul & I. C. F. Stewart (Eds.), *The red sea: The formation, morphology, oceanography and environment of a young ocean basin* (pp. 1–28). Springer. https://doi.org/10.1007/978-3-662-45201-1_1

Robitzch, V., Kattan, A., Dunne, A., & Coker, D. J. (2023). Saudi Arabia case study. In M. Wolff, S. C. A. Ferse, & H. Govan (Eds.), *Challenges in tropical coastal zone management: Experiences and lessons learned* (pp. 73–93). Springer International Publishing. https://doi.org/10.1007/978-3-031-17879-5_6

Romaniv, O., & Yarmolyuk, D. (2021). *The red sea as tourist destination*. Baltija Publishing. https://doi.org/10.30525/978-9934-26-077-3-5

Savary, R., Barshis, D., Voolstra, C., Cardenas, A., Evensen, N., Banc-Prandi, G., Fine, M., & Meibom, A. (2021). Fast and pervasive transcriptomic resilience and acclimation of extremely heat-tolerant coral holobionts from the northern Red Sea. *Proceedings of the National Academy of Sciences, 118*, e2023298118. https://doi.org/10.1073/pnas.2023298118

Shellem, C. (2020). *Assessing the potential impact of a mass coral bleaching event on Red Sea fisheries* [thesis]. https://doi.org/10.25781/KAUST-Q7Y7A

Spaet, J. L. Y., & Berumen, M. L. (2015). Fish market surveys indicate unsustainable elasmobranch fisheries in the Saudi Arabian Red Sea. *Fisheries Research, 161*, 356–364. https://doi.org/10.1016/j.fishres.2014.08.022

Turner, R. K. (2015). *Coastal zones ecosystem services*. Springer. https://doi.org/10.1007/978-3-319-17214-9

van de Water, J. A., Tignat-Perrier, R., Allemand, D., & Ferrier-Pagès, C. (2022). Coral holobionts and biotechnology: From blue economy to coral reef conservation. *Current Opinion in Biotechnology, 74*, 110–121. https://doi.org/10.1016/j.copbio.2021.10.013

Wahlig, P., & Sommer, N. (2021). On the collection and use of fish scales for fisheries management: Pros, cons, and potential solutions. *Journal of Student Research, 10*(4), 2410. https://doi.org/10.47611/jsrhs.v10i4.2410

Walter, R. C., Buffler, R. T., Bruggemann, J. H., Guillaume, M. M. M., Berhe, S. M., Negassi, B., Libsekal, Y., Cheng, H., Edwards, R. L., von Cosel, R., Néraudeau, D., & Gagnon, M. (2000). Early human occupation of the Red Sea coast of Eritrea during the last interglacial. *Nature, 405*(6782), 6782. https://doi.org/10.1038/35011048

White, A. T., Aliño, P. M., Cros, A., Fatan, N. A., Green, A. L., Teoh, S. J., Laroya, L., Peterson, N., Tan, S., Tighe, S., Venegas-Li, R., Walton, A., & Wen, W. (2014). Marine protected areas in the coral triangle: Progress, issues, and options. *Coastal Management, 42*(2), 87–106. https://doi.org/10.1080/08920753.2014.878177

Zakai, D., & Chadwick-Furman, N. E. (2002). Impacts of intensive recreational diving on reef corals at Eilat, northern Red Sea. *Biological Conservation, 105*(2), 179–187. https://doi.org/10.1016/S0006-3207(01)00181-1

Chapter 3
The Green Transformation of the GCC: A Path to Sustainable Finance

Elhachemi Gherbi and Habib Ouni

Abstract This chapter explores the strategic transition of the Gulf Cooperation Council (GCC) nations—Oman, Bahrain, Saudi Arabia, Kuwait, Qatar, and the United Arab Emirates—from economies that depend on fossil fuels to those prioritizing renewable energy and sustainable finance. It highlights key policy measures that support renewable energy, enhanced energy efficiency, and sustainable financial systems, all vital for addressing climate change and attaining long-term environmental sustainability. To foster economic diversification while encouraging environmental stewardship, GCC countries are directing investments toward green finance instruments such as green bonds and sustainability-linked loans. The chapter examines the increasing reliance on renewable energy sources like solar and wind to enhance energy security and promote job creation in the green energy sector. Key initiatives, such as solar parks and carbon capture projects, are underscored as essential elements of this shift. Furthermore, the analysis explores the challenges faced during this green transition, including technological limitations, regulatory hurdles, and societal barriers. Despite these difficulties, it highlights the importance of international collaboration, innovation, and forward-thinking policymaking in positioning the GCC as leaders in global sustainability efforts. The discussion also stresses the role of green finance in supporting this change, suggesting a promising future for sustainable development in the region.

Keywords Green transformation · Sustainable finance · Renewable energy · Economic diversification · Green finance mechanisms · Climate change mitigation

E. Gherbi · H. Ouni (✉)
Faculty of Business, Higher Colleges of Technology, Al Ain, United Arab Emirates
e-mail: houni@hct.ac.ae

E. Gherbi
e-mail: egherbi@hct.ac.ae

H. Miniaoui and M. Bellalah (eds.), *Green Growth Opportunities and Sustainable Finance in the Gulf Cooperation Council Region*, Gulf Studies 24,
https://doi.org/10.1007/978-981-96-8287-4_3

3.1 Introduction

Bahrain, Kuwait, Oman, Qatar, Saudi Arabia and United Arab Emirates are all actively striving toward two primary objectives: responsible environmental governance and sustainable development. Traditionally, these nations have depended significantly on energy production and export revenues as their principal means of generating wealth. This dependence has been driven by the exploitation of fossil fuels, which have underpinned their economies. Now is the perfect moment for these countries to move beyond greed and embrace renewable energy sources along with sound financial practices.

The GCC's policy initiatives that push for renewable energy and sustainable financial systems evince that their goals are oriented toward mitigating the future risks of climate change and building long-term environmental sustainability. Acknowledging that the time for greening their economies and cutting back on greenhouse gas emissions is now, GCC member states have proposed a variety of ambitious undertakings that encompass renewable energy projects, energy efficiency programs, and the creation of sustainable structures.

The driver of the environmentally sustainable financing agenda is the fact that the GCC's commitments to economic diversification and ecological preservation are mutually inclusive. Through exploiting the green clean energy sources, such as solar and wind energy, the GCC countries aim to secure economic growth, reduced use of the finite fuel resources, and job creation in the green energy sector. This shift not only positively contributes to energy security as well as the safety and resilience of the system, but it also stimulates innovation, technological progress, and human capacity development.

Furthermore, the GCC adoption of green finance methods like green bonds, sustainability-linked loans, and ESG investments has given a glimpse of the paradigm shift toward sustainable investment which focuses on preservation of environmental standards. These capital mobilization structures are not only seeding capital for environment preservation, but also their benefits encompass allowing businesses and enterprises to follow cleaner, greener practices, which in turn optimizes their financial chances.

The GCC green financing transition is a vital aspect of the region's overall sustainability vision and goal of climate resilience. To that end, through the process of environmental concerns reaching financial decision-making corridors, GCC states are moving sustainability principles up to the level of industries and sectors such as energy, transportation, agriculture, and urban development. Through this broad approach, growth which is not at the cost of environmental degradation but rather is in line with ecological integrity and social well-being is achieved.

While the upsurge of green finance and renewable energy worldwide is usually characterized by acceleration, GCC governing companies that are responsible for this will affect the region's future economically. Through embrace of innovation, collaboration promotion, and policy development using forward thinking, the countries in the GCC may act as forerunners in global transformation to a low carbon, sustainable future that could be an inspiration for green development which echoes far beyond their borders.

3.1.1 *Promotion of New Energy Sources*

GCC countries are actively promoting the use of alternative energy sources, particularly renewable energy, in order to diversify their energy portfolio and reduce their reliance on fossil fuels. The main objective of this project is to promote the use of solar energy by making use of the abundant sunlight in the region. Most countries within the Gulf Cooperation Council have initiated significant solar projects, which include the creation of large solar parks and the placement of solar panels on building rooftops. Among these, the Mohammed bin Rashid Al Maktoum Solar Park in Dubai, UAE, stands out as one of the largest solar parks in the world. By 2030, it is anticipated to reach a capacity of 5000 mW.

In addition to solar power, the GCC nations are exploring a range of renewable energy options, including wind, geothermal, and biomass (Hafner et al., 2023). Oman has recently announced its intention to initiate projects centered on wind and solar energy, while Saudi Arabia is actively investigating the possibilities of geothermal energy. These efforts are designed to increase the variety of energy sources, reduce greenhouse gas emissions, and improve energy stability.

3.1.2 *Reduction of Dependence on Fossil Energy*

The Gulf Cooperation Council (GCC) nations are actively working to reduce their reliance on fossil fuels, especially oil and gas. A key approach to achieving this goal is through boosting energy efficiency and encouraging conservation efforts. This includes optimizing the efficiency of systems related to power generation, transmission, and distribution, along with promoting energy-saving technologies and practices across different sectors, including industry, buildings, and transportation.

An alternative approach involves transitioning to cleaner options like natural gas, known for producing lower greenhouse gas emissions than coal and oil. Countries like Qatar and the UAE, which have abundant natural gas reserves, are investing in infrastructure and technology to improve the use of natural gas across power generation, transportation, and various other sectors.

Furthermore, the countries of the GCC are exploring carbon capture and storage (CCS) technology as a means of reducing emissions from fossil-fuel-dependent power plants and industries. The CCS process involves capturing carbon dioxide emissions and safely storing them beneath the Earth's surface to prevent their escape into the atmosphere.

The GCC nations, recognized for their rich oil reserves, are experiencing a significant transformation as they embrace environmentally sustainable practices, leading to the emergence of a green finance sector. This shift is driven by the necessity to move economies beyond fossil fuels and the pressing global demand to address climate change, offering opportunities for sustainable development within the region. Furthermore, politicians and international institutions are exerting growing pressure to address climate change, in accordance with the goals set forth in the Paris Agreement.

3.2 Investing in Renewable Energy in the GCC

Timucin Engin's analysis reveals that the Gulf Cooperation Council (GCC), under the leadership of the United Arab Emirates (UAE), has been actively allocating investments toward renewable energy sources, with a specific focus on solar power, in recent times. Therefore, there has been an expanding and vibrant market for renewable energy sources. Since then, the Masdar Company has been instrumental in driving clean energy and sustainability efforts in the UAE and across the globe. Today, the UAE's renewable energy sector is expanding rapidly, with a goal of achieving 14.2 GW of renewable energy capacity by 2030. Despite the green bond market's meteoric rise over the past 5 years, the GCC's ecological finance industry is quite young. Al-Sarihi and Mansouri (2022a, 2022b) argue that throughout history, the region has mostly concentrated on its crucial hydrocarbon sectors, which have been the primary catalysts for the expansion of its Gross Domestic Product (GDP) since the 1930s. In March 2017, the National Bank of Abu Dhabi (now First Abu Dhabi Bank) issued the inaugural green bond in the area, amounting to $587 million. Masdar, located in Abu Dhabi, is among the limited number of organizations that have been able to get into the green loan market. In September 2018, it successfully obtained a green revolving credit facility (RCF) worth $75 million for its sustainability initiatives. Yet, there have been no instances of GCC issuers issuing green sukuk. Financing for climate change initiatives in the area is now concentrated on a handful of large-scale projects supported by climate funds such as the Green Climate Fund, as well as banks, governments, and international organizations.

3.3 Accelerating Renewable Energy Capacity in the GCC

The GCC area's 867 MW of RE capacity at year-end 2018 was equivalent to 0.6% of its total electrical capacity. The United Arab Emirates has the greatest percentage of renewable energy capacity at 68%, with Saudi Arabia, Kuwait, and Qatar having 16%, 9%, and 5% respectively.

3.3.1 Challenges Facing the GCC in the Adoption of Renewable Energy

In the process of clean energy programs for the Gulf countries, there are many obstacles to diversifying the technologies for renewable energy. With their technical, cultural, and institutional aspects, these different challenges define the right approach that needs to be followed to cope with them properly.

One of the key technical issues that hampers the growth of renewables is the unreliability of supply. With the concentration of wind and solar power, which are affected seasonally or under changing weather conditions, stability of supply remains a main barrier to continuous and predictable generation of energy. Furthermore, an unfavorable environment, e.g., dust, excessive heat, or high humidity, along with other factors, increases the list of challenges that renewable energy infrastructure must deal with.

In addition, the remote locations of renewable energy sources make the development of distributed energy systems and power grid connection a difficulty. Connection costs, which are often high and difficult to come by, as well as limited space for the siting of renewable energy technologies, are primary challenges to their mass deployment.

The cultural dimension also influences attitudes to renewable energy adoption in the GCC. Residents live today with the ready availability and low unit costs of conventional energy as well as the customary culture of high consumption, which has been supported by historically low power costs subsidized by the state. The reliability of abundant cheap conventional energy supply has resulted in a lack of interest for most people in replacing them with renewable power sources that carry the stigma of an expensive alternative.

An additional barrier is a lack of knowledge concerning renewable technologies among the public at large. This, in its turn, further complicates the extensive promotion of these renewable technologies. The fact that there is no appealing narrative that will spark the joy of adopting renewable energy lengthens the cultural transition to sustainable energy use.

It is extremely challenging for the GCC states to adopt renewable energy due to institutional obstacles such as an absence of consistent energy policy and governance issues in renewable energy regulatory frameworks. Without a constituted authority to oversee the deployment of renewable energy sources, the regulatory environment and coordinating efforts are disorganized.

Some feel that the local power sector has failed to make the most of the research and development budget or the funds sent to it from other countries in the form of renewable energy sources and expert knowledge. Research and development (R&D) funding has been a focus for the Masdar Institute and the Qatar Environment and Energy Research Institute. The GCC still lacks the necessary infrastructure to develop renewable energy solutions tailored to its unique environment.

Addressing these complex issues requires a cooperative approach involving policymakers, industry regulators, and members of civil society. By creating a conducive regulatory framework, increasing public awareness, and committing resources to local research and development, the GCC nations can effectively overcome the obstacles hindering the transition to renewable energy and promote opportunities for a sustainable energy future.

3.4 Green Finance's Role in Funding Renewable Projects in the GCC

The term "green finance," which The GCC countries introduced, pertains to the investment of capital in companies that promote actions aimed at reducing environmental impact, improving quality, and optimizing the use of current resources (2020). Since its introduction in the 1980s, "green finance" has experienced a significant increase in global interest. The green energy projects within the GCC could greatly benefit from this trend.

This area has been seeing substantial advancements thanks to the green growth movement and the transition to a low-carbon economy. Adding additional goals for resilience and sustainability to the infrastructure is a sign of success. With these goals in mind, there is a pressing need for money as well as innovative environmental financing solutions. The United Arab Emirates and Saudi Arabia are leading the charge since they control most of the region's energy markets. The UAE Banks Federation (UBF) reports that six large UAE banks—FDB, ADCBC, ENBD, DIB, Mashreq, and ADIB—have committed to donating more than AED 190 billion (about $51.8 billion) to environmental projects by 2022. Sources of renewable energy, waste-to-energy programs, and companies dealing in green technologies will all benefit from this financing.

The United Arab Emirates pledged in its 2016 Paris Agreement nationally determined contribution (NDC) to boost the share of renewable energy in its total primary energy mix from 21% to 27% by 2021. In addition to reducing gas flaring, the UAE has promised to implement carbon capture and storage, modernize tariffs, set efficiency requirements for buildings and appliances, and control the price of fossil fuels. Table 3.1 provides an overview of the sustainability and renewable energy goals for each country. For example, the Energy Strategy 2050 of the United Arab Emirates aims to diversify the energy mix to embrace 50% renewable sources, boost energy efficiency for households and businesses by 40%, and reduce carbon dioxide emissions from power generation by 70%.

In fact, in 2019, the UAE registered the highest growth in renewable energy investments, recording a remarkable 1223% increase compared to 2018. Dubai plans to increase the amount of power produced by solar panels from 7% in 2020 to 25% in 2030 and 75% in 2050, as seen in the city's integrated clean energy policy for 2050.

A thorough effort to diversify the Saudi economy and reduce the country's carbon footprint are central tenets of the Saudi Arabian Nationally Determined Contribution

Table 3.1 Share of renewable energy target in the Gulf Cooperation Council region for 2030, by country: a comparative overview

Country	Renewable energy capacity target	Actual share of renewable energy electricity capacity (2022)
Bahrain	10% by 2030	0.14%
Kuwait	15% by 2030	0.48%
Oman	20% by 2030	6.38%
Qatar	20% by 2030	7.22%
Saudi Arabia	50% by 2030	0.53%
UAE	30% by 2030	16.64%

Source: International Renewable Energy Agency (IRENA)

(NDC). Improvements in water and wastewater management, increased use of natural gas, carbon capture, energy efficiency, and sustainable energy are all part of this program's intended outcomes. Gas flaring will be further reduced as a result. As part of its Vision 2030 plan, the Saudi government has pledged to increase the output of renewable energy from 3.45 GW in 2020 to 9.5 GW in 2023. The ultimate goal is to acquire 30% of energy from sustainable sources by the year 2030. The government of Saudi Arabia has committed to investing \$30–\$50 billion more in renewable energy by the year 2030. The other objective is to cut peak demand by 14% and power consumption by 8% by 2021 (Samargandi et al., 2024).

Good, but modest, environmental protection strategies are included in the four nations' Nationally Determined Contributions (NDCs). Plans to reduce the environmental effect of human activities and legislation to encourage adaptation and mitigation are all part of these projects. By 2030, renewable energy is targeted to provide 15% of Kuwait's energy needs. The ambitious goals set for Oman and Bahrain are 10% by 2035 and 5% by 2025, respectively. The target of producing 200–500 MW of solar electricity by 2020 is part of Qatar's strategy to diversify its economy and lessen its reliance on oil.

3.4.1 Global Green Bond Market

Given the increasing global concern and the urgency of addressing climate change, it is unsurprising that the green bond market has witnessed substantial expansion in the last 5 years. In 2018, despite higher interest rates and a decrease in new bond offerings, the green bond market surpassed the previous record established in 2017, reaching a total of \$167.3 billion, which represents a 3% rise compared to the previous year's \$155 billion.

3.4.2 *The Rise of Green Finance in Real Estate*

Several factors are pushing green financing forward in the GCC, not the least of which is the expected increase in renewable energy capacity there. The real estate industry received nearly the same amount of funding as all other global green bond issues in 2018. The energy industry and the buildings sector were the market leaders, receiving 31% and 28% of the total funds, respectively. As of the end of November 2018, a total of $126 billion had been allocated to low-carbon building assets and projects, according to the January 2019 study "Financing Low Carbon Buildings with Green Bonds" published by the Climate Bonds Initiative. Similarly, green construction projects received 65% of First Abu Dhabi Bank's green bond income, followed by projects including solar plants (14%), and district cooling (14%). Even though 70% of homes in the GCC area do not have enough insulation, the residential sector is responsible for around 50% of the region's overall power usage. As a result, energy efficiency standards, newly developed energy-saving technologies, and ecologically friendly building practices account for a large chunk of the region's efforts to reduce greenhouse gas emissions.

3.4.3 *Potential Growth of the Green Bond Market*

The creation of a GCC project bond market is crucial due to the GCC's emphasis on the property sector and the large number of ongoing renewable energy projects. Local and international banks in the region have a history of being flexible and offering favorable terms and rates for large-scale transactions. There has been little impact from vanilla bond markets, but that is beginning to change.

There will likely be an increase in the size of the green bond markets because of the capital markets' involvement in financing the recently announced projects. Refinancing via the financial markets may be possible for some green projects once they are operational, for instance through the issuance of green bonds.

Two recent instances of project finance arrangements that involved bond market refinancings occurred in the local area: one in 2012 involving an $825 million pre-existing debt and another in 2013 involving the $2.3 billion Ruwais Power Co. PJSC (Shuweihat 2). The first of their type, these bonds were issued by an independent Middle Eastern electricity and water firm. Emirates Sembcorp Water & Power Co. recently restructured some of its debt by issuing a $400 million project bond. When issued, both agreements were rated 'A-' by S&P Global Ratings.

3.4.4 Islamic Green Finance: A Growing Market

The green sukuk market has arisen as a response to the growing concern about climate change among both traditional and non-traditional investors. These investors are looking for new and creative methods to finance environmental projects. Green sukuk are Sharia-compliant Islamic financial instruments that can be used to invest in environmentally friendly assets such as solar parks, biogas facilities, wind power projects, and renewable transmission and infrastructure.

The Islamic finance industry globally already holds $2.1 trillion in assets, and S&P worldwide ratings predicts that this will increase by 5% over the next 2 years. Historically, investors from the GCC have been instrumental in driving the global sukuk market. We are expecting the first green sukuk issue in the area to happen soon.

The first green sukuk according to Leong et al. (2024), were issued by Tadau Energy of Malaysia in July 2017, raising $58 million for a solar power project. Since then, the industry has grown. The market's asset value is $2.1 billion as of September 2018. The green sukuk market has recently shifted its focus to Indonesia and Malaysia. However, in the future, we anticipate that issuances will increase in the GCC area.

Green sukuk are a new sort of Islamic debt financing that might increase diversity within the system while also bringing conventional and Islamic finance closer together. The growth of the green sukuk market depends on the availability of diverse funding sources for GCC issuers. Green sukuk may attract conventional investors by satisfying a need for fixed-income options and providing a high level of transparency on the distribution of profits. Using Sharia-compliant funding for environmental projects is one strategy to entice additional investors to the green finance industry. Major issuers of environmentally friendly bonds will most likely include utilities, municipalities, and perhaps real estate companies.

Table 3.2 below outlines the potential growth of the green bond market and the rise of Islamic green finance:

3.4.5 Green Finance as Emerging Trend and Future Outlook in the GCC

On the one hand, green finance and real estate in the GCC will join together on the path toward a sustainable future. With the governments, investors, and stakeholders nowadays frantically shifting the focus toward environment sustainability, we foresee continued green financial initiatives specifically within the real estate sector. However, the impetus has further to go through the latest financial instruments like the green bonds and the green sukuk that offer enviable opportunity for earnings to traditional and Islamic investors alike. Taken together with its target renewable energy goals and its aim of cutting down carbon emissions, the GCC's commitment

Table 3.2 Growth prospects in green finance and Islamic green bonds

Key points	Details
Potential growth of green bond market	– Establishing a project bond market inside the GCC is essential for securing funding for renewable projects and promoting growth in the real estate sector
	– Traditionally, both local and international banks have been open to conducting significant transactions, but the vanilla bond markets have had a limited impact
	– Capital markets could have a greater influence in funding projects and promoting the growth of green bond markets
Examples of project finance transactions	– In 2013, Ruwais Power Co. PJSC (Shuweihat 2) obtained a project bond to restructure a debt of $825 million
	– Emirates Sembcorp Water & Power Co. obtained new financing by issuing a $400 million project bond to replace its current debt
Islamic green finance: a growing market	– Increasing interest in innovative green finance among Islamic and conventional investors
	– Rise of green sukuk market for Sharia-compliant investments in renewable energy and environmentally friendly assets
	– $2.1 trillion in assets in the global Islamic finance sector, with a predicted 5% growth rate over the next 2 years
Green Sukuk Market	– In July 2017, Tadau Energy raised $58 million for a solar power plant through its first issuance in Malaysia
	– With an emphasis on Indonesia and Malaysia, the market has expanded to $2.1 billion in assets as of September 2018
	– Anticipated emergence of issuances from the GCC region, leveraging the robust presence of the Islamic financial industry
Potential impact of green sukuk	– The objective is to enhance the diversification of the Islamic financial industry and facilitate the integration of conventional and Islamic finance
	– Access to a broader pool of capital, attracting conventional investors and expanding the investor base in green finance
	– High level of certainty regarding proceeds allocation and addressing fixed-income supply gap for green investors

International Renewable Energy Agency (IRENA)

to a low-carbon future makes it the most probable area for sustainable real estate development for the world to marvel. Green finance can be used as an instrument of the solution to the existential environmental threats, but the development will be a prolonged effort and the achievement will require persistence.

Recent numbers show green finance increasing within the GCC at a high pace, indicating that there is a vital shift toward sustainability in property development in the region. The Middle East's green bond issuances reach $16.7 bn for 2024: S&P Global.

This manifestation not only highlights the fact that investors and developers now report the inclusion of environmental concerns in their real estate development strategies but also reflects the increasing adoption of environmental factors into property development. Moreover, there are forecasts that the green housing market in the region could experience an upwards trend, as a result of which green bond issuance for building projects is projected to rise by more than 60% by 2025 and hit about $4 billion. In concurrence with the intensified actions by governments and other real estate constituents to advance energy efficiency strategies and put in place sustainable development for buildings, the real demand for green finance is likely to be on a surge in the Middle East and North Africa Region, leading to a sustainable and environmentally competent built environment.

3.4.6 The UAE's Green Energy Revolution: Advancing the Use of Renewable Energy and Promoting Sustainable Development

As a result of its leaders' acute awareness of climate change's dire effects, the United Arab Emirates is driving innovative efforts to increase energy efficiency and use renewable energy sources. In line with the Paris Agreement, they have launched the first Middle Eastern Net Zero by 2050 Strategic Initiative to foster the growth of innovative thinking, eco-friendly businesses, and green jobs.

The United Arab Emirates has come a long way in achieving economic and energy diversification. Masdar, a renewable energy corporation, formed a notable collection of clean energy assets in 2022 in partnership with state-owned enterprises such as the Abu Dhabi National Oil Corporation, Taqa, and Mubadala. This initiative commenced with a committed renewable energy capacity exceeding 23 gigawatts. At the present, the UAE's economic activity is only boosted by approximately 30 percent from oil and gas exports.

3.4.7 Domestically, the UAE Is Rapidly Increasing Its Use of Clean Energy

Two of the world's biggest solar parks—Dubai's Mohammed bin Rashid Al Maktoum Solar Park and Abu Dhabi's Noor Solar Park—are located in the United Arab Emirates. The Noor Abu Dhabi solar park is expected to decrease the UAE's carbon emissions by one million metric tons per year, while the Mohammed bin Rashid Al Maktoum Solar Park is projected to provide electricity to 800,000 homes by 2030.

In October 2023, the UAE initiated its inaugural wind power initiative, featuring a project with a capacity of 103.5 megawatts (MW) spread over four locations. This program is designed to provide energy for over 23,000 residents in the UAE while cutting carbon dioxide emissions by 120,000 tons annually.

A regional frontrunner in nuclear energy, the United Arab Emirates is advancing a clean nuclear energy initiative aimed at fulfilling 25% of its electricity requirements. To address the ongoing demand for oil and gas, the UAE is adopting extensive carbon capture technology, marking it as a groundbreaking effort within the region.

As outlined in the National Hydrogen Strategy 2050, the UAE is ramping up the production of both blue and green hydrogen, aspiring to emerge as a top hydrogen producer by 2031. In alignment with economic requirements and environmental objectives, the UAE Energy Strategy 2050 seeks to create an energy mix that incorporates renewable, nuclear, and clean energy sources.

According to the plan, renewable sources will account for 44%, natural gas for 38%, clean coal for 12%, and nuclear power for 6%. In 2023, the United Arab Emirates revised its National Energy Strategy, outlining definitive plans. Among the goals set forth are the following: achieving carbon neutrality by 2050; raising the proportion of clean energy in the overall energy mix to 30% by 2031; generating 50,000 new jobs in the green sector by 2030; tripling renewable energy capacity to 14 GW by 2030; and overall creating 50,000 new jobs in the green sector.

3.4.8 Renewable Energy in UAE

The United Arab Emirates, a leader in renewable energy innovation, has a deep-rooted connection to the petroleum sector. In 2023, the country announced a significant investment of $54 billion focused on energy and renewable resources, in accordance with its updated National Energy Strategy. This financial commitment will be executed over the next seven years to meet the increasing demand for electricity.

Solar power, which has been the subject of intensive study for more than 15 years, is now available at the very low cost of 1.35 cents per kilowatt-hour. Three of the world's biggest solar power facilities are located in the United Arab Emirates (UAE), and the country's capacity to generate solar electricity is expanding rapidly.

Taking 200,000 vehicles off the road would be the same as reducing the United Arab Emirates' carbon emissions by one million metric tons per year, according to a study by the Noor Abu Dhabi solar park. Enormous enough to cover 800,000 houses' worth of electricity needs by 2030, the 4000-acre Mohammed bin Rashid Al Maktoum Solar Park in Dubai will be the biggest solar park in the Middle East. To improve the UAE's energy mix, Masdar's wind program is working to connect wind power to the grid, which will allow for further large-scale projects. More than $20 billion has been pledged by Abu Dhabi to renewable energy projects through Masdar, with an emphasis on energy efficiency, desalination, carbon management, water consumption, and renewable energy technology.

3.4.9 Nuclear Energy

To achieve its goal of net-zero emissions, the United Arab Emirates is building a nuclear energy sector that is not used for military purposes. It aspires to be the first Middle Eastern country to produce nuclear power without emitting carbon. The UAE is expected to get 14 gigawatts (GW) of eco-friendly power from nuclear power and renewable sources by 2030. In line with the goals outlined in the UAE's Energy Strategy 2050, this shows the country's commitment to diversifying its energy sources and reducing its reliance on fossil fuels.

In March 2024, the Emirates Nuclear Energy Corporation (ENEC) marked the completion of the Barakah Nuclear Energy Plant by successfully connecting it to the UAE's transmission system. With the addition of Unit 4, the national grid will receive an extra 1400 megawatts of clean electrical capacity, enhancing the UAE's efforts to strengthen energy security and grid stability through the production of emission-free electricity (Mahama and Mumin, 2023a, 2023b).

In December 2009, a bilateral agreement called the "123 Agreement" was established between the United States and the United Arab Emirates to promote cooperation in peaceful nuclear energy. This agreement aims to enhance standards, safety, and security regarding nonproliferation while creating a legal framework for the exchange of nuclear energy technologies between the two nations. US officials and nonproliferation specialists have commended the pact for its robust commitment to operational transparency, safety, and security.

The enactment of various treaties and legislation, including the UAE Nuclear Law ratified in October 2009, illustrates the UAE's dedication to a non-aggressive approach in its nuclear energy ambitions. A significant factor in advancing the Barakah Nuclear Energy Plant, managed by the Emirates Nuclear Energy Corporation, has been the pivotal role of the 123 Agreement.

To adhere to internationally recognized standards, the International Atomic Energy Agency (IAEA) and the World Association of Nuclear Operators (WANO) have carried out more than 42 missions and reviews of the UAE Program, the Barakah facility, and the operations team. Multiple American companies have participated in the United Arab Emirates' nuclear energy program.

- Westinghouse, based near Cranberry, Pennsylvania, is one such company. Westinghouse has provided crucial components, instrumentation, and control equipment, in addition to design, technical, and engineering support services, as part of a consortium headed by Korea Electric Power Corporation.
- The United Arab Emirates has received advice from the Virginia-based Lightbridge Corporation on how to set up a nuclear energy program that complies with international standards for design, development, and administration.
- In October 2008, the United Arab Emirates' nuclear program granted a 10-year contract to CH2M Hill, a Colorado-based engineering and construction firm.
- During the planning phase, the renowned Pennsylvania-based engineering and consulting company Paul C. Rizzo Associates was involved in site selection and engineering tasks.

3.4.10 Minimizing and Sequestering Deforestation

A growing body of opinion holds that in order to fulfill our climate obligations, widespread deployment of carbon capture technologies is essential.

In 2016, Abu Dhabi National Oil Company (ADNOC) finished the first large-scale carbon capture utilization and storage (CCUS) facility in the region; it captured 800,000 tons of carbon dioxide. By 2030, the UAE intends to have doubled its capacity. The United Arab Emirates is heavily developing its mangroves. Reduced vulnerability to coastal erosion, more biodiversity, and greater carbon sequestration per hectare than rainforests are the three main advantages of mangroves. The UAE plans to plant 100 million more mangroves by 2030.

The carbon intensity of UAE-produced crude-grade Murban oil is significantly lower than the industry standard. Over the next 10 years, ADNOC, an oil and gas firm, plans to halve its carbon impact. The first national oil corporation to achieve this feat, ADNOC gets all of its grid electricity from nuclear and solar power. One thousand government services and developments in six critical areas—electricity, urban planning, transportation, infrastructure, and communications—are at the heart of Dubai's "Smart City" agenda. The Emirates Green Building Council launched the initiative in 2013 with the goal of lowering the UAE's carbon footprint by implementing energy efficiency retrofits into existing structures. As part of its Estidama initiative, the Abu Dhabi Urban Planning Council is working to ensure that all projects and operations within the emirate adhere to cutting-edge environmentally friendly practices. Among the policy's declared objectives are the improvement of transportation networks, the expansion of green spaces, and the maximization of energy efficiency in both buildings and communities. "Estidama" means "sustainability" in Arabic.

When it is up and running, a waste-to-energy plant in Abu Dhabi may use the energy it produces from organic trash to power 20,000 homes.

To create hydrogen, an environmentally friendly fuel with a clean burn, various energy sources such as nuclear power, biomass, and renewable power sources like sun and wind may be combined. It has several potential applications, but its most obvious ones are in the transportation and residential sectors. Because of its abundant natural gas reserves, the United Arab Emirates can manufacture blue hydrogen. In addition, the country's plentiful sunshine might be used to create environmentally beneficial hydrogen. Unveiled at COP26 in November 2021, the Hydrogen Leadership Roadmap is an all-encompassing national plan. As a whole, it aims to help the UAE's low-carbon businesses thrive, get the nation closer to its zero-carbon targets, and set it up to become a leading hydrogen exporter.

Joined in January 2021 by the United Arab Emirates, ADNOC, Mubadala, and ADQ formed the Abu Dhabi Hydrogen Alliance. The main objective of this partnership is to encourage the use of low-carbon green and blue hydrogen in growing worldwide markets, with the aim of establishing a substantial green hydrogen economy in the UAE. Dubai made history in October 2017 when it opened the region's first hydrogen refueling station.

3.4.11 The Green Revolution: Saudi Arabia's Shift to Renewable Energy

The Kingdom of Saudi Arabia is rapidly transitioning away from its long-standing reliance on exports of crude oil and toward renewable energy sources. A major step toward the monarchy's larger goal of economic diversification, contracts for 700 MW of wind and solar power plants are set to be awarded in September 2024.

First, in the northern parts of Saudi Arabia, there will be solar power facilities with a capacity of 300 MW in Al Jouf and wind power projects with a capacity of 400 MW in Tabuk as part of this massive initiative. The Minister of Energy, Industry, and Mineral Resources, Khalid A. Al-Falih, has said that an auction is planned for April 2025 and that both local and foreign enterprises are welcome to take part in the competitive process.

The attractive terms of the renewable contracts, which seek to achieve the most economical means of producing power from these sources, are a crucial part of this endeavor. Being the first of its kind in the area, these projects will serve as a model for Saudi Arabia's private-public cooperation program. To manage and carry out this massive project, the Energy Ministry has established a brand-new Renewable Energy Development Office.

The initial phase of Saudi Arabia's National Renewable Energy Program (NREP) is represented by this project, which is projected to require an investment of $30–$50 billion and aims for a renewable power capacity of 9.5 GW by 2023. Minister Al-Falih has emphasized the wealth of the country's renewable energy resources. Effectively utilizing these assets brings it nearer to fulfilling the Sustainable Development Goals and strengthens its status as the leading energy supplier globally. This will accelerate the pace of economic transformation.

3.4.12 The Kingdom's Renewable Energy Renaissance

A groundbreaking strategy aimed at strengthening economic development is being led by the Ministry of Energy of the Kingdom of Saudi Arabia through the integration and connection of various energy sectors. This initiative is designed to stimulate economic expansion and improve both domestic and global economies, particularly emphasizing traditional energy sources, renewable energy, and the potential use of nuclear energy for non-military purposes.

The main goal is to integrate renewable energy initiatives with the plentiful hydrocarbon resources found in Saudi Arabia. The unique geographic location and climate of the nation provide significant advantages, making renewable energy both economically viable and a major factor in efforts to diversify energy sources. By leveraging its leading role in the global oil market, Saudi Arabia is expanding into alternative energy sectors, thereby enhancing its position even more.

At the heart of this initiative lies the National Renewable Energy Program, which aligns with the government's commitment to the development of renewable energy in the Kingdom and with Saudi Vision 2030. The goal of the program is to enhance the use of sustainable energy sources within Saudi Arabia, creating a balanced integration of local energy resources while meeting emission reduction commitments.

The Ministry of Energy is carrying out this initiative to diversify the energy sources for power generation, particularly focusing on increasing the shares of natural gas and renewable energy. By the year 2030, it is projected that renewable energy sources will make up around 50% of the power generating energy mix, hence encouraging an ideal, productive, and economically advantageous combination of energy sources. This transition entails shifting from liquid fuels to natural gas.

In addition, the Ministry is promoting the development of the renewable energy industry by establishing a competitive domestic market that incentivizes private sector investments and collaborations between the public and private sectors. The Kingdom has successfully attained globally unparalleled affordable costs in solar and wind projects, underscoring its dedication to growing renewable energy on a global scale.

3.4.13 Expansion of Green Finance and Sustainable Development in the GCC: A Promising Future

3.4.13.1 UAE

To support their environmental objectives, the governments of the GCC nations have laid the framework for a green financing sector. The UAE approved new laws in October 2018 to establish a framework for the issuance of sovereign bonds and sukuk, with the objective of developing its main and secondary green financial markets. The government of Dubai has made a significant pledge to spearhead green financing in the region. The goal of the Dubai Electricity and Water Authority's (DEWA) Dubai Green Fund is to make a big difference by giving investors preferential financing and funding eco-friendly projects. Its goal is to control environmental assets with a value of $27.2 billion.

It is Abu Dhabi's goal to issue more green bonds than any other city in the area. The renewable energy firm Masdar, which is based in the emirate, reached an agreement with four regional and foreign banks in September 2018 to establish a $75 million green revolving credit facility (RCF). There had never been a facility like this in the Middle East before. In line with the Green Loan Principles set out by the Loan Market Association, this RCF seeks to offer financial backing for both current and future investments in Masdar's global technology and eco-friendly property developments. Further, in an effort to promote sustainable investment, the Abu Dhabi Securities Exchange has joined the Sustainable Stock Exchanges initiative run by the United Nations.

Though there have been notable advancements, the green finance market in the GCC area is still in its early phases and lacks institutional investors like pension funds and cross-border lending, which are common in more developed capital markets. We argue that the substantial financial resources needed to achieve the sustainability goals of the region might be generated through the combination of conventional green bonds, green and conventional sukuk, and other similar instruments. To attract a broader range of investors, both Islamic and otherwise, and to reduce the cost of capital for international financing, these new and old green finance strategies might be the way to go. Many GCC utilities are considering green bond offerings, but they will likely need a price structure comparable to traditional bonds before moving further.

3.4.13.2 Qatar

Green Financing, also known as Clean Energy Financing or Eco-Friendly Financing, primarily seeks to assist small and medium-sized enterprises (SMEs) that are engaged in the production of eco-friendly goods or services or in the development of sustainable infrastructure projects, with the objectives of lowering energy expenses, environmental impacts, and carbon emissions. Energy efficiency and carbon emission reduction are two of the goals of the many programs that will benefit from these funds. Included in these endeavors are cleantech, water treatment, recycling, renewable energy, and waste management.

The Qatar Central Bank (QCB) intends to execute strategic measures regarding green finance as a component of its Second Strategic Plan for Financial Sector Regulation 2017–2022. These efforts involve organizing policy initiatives to broaden and enhance financial goods and services through diversification and innovation. The primary objectives of QCB are to streamline the process of issuing green bonds, strengthen collaboration with Qatar Development Bank to promote economic diversification through green finance, and encourage regulators to include environmental and social factors in corporate governance regulations. The Twelfth Financial

Stability Report by QCB emphasized that one of its banks has issued green bonds, while other banks have incorporated environmentally friendly methods into their business operations.

In its quest to achieve its sustainability goals, the Qatari government has effectively implemented various renewable energy initiatives and projects. The financial sector has undergone a notable transformation with the embrace of eco-friendly practices, particularly in areas like banking, lending, and microfinance. In 2019, the total value of the global green bond market reached £258 billion. It is anticipated that by 2030, economies worldwide will invest $90 trillion to achieve global climate and sustainable development objectives. London serves as a vital hub for the promotion of green finance. During the first quarter of 2019, India emerged as the second-largest market for green bonds, with a total transaction value of $10.3 billion.

Qatar is unwavering in its commitment to fostering sustainable global development through initiatives in green finance. To combat climate change, the nation has pledged to lower both air pollutants and carbon emissions. The Ministry of Municipality and Environment collaborates closely with international organizations focused on environmental and climate issues to address these challenges. As a testament to its commitment to safeguarding the environment, Qatar has officially ratified significant agreements such as the Paris Agreement, the Kyoto Protocol, and the UN Framework Convention on Climate Change. In alignment with the objectives of Qatar Vision 2030, Doha Bank has introduced green financing programs, which include loans for eco-friendly vehicles.

Qatar is diligently pursuing the goal of sustainability. This effort includes attaining Level 3 'Optimization' status in the Airports Council International's Airport Carbon Accreditation program at Hamad International Airport, advocating for compressed natural gas as a fuel for transportation, and establishing the Jetty Boil-Off Gas Recovery Facility to minimize carbon dioxide emissions. Additionally, Qatar has introduced a Green Mortgage program for home loans, providing benefits such as reduced interest rates and exemptions for energy-efficient homes.

In response to the COVID-19 epidemic, Naeem et al. (2021) argue that Qatar has experienced a significant change in its focus, with a greater emphasis on incorporating sustainability into investment and economic initiatives. Despite facing difficulties, green finance continues to be a very attractive investment option, as seen by the yearly global green bond market exceeding $100 billion. Qatar is developing its green finance frameworks to reduce negative environmental and social effects, with a particular emphasis on conserving water, improving energy efficiency, and mitigating climate change. Qatar's green finance industry is set to grow even more despite the obstacles posed by the pandemic. This growth is supported by government efforts and the strong dedication of financial institutions to sustainability.

3.4.13.3 Bahrain, Kuwait, and Oman

Bahrain, Kuwait, and Oman, like many countries globally, are progressively prioritizing sustainable growth and minimizing their carbon emissions. An essential

Table 3.3 Green finance initiative distribution in percentage

Countries	Green finance initiative distribution in percentage (%)
Bahrain	30
Kuwait	40
Oman	30

Source: International Renewable Energy Agency (IRENA)

element of this transition is the advancement of green financing projects. Green finance encompasses financial instruments and services that provide backing for projects that promote sustainability and environmental friendliness. This review will examine the green finance development initiatives in Bahrain, Kuwait, and Oman. The primary focus will be on the advancement of alternative energy sources and the decrease in reliance on fossil fuels.

The government provides incentives and subsidies to promote the use of solar electricity among enterprises and families. The wind energy potential on land is constrained and so governments are investigating the possibility of utilizing offshore wind farms as a viable and sustainable source of electricity. The nations are investigating the feasibility of transforming trash into energy, a potential solution to decrease reliance on fossil fuels. Furthermore, they are adopting energy efficiency measures across several sectors to curtail overall energy use. Furthermore, governments have implemented policies to incentivize the use of renewable energy sources and decrease dependence on fossil fuels.

Bahrain, Kuwait, and Oman are actively advancing the promotion of green financing and shifting toward sustainable energy sources, as shown in Table 3.3. By advocating for the use of alternative energy sources and decreasing reliance on non-renewable resources, these nations are actively supporting worldwide initiatives to address climate change and attain sustainable development. Sustained investment in green financing initiatives will be essential for attaining their goals in renewable energy and mitigating carbon emissions.

3.5 Renewable Energy and Sustainable Development: GCC Case Studies and Success Stories

In the march toward renewable energy and sustainable development, the GCC countries have witnessed some achievement and success of projects and initiatives. These case studies therefore provide a real picture of the way renewable energy technologies and sustainable measures are actually put into operation in our region, highlighting the success stories of green transition. Here are some notable examples:

1. Masdar City, Abu Dhabi, UAE: Masdar City is a shining beacon of sustainability, which was lit by the compelling action of the UAE to lead sustainable construction of cities. Unveiled in 2006, this innovative program represents

a first-of-its-kind eco-city, whose founding principle is predicted zero-carbon electricity generation derived exclusively from renewable energy. Masdar City organically unites state-of-the-art technologies, renewable energy sources, and urban planning progress in a buzzing redefinition of sustainable living in the world. Besides having walking-friendly streetscapes, low-maintenance public transportation modes, and extensive application of renewable energy, Masdar City is a living lab for green developments and, at the same time, a global hub for sustainable urbanism.

2. Shams Solar Power Station, UAE: The Shams solar power station, which is situated in the center of the sunlit Al Dhafra region, unveils the huge renewable resource capacity that our region is proud of. With Shams being one of the largest concentrated solar power (CSP) plants in the world, a stipulated amount of energy from the sun is employed whenever the desire for clean and secure electricity arises for a number of homes. The installed capacity of 100 MW in the Shams area not only describes a success in solar power production in the region's hot desert climate, but it also certifies the possibility of renewable power usage there. Its cutting-edge CSP system, together with its energy storage feature, provides reliable power without emission, thus aiding the fight against climate change.
3. King Abdullah City for Atomic and Renewable Energy (K. A. C. A. R. E), Saudi Arabia: K. A. C. A. R. E, an arm of the Saudi Arabian government dedicated to atomic and renewable energy, represents the country's strong vision of diversifying its energy mix and no longer limiting itself to old, fossil fuel sources. Set up in 2010, K. A. C. A. R. E represents an important catalytic force in the process of achieving the Kingdom's renewable energy objectives by influencing strategic planning, project building, and technology development. The establishment of the Renewable Energy Project Development Office (REPDO) played a significant role in increasing access to renewable energy, as utility-scale renewable energy projects such as solar and wind farms have been procured to the extent that the goals for renewable energy are becoming realizable in Saudi Arabia. K. A. C. A. R. E. is a method by which the Kingdom intends to achieve its goal of building an energy future that is both resilient and sustainable.
4. Durrat Al Bahrain, Bahrain: Durrat Al Bahrain, as a project whose roots are planted in sustainability and ecological development, will reshape the understanding of eco residing in the Kingdom of Bahrain. Durrat Al Bahrain, a perfect blend of man-made islands, is its signature element. Eco-friendly layouts of energy-efficient buildings and sustainable water management systems allow nature and the urban environment to co-exist in a harmonious manner. The nature of this significant operation is an example of environmentally friendly housing construction, highlighting sustainability values while improving the amenities for the life of the inhabitants. As a manifestation of the desire to fulfill the goal of sustainable urban planning as well as achieve clean design, Durrat Al Bahrain initiates the trend of sustainable development projects across the GCC.
5. Kuwait's Renewable Energy Pilot Projects: Kuwait has launched several pilot schemes to consider alternative energy sources. This approach includes solar PV panels and wind energy projects, which were implemented by Kuwait as efforts

toward the diversification of its energy mix as well as the reduction of exports of fossil fuels. Although they are still in their early stages, these testbeds are the crucial stepping stones toward the destination to which Kuwait is heading, and embracing renewable energy technology and environmental safeguarding are the real aims.

What these cases tackle is the use of renewable energy and sustainable development through the enhanced technologies and improved procedures that are happening in the region. Through the exchange of experience, good practices, and cutting-edge technologies the sustainable development projects end up preparing a model of a green, sustainable future in which people will be able to live, get an education, and be safe.

3.5.1 *Achievement in Green Technology and Sustainable Manufacturing*

The transformative impact of solar power is increasingly recognized across the Gulf. While it is a natural choice for domestic energy supply, the region is also placing greater emphasis on solar technology manufacturing. From Bahrain's development of a 100-MW solar park atop a landfill to initiatives by companies like Arla Foods to transition to greener electricity, Gulf states are actively advancing solar technology adoption. A key factor in this progress is collaboration between the public and private sectors, particularly in smaller, highly urbanized nations where limited land availability makes corporate-hosted solar installations essential for achieving government sustainability goals.

Using cloud technologies to enhance efficiency is a key component of sustainable initiatives. Integrating data capture and analysis tools into the energy infrastructure of both private and public organizations enables better monitoring of energy consumption and more effective optimization of manufacturing processes. For instance, Mondelēz International has implemented a cloud-based system to track water, electricity, and natural gas usage at its Bahrain facility in real time. This data can be easily compared with sustainability benchmarks, leading to measurable improvements—CO_2 emissions from the factory dropped by 65% in 2020 compared to 2018, while total waste was reduced by 70% over the same period.

Green technologies will play a vital role in the global push for net-zero emissions. The manufacturing sector, particularly in the GCC, is fostering key opportunities for technological and sustainable innovations that will drive this transition, recognizing the urgent need for climate action. Beyond technological advancements, the path to sustainability will be shaped by collaboration across all levels of society, spanning borders and continents. This is not just an economic necessity—it is essential for the stability and progress of our communities. The future of manufacturing in the GCC is promising, with green innovations leading the way forward.

3.6 Challenges and Barriers to Sustainable Finance and Renewable Energy Growth in the GCC

Overcoming the diverse difficulties and obstacles encountered in the way of sustainable finance and renewable energy growth in the GCC region needs a more complicated assessment, which is a combination of financial, technical, regulatory, and social factors.

1. Financial constraints: Firstly, the GCC countries need to notice that renewable energy and sustainable infrastructure face financial obstacles and they should address the problem. Although it is true that the energy price tags have gone down, the initial investment is still way too expensive. Even though the availability of financially sustainable infrastructural networks and investments in green projects with surmountable financial risks impede the process of scaling up these green initiatives, the government is actively taking steps to eliminate barriers and risks.
2. Technological limitations: The region struggles with persistent technological shortcomings, especially in the area of energy storage and grid integration. The typical characteristics of renewables, namely the unpredictable nature of sources and systems like solar and wind, create a need for reliable energy storage mechanisms that will contribute to the stability and consistency of the grid. The development and implementation of scalable energy storage technologies are viewed as the most critical technical riddle the Gulf state leaders are yet to solve.
3. Regulatory issues: While renewable energy and sustainability become matters of discussion, regulatory barriers grow into the biggest challenges for the expansion of renewable energy and sustainability projects. A building permit process which is full of problems, differences in the regulation policies, and ill-defined regulations can be counted among the difficulties facing investment in and development of projects. Harmonization of regulations as well as policy coherence pose insuperable problems which lead to unfavorable investment conditions and, finally, extremely limited growth of renewable energy.
4. Social and cultural factors: Interpersonal ties are a crucial factor and account for a large percentage of the reasons why people adopt renewable energy and sustainable practices in GCC societies. The established mindset that green energy is not reliable or is inferior, and reinforced mythological norms about energy consumption, wind up being a barrier to greater implementation of green technologies. Important tactics intended to raise awareness, unveil behavioral and structural obstacles, and change corresponding societal mindsets are critical to overcoming the persistent socio-cultural barriers.
5. Infrastructure and grid challenges: This means that renewable energy transition will demand sizable modernization of the initially built systems and networks to be able to handle the new features which are related to the irregularity of energy resources and distributed generation. The lack of an undercapacity connection, outdated grid infrastructure, and the perennial grid stability problems are the main hurdles that stand in the way of the effective connection into the prevailing energy systems of the large-scale renewable projects. It is necessary to invest

swiftly in grid modernization and expansion to sidestep these bottlenecks of existing infrastructure.

6. Resource constraints: In addition to having numerous resources of sun and wind, the GCC region is faced with grave challenges of water scarcity and land inadequacy that considerably thwart the reliance on renewable energy, as well as other complex environmental preservation policies and initiatives. The varied factors complicating the siting decisions, notably the necessity of finding locations with ease of access to water for urban communities as well as those lacking water availability and addressing the water-dependent cooling needs of renewable energy projects provide formidable constraints. Strategic land-use monitoring and planning are essential to overcome these limitations successfully.

The encountering of these acute difficulties has demonstrated that a single, integrated, holistic effort among all stakeholders is necessary to lift the barriers that are hindering the green transformation of the GCC. The collaborative efforts, anchored on strong policy legislation, innovative financial options, and stakeholder engagement, have to be at center stage to overcome the obstacles and chart a sustainable and resilient energy future.

3.7 Renewable Energy and Sustainability Policy and Regulatory Framework

Regulation, legislation, and policymaking structure of renewable energy and sustainability in GCC countries serve as an imperative factor determining the wise way ahead. In this portion, the paper touches on the complex but interesting policy frameworks, which include recent policy reforms, government grants, and regulations intended to propel the adoption of green energy and foster a guardianship philosophy among the masses. Table 3.4 displays the policies set by the GCC countries in terms of renewable energy strategy.

Table 3.4 Overview of key renewable energy policies in the GCC

Policy instrument	Description
UAE Energy Strategy 2050	Sets ambitious targets for renewable energy integration, energy efficiency enhancement, and carbon emissions reduction
GCC Framework for Climate Change and Environment	Fosters regional cooperation on climate action and environmental conservation
GCC Unified Environmental Law	Harmonizes environmental regulations and enforcement mechanisms across member states
GCC Renewable Energy Policy Framework	Demonstrates collective commitment to advancing renewable energy deployment and energy diversification

UAE Policy Landscape The UAE has been through the most noticeable policy change in recent years, with further initiatives adopted to motivate a greener approach focusing on renewable energy and sustainable living agendas. The Vision 2021 of the UAE, which contains the twin goals of economic and social development in a sustainable energy transition and ecology, provides a concise roadmap for the country's way forward. However, the UAE Energy Strategy 2050 is the key guide that is at the center of the renewable energy aspirations of the nation, pledging to integrate energy sourced from renewables, enhance energy efficiency, and reduce carbon emissions. The UAE National Climate Change Plan 2017–2050, which is a crucial part of the government's tax policy, serves to reinforce the government's resolve to combat climate-related challenges through appropriate mitigation and adaptation steps.

GCC Policy Harmonization Although the heterogeneity of policy approaches in the GCC member states is well recognized, the thrust toward policy harmonization and standardization is gradually hitting high gear. The framework "GCC Framework for Climate Change and the Environment" represents a major watershed in the creation of a structured environment among countries for the fight against climate change and environmental conservation. Moreover, 2019 was the date of the establishment of the GCC Unified Environmental Law that served as the basis of the harmonization of environmental laws and enforcement mechanisms at member states level, thus boosting a facilitating regulatory framework that is making way for green projects. The creation of a GCC Renewable Energy Policy Framework enhances the region's further action to accelerate renewable-energy development and energy diversification.

Government Incentives and Subsidies The governments of the GCC have played a vital role in influencing policies encouraging renewable energy adoption and the wide scale utilization of environmentally friendly practices. Strong FIT (feed-in tariffs), procurement guarantees, and RECs (renewable energy certificates) have promoted private sector investment in renewable energy projects. And lastly, the targeted subsidies distributed for energy-efficiency technologies have been an option, e.g., LED lighting and energy-efficient fridges to motivate consumers and make the required improvements.

Regulatory Mechanisms and Market Reforms The regulatory instruments have a hand in determining the habitat of renewable power implementation as well as greening efforts. The enactment of independent regulators, for instance UAE's Federation Electricity and Water Authority (FEWA) and the Dubai Electricity and Water Authority DEWA), shows how much they care about establishing a transparent regulatory body that is accountable and has oversight of the market. Renewable energy deployment will be boosted by reforms implemented in local markets, such as the ones designed to empower interconnection and beneficial grid modernization through ramps of renewable energy inclusion, generating a very energy secure and resilient scenario.

Public Awareness and Education Through organizing campaigns, the knowledge of conservation of energy and sustainability will be communicated and will enlighten the people on the use of these resources. Thus, environmental stewardship will be embedded as one of the most important pillars for installing green behaviors in all aspects of life. It further sheds light on how the GCC countries and other nations are actively encouraging participation in activities which have conservationism and sustainability as their main objectives through communities, schools, and the business sector.

1. Community engagement initiatives: Local campaigns to make sustainable energy more popular and eco-friendly initiatives literally rally the community around change. The UAE is where the grassroots organizations and non-governmental groups head the initiatives of teaching and spreading community knowledge on the benefits of adopting renewable energy at home as well as embracing environmental conservation habits. For instance, widely known events, like Sustainability Week, offer a forum for interactivity and networking among community members, policymakers, and industry facilitators as well as a tangible path to promote sustainable development objectives.
2. School curriculum integration: Incorporation of renewable energy as well as sustainability lessons in the school curriculum is a strategy that can actively generate the next generation of nature-loving youth and embed the concept of an environmental culture early in lives. The education authorities in the GCC countries are establishing multi-faceted renewable energy and sustainability streams within the national curriculum, covering social sciences just like science and technology, as well as geography. By means of the students' learning with their hands, for example, when organizing field trips to green energy facilities and class activities designed to be interactive, students are able to understand complicated environmental concepts and value the importance of leading a green life.
3. Corporate sustainability programs: Business and industries, being highly influential, play an instrumental role in initiating sustainable activities within the communities within their boundaries. Sustainable programs at corporate level have become more apparent, ranging from energy efficiency initiatives, reducing carbon footprint, and other similar strategies. Companies in the GCC have adopted these programs as well. Major corporations capitalize on their large user bases to share with them the renewable energy aspirations as well as the economic benefits linked to environmentally responsible business ownership. Being targeted in collaboration allows public-private partnerships to achieve the largest impact of sustainability efforts, because innovative interventions and solutions can be scaled and spread out into the environment.
4. Mass media campaigns: Mass media platforms serve for much-needed information-distributing and opinion-building about renewable energy and sustainability. Such campaigns on these umbrella platforms often play pertinent roles in reaching people with varied backgrounds and in creating behavioral changes toward more green living. Forwarding the message with the application of multimedia networks such as videos, infographics, and social media

campaigns, aids in telling narratives about innovative renewable energy as well as the mitigation of climate change and the collective need for the environment's preservation to protect planet earth for the subsequent generations.

3.7.1 International Collaboration and Partnerships

The joint action and association of the GCC countries and international organizations seem to us the most powerful and efficient tools to create unity and speed up changes for the same cause. This part explains the complex aspect of international cooperation, which involves skills and knowledge sharing, and technology transfer, together with the financial institutions that are being set in place to further advance innovation and green projects.

Bilateral and Multilateral Agreements The GCC countries proactively undertake bilateral and multilateral agreements with foreign governments and international organizations, as a basis to build up collaborative frameworks for advancing renewable energy and sustainability agendas. These treaties ease the sharing of skills, expertise, and best practices within the community as well as promoting collaborations and joint research and development initiatives. Notable is the readiness of the UAE to go into specific strategic partnerships with outstanding renewable energy leaders from Germany, USA, and Japan, which were very effective and paved the way for more bilateral cooperation.

Global Platforms and Forums The GCC countries‘ participation in international platforms and meetings not only brings the capabilities that they possess to worldwide recognition but also enhances the recognition of their contributions to the issues of renewable energy and sustainability in the international arena. In addition to two forums, the World Future Energy Summit and the United Nations Climate Change Conferences, that bring diverse stakeholders—decision makers, representatives of industry, and civil society—onto the common platform, dialogue and knowledge sharing is realized. Since the countries in this region actively involve themselves in these platforms, they are able to pool their expert knowledge and resources as well as address the challenges that are common to the region as a whole and network opportunities for sustainable development.

Technology Transfer and Capacity Building International cooperation helps the technology transfer and the capacity-building initiatives which are shown to benefit the renewable energy infrastructure of the GCC countries. Technical collaboration with technologically advanced countries ensures that the local community is abreast of the latest renewable energy technologies; the capacity-building programs ensure that the local community has requisite knowledge and skills to deploy, run, and maintain renewable energy systems effectively. Joint R&D projects initiate innovation and the sustainable solutions are developed in accordance with the social and environmental setting of the GCC region.

Financial Support and Investment Opportunities International cooperation also emphasizes financial support and investment procedures that have been made or are yet to be made to foster renewable energy and sustainable development in the GCC countries. Multilateral development banks, charitable organizations, and international funding agencies, who act as primary players, offer variable funds, grants, and technical assistance aimed at realizing the commencement and continuation of renewable energy projects, energy-efficient practices, and sustainability initiatives. Through public-private partnerships and innovative financing methods, more capital can be created, which in turn offers a basis for private investment to develop renewable energy and infrastructure projects in this area much faster and on a larger scale.

3.7.2 Future Directions and Opportunities

Knowing which direction developments will take in renewable energy and sustainability in the GCC region is the foundation for charting the course toward a green and resilient future. This part of the paper focuses on new trends, finds scope for innovation, and, finally, reveals the fields that may still be possible for the future development of the green energy market in a context of industry dynamics.

1. Accelerating renewable energy deployment: The GCC is set to drive the acceleration of the renewable energy technologies by means of its lofty ambitions, the well-established frameworks for regulation, as well as the fact that the technologies of renewable resources are getting cheaper every day. With clean energy sources' competitiveness furthered by improved costing, ramping up solar, wind, and other renewable capacity becomes even more of an appealing push. Undertakings aimed at simplifying the approval process for projects, increasing the grid integration of renewables, and encouraging the cooperation of government and the private sector will be effective catalysts of renewable energy expansion and contribute to the diversification of the energy mix.
2. Investment in research and innovation: Spending money on research and innovation is a very meaningful step toward setting free the whole capacity of renewable energy and sustainability in the GCC zone. Additional funding for technology research institutions and various incubators and innovation centers will be the impetus for growing technologies and improving the storage methods as well as smart grids. Interdisciplinary research projects with the participation of academia, industry, and government officials with new expertise will work on materials science and engineering, energy management, and policy formulation of the entire value chain.
3. Integration of digital technologies: Employment of digital technologies including AI, blockchain, and Internet of Things (IoT) provides a chance for a remarkable development of renewable energy systems. This could help with greater reliability, efficiency, and resilience. Using high technology predictive modeling and advanced analytics, one can achieve effective energy forecasting, demand side

management, and optimization of the grid, which will ease incorporation of variable renewable energy sources into the grid without impact on the other sources. Smart energy management systems stimulate consumers' ability to tune in and manage their energy use down to the moment, thus promoting the growth of an energy efficiency and conservation environment.

4. Ecosystem for green finance and investment: Designing an environment that includes green financial instruments and investments is of the essence, as it will allow the allocation of capital toward green development. GCC nations can shift the paradigm to become a global center for green finance owing to their strategic location, financial expertise, and the strive toward sustainability. Established creative funding strategies like green bonds, green sukuk, and green impact funds will meet the demand of both articulate local and foreign investors seeking profit and social impact with their assets.
5. Circular economy and resource efficiency: Implementation of the principles of the circulating economy and resource efficiency will be the core factors in achieving lasting success in the GCC region's sustainability objectives. To decrease local impact, save energy resources, and minimize waste production, strategies such as circular business models, waste-to-energy technologies, and sustainable management practices will be implemented. This transition will be grounded in regenerative eco-design, product lifecycle optimization, and sustainable consumption patterns that promote an economy balancing economic growth with environmental preservation.

3.8 Conclusion

GCC nations are tackling the interconnected challenges of environmental degradation and economic resilience across sectors like energy, transportation, and urban development. The findings of this chapter highlight the critical role of green transformation in reshaping the region's economic and environmental landscapes. By incorporating sustainability principles, this comprehensive approach ensures that economic growth progresses without worsening climate-related risks, fostering a future where ecological health and economic development mutually reinforce each other.

A central aspect of this transformation is sustainable finance. The region is using innovative financial tools such as green bonds, sustainability-linked loans, and ESG investments to build robust financial systems that can withstand global shocks. These new financial instruments not only attract investment but also incentivize companies to adopt cleaner, more sustainable practices. Green finance enables GCC countries to support large-scale renewable energy projects, such as solar parks and carbon capture initiatives, while ensuring long-term economic stability.

The implementation of these financial mechanisms highlights the connection between financial stability and environmental sustainability, demonstrating that economic growth can occur without harming the ecological environment. This chapter focuses on the importance of integrating GCC sustainability goals, especially at a time when the world is working to address climate change. The GCC countries are embedding sustainable development principles into key sectors such as energy, transportation, and urban development to address the dual challenges of economic resilience and environmental degradation. This approach ensures that economic development can proceed alongside the preservation of ecological well-being, whether there is rapid accumulation of financial wealth. Additionally, sustainability initiatives in renewable energy and green finance provide reliable opportunities for collaboration and progress in enhancing energy security, job creation, and technological innovation. These are central to a broader international effort to fight climate change and are an opportunity to create more resilient and sustainable infrastructure. Other examples are the building of the Mohammed bin Rashid Al Maktoum Solar Park in the UAE and wind energy projects in Saudi Arabia. They clearly reflect the focus of the region on the adoption of renewable energy. More importantly, they position the GCC as the international forerunner in the rapid global transition toward a low-carbon economy. This chapter also highlights the importance of addressing issues such as regulatory barriers, technological limitations, and social acceptability of renewables for all the combined forces of policymakers, industrialists, and civil society in the GCC. That is, under strong regulatory frameworks, with proper communication strategies for public awareness, and inclusiveness in local research and development investments, the GCC nations will be able to accelerate the green transition with the opening up of new avenues for sustainable growth. The GCC's path toward sustainable finance and green growth is a contribution beyond a merely regional initiative to the fight against climate change. Through the GCC, there is forming a system that unites economic resilience with ecological sustainability via new financial mechanisms and progressive policies. That transformation tends to put the GCC in front with regard to sustainability, with creation space for a sustainable inclusive future. The chapter underscores how important is the shift of the whole GCC to green practices not only for global challenges but also for energy security and long-term growth for the region and beyond.

References

Al-Fares, H. (2019). *Sustainable finance in the Arabian Gulf: A challenge and opportunity*. Emerald Publishing Limited.

Al-Sarihi, A., & Mansouri, N. (2022a). Renewable energy development in the Gulf Cooperation Council Countries: Status, barriers, and policy options. *Energies, 15*, 1923.

Al-Sarihi, A., & Mansouri, N. (2022b). *Sustainable finance in the Gulf Cooperation Council Countries: Challenges and opportunities*. Springer.

Hafner, M., Raimondi, P. P., & Bonometti, B. (2023). *The energy sector and energy geopolitics in the MENA region at a crossroad: Towards a great transformation?* Springer Nature.

Khalid, R., & Alshami, H. S. (2020). *Sustainable finance and development in the GCC Region: Policies, practices, and prospects*. Palgrave Macmillan.

Leong, W., Heng, L., & Leong, Y. (2024). Malaysia's renewable energy policy and its impact on ASEAN countries. *Advances in Science, Technology and Engineering Systems Journal, 9*, 67–73.

Mahama, A. J., & Mumin, M. A. (2023a). *Renewable energy resilience*. IGI Global.

Mahama, A. J., & Mumin, M. A. (2023b). *Renewable energy resilience: Strategies for sustainable finance in the GCC*. IGI Global.

Naeem, M. A., Farid, S., Ferrer, R., & Shahzad, S. J. H. (2021). Comparative efficiency of green and conventional bonds pre- and during COVID-19: An asymmetric multifractal detrended fluctuation analysis. *Energy Policy, 153*, 112285.

Samargandi, N., Islam, M. M., & Sohag, K. (2024). Towards realizing vision 2030: Input demand for renewable energy production in Saudi Arabia. *Gondwana Research, 127*, 47–64.

Chapter 4
Good Governance, Democracy, and Economic Resilience in the Gulf Region: A Post-Pandemic Analysis

Fatih Kırşanlı

Abstract This chapter concentrates on a post-pandemic political economy analysis of the Gulf region. It examines the institutional quality (i.e., corruption and participatory democracy scores) of the Gulf Cooperation Council (GCC) members (Bahrain, Kuwait, Oman, Qatar, the Kingdom of Saudi Arabia, and the United Arab Emirates (the UAE)) by looking at the Transparency International and Varieties of Democracy (V-Dem) data sets. The data reveals that the region was resilient during the pandemic due to the policies implemented by each government; however, there is an acute need for institutional improvement, particularly for democratic governance at the local and national levels. The chapter examines the fiscal breakeven oil prices for each country. The findings highlight the inevitability of an increased pace of diversification, considering the present dependence on oil. The analysis underlines that GCC countries performed relatively well compared to the rest of the world due to their financial resources. However, countries with higher-quality institutions before COVID-19 (e.g., the UAE and Kuwait) were more resilient to the negative repercussions of the pandemic. Finally, the chapter examines regional and country-specific policies implemented during the pandemic to improve institutional quality and strengthen economic resilience in the long term.

Keywords Institutions · Economic resilience · Democracy · Political economy · GCC

JEL Codes E02 · F50 · N25 · N45

F. Kırşanlı (✉)
Department of Economics, College of Political Science, Social Sciences University of Ankara, Ankara, Türkiye
e-mail: fatih.kirsanli@asbu.edu.tr

H. Miniaoui and M. Bellalah (eds.), *Green Growth Opportunities and Sustainable Finance in the Gulf Cooperation Council Region*, Gulf Studies 24,
https://doi.org/10.1007/978-981-96-8287-4_4

4.1 Introduction

The COVID-19 pandemic created an unprecedented global environment, politically, socially, and economically. When this unexpected crisis began in China in late 2019, it disrupted the global economy. It led to lengthy public isolation through social distancing, closure of educational and communal institutions, and travel restrictions. The way businesses, education, tourism, transportation, health services, and telecommunications are conducted has altered in a way never seen before in human history. The cancelation of big events and international flights, and the transition to online learning, were the main consequences of the pandemic. The global economy shrank, inflation increased due to supply chain bottlenecks, public debts increased because of financial support provided to households and businesses, and the stock markets collapsed for a long time.

Total global output shrank by 3.0% in 2020, while contractions were 4.4% for advanced nations and 1.9% for emerging markets and developing economies (International Monetary Fund, 2022). Reduction in total global output combined with future projections of low oil demand, higher supply because of technological breakthroughs and the US production levels, and the necessity of government intervention with stimulus packages altered the paradigm of the world and made it more vulnerable to shocks (Dale & Bassam, 2018). Along with these counted shifts, the pandemic also caused the loss of recent gains such as women's labor force participation, which decreased (Young, 2023b) and exacerbated gender inequality.

After the initial shock in 2020, the post-pandemic environment steadily recovered in Asia, followed by the MENA region. Due to its power, financial stability, and lower population, the GCC members amplified their dominance over the rest of the MENA region, since the countries outside the Gulf had economic and financial fragilities (Hanieh & Rafeef, 2022). GCC members have, though, also been affected by the damaging aftershocks of the pandemic. Since GCC countries meet their budget requirements mostly from oil, the pandemic aggravated the late 2014 oil shock after less than a decade. Figure 4.1 below shows the volatility of oil prices, with three major shocks since the 1990s; the first was after the 2008 mortgage crisis, the second in late 2014, and the most recent the pandemic. The precariousness of oil prices caused budget deficits in many countries including GCC members, pushing them to implement economic diversification policies.

Through their social safety nets (e.g., Sovereign Wealth Funds (SWFs)) and highly financialized economies, GCC members had mechanisms to stabilize their economies when the pandemic hit. Through this financial and business resilience, the Gulf countries kept going through the pandemic's demand, supply, and distribution uncertainties. The search for diversification, which was already underway before the pandemic, intensified following the pandemic's economic effects, with increased supply and lowered demand in the oil market (Young, 2023b). The need for diversification due to external shocks, the last of which is the pandemic, may also be seen from the debt structure of the GCC members. The total debt of US$70 billion verifies that there is a sound capital market for all GCC members, including Bahrain

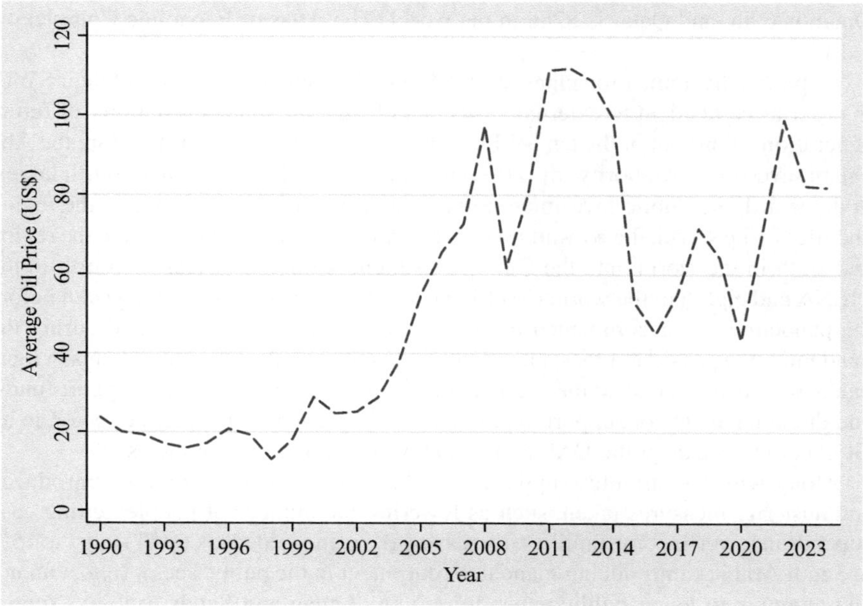

Fig. 4.1 Average oil price (per barrel). Source: FRED Economic Data—St. Louis FED

and Oman, which had relatively lower credit ratings, yet that compels the region into a debt habit that relies on additional borrowing to balance the budgets (Barclays, 2021). Hence, GCC countries strived to diversify their economies, and these efforts are expected to drive the MENA region's growth (Fitch Solutions, 2023).

Economic diversification is the precondition of economic development and resilience, but each council member has had different paths for implementation since the 1970s. Despite plunging oil prices in 2014, weak macroeconomic indicators, fiscal and monetary deterioration, current account problems, and a sharp decline in the private sector, oil revenues are still at the core of economic growth in the Gulf (Fattouh & Anupama, 2021). A satisfactory performance requires further implementation of complex structures of technology, advanced knowledge, and know-how in the long run, as diversification means diversifying export packages by moving away from one single good (Beutel, 2021).

Since the 1970s, and especially after the collapse of oil prices from 2014 until 2016, the GCC members have implemented structural energy reforms that vary depending on the size of the country, its resources, and its geopolitical view (Moerenhout, 2021). The COVID-19 pandemic accelerated the GCC members' transformation dynamics and diversification channels. The Economic Complexity Index ranks countries based on the diversification and complexity of their exports. The GCC countries are doing relatively better than their counterparts in the rest of the MENA region, as Saudi Arabia is 38th, Bahrain is 43rd, Kuwait is 55th, the UAE is 56th,

Oman is 69th, and Qatar is 82nd in the world (The Atlas of Economic Complexity, n.d.).

As part of diversification plans, the GCC members enlarged the size of their SWFs in order to be resilient to external shocks and hedge the financial stability of future generations. Four out of the ten SWFs in the world are from the Gulf region; the Abu Dhabi Investment Authority (the UAE) is the largest in the MENA and fourth largest in the world, the Public Investment Fund (Saudi Arabia) is the second in the region and fifth in the world, the Kuwait Investment Authority (Kuwait) is third in the region and sixth in the world, and the Qatar Investment Authority (Qatar) is fourth in the MENA and eighth in the world (SWFI, n.d.). SWF investments, existing even before the pandemic, are accompanied by other general policies implemented during the pandemic. Suspension of fees, electricity payments in industrial areas, loans in local banks with less interest, some personal checks to local banks, SMEs support funds, and direct payments of support to low-income employees in the private sector, in all GCC countries except the UAE and Oman, were among those policies.

Along with the stimulus support, there have been some restrictions introduced and austerity measures taken, such as lowering the number of people getting cash stipends in lower-income families, increasing the value-added tax (VAT) (e.g., to 15% in Saudi Arabia), introducing mandated retirement in the public sector (e.g., Oman), and attempts to lower public expenditure (e.g., Oman and Saudi Arabia) (Young, 2023b). Further, countries chose different paths for pricing reforms. For example, Bahrain, Kuwait, and Saudi Arabia chose ad hoc pricing. Oman, Qatar, and the UAE have adjusted price settings according to the regional and international movements of oil prices (Moerenhout, 2021). In Oman, the government introduced a repayment plan for borrower SMEs rather than providing checks. Nonetheless, the pressure of social spending despite declining government revenues creates a structural weakness for GCC members, and the pressure was being felt by the ruling governments even before COVID-19. Thus, fiscal and monetary spending need to be readjusted and contracted (Young, 2023b), and they need to be supported with higher institutional quality and democratic openness.

Foreign employment in the Gulf is also affected by the dramatic shift created by the pandemic, which jeopardized hiring commitments and remittances flow. The governments conceive through the lenses of their Visions (e.g., Saudi Arabia Vision 2030, the UAE Vision 2040), and how to lower foreign employment and nationalize the labor force to sustain economic development for the locals. Regarding that path, the UAE has more flexibility due to its smaller national population and relatively fewer public sector payment commitments. The UAE can maintain this tie of foreign employment for the foreseeable future. Expatriates play a key role in the diversification efforts of the GCC nations and in the institutional quality arrangements and democratization processes, which will be discussed later in relation to the *kafāla* (sponsorship) system.

Another important implication of the pandemic is the existence of Gulf capital in other regions in terms of military spending or humanitarian aid, such as within the MENA region or in Africa and Pakistan. An increasing number of countries are looking for GCC investment, defense, or humanitarian assistance due to the

Gulf's economic growth potential and its geostrategic proximity to Europe, Asia, and Africa. Nonetheless, when fiscal and monetary pressures rise in the region, the external funding of these countries may be at risk simultaneously (Young, 2023b).

The following sections focus on individual members of the GCC and examine the policies executed during the pandemic, along with a political economy analysis of their fiscal breakeven oil prices, and their corruption and democracy scores. Oil price and institutional quality analysis reveal that the region has gone a long way in terms of diversification, regional politics, and institutional order transformation. Yet, it still has fragilities that the governments must address. Considering the ramifications of the pandemic, the country-specific analysis will help comprehend the overall performance of the GCC region through programs and reforms, as well as its governance and resilience. After examining each country separately, the final section concludes.

4.2 Bahrain

Bahrain was the first country in the Gulf where oil was discovered, in 1931 (Aljazeeri, 2021). Despite this, it is affected by oil price fluctuations due to budget concerns. The debt-to-GDP ratio of Bahrain was 101% just before the pandemic in 2019. Since then, albeit only slightly, there have been recoveries in the debt structure of other members of the Gulf region, yet Bahrain's debt is increasing annually and is expected to reach about 141% in 2029 (Statista, n.d.-a). Under this debt burden, Bahrain needs the highest level of oil price to meet its fiscal needs. Debt that GCC members incur to sustain their budget will constrain future fiscal policies. The total funding needs to be 11.4% of Bahrain's GDP by 2024, meaning that borrowing will continue (Barclays, 2021). Figure 4.2 below illustrates Bahrain's fiscal breakeven oil price levels. According to the IMF projections, Bahrain requires more than US$120 per barrel of oil in 2024 and 2025. At the time of writing oil prices are around US$83, and the Bahraini budget will be in massive deficit in upcoming years if oil prices continue to be around the US$80–85 range.

Before the pandemic, Bahrain launched its Vision 2030 campaign. The Vision aims to reduce oil dependency by 2030 and achieve sustainable development goals. It explicitly aims for a 700-MW renewable energy generation capacity. It also aims to produce 5% and 10% of the country's energy from renewable resources by 2025 and 2035, respectively, with electricity being generated by natural gas until 2030 (Al-Sarihi et al., 2023). As a result of diversification processes, Bahrain's goal is to have renewable energy sustained by low oil dependence; however, considering Bahrain's need for above US$120 per barrel of oil, the goals seem ambitious, at least to reach by 2030.

Among other policies initiated during the pandemic, Bahrain installed 535 solar panels in government institutions to produce clean energy (Al-Sarihi et al., 2023). The country aims to increase the installation of solar panels in the future. This transition increases the availability of clean energy and new methods of financing. COVID-19 directly affected the Bahraini economy in terms of economic growth and

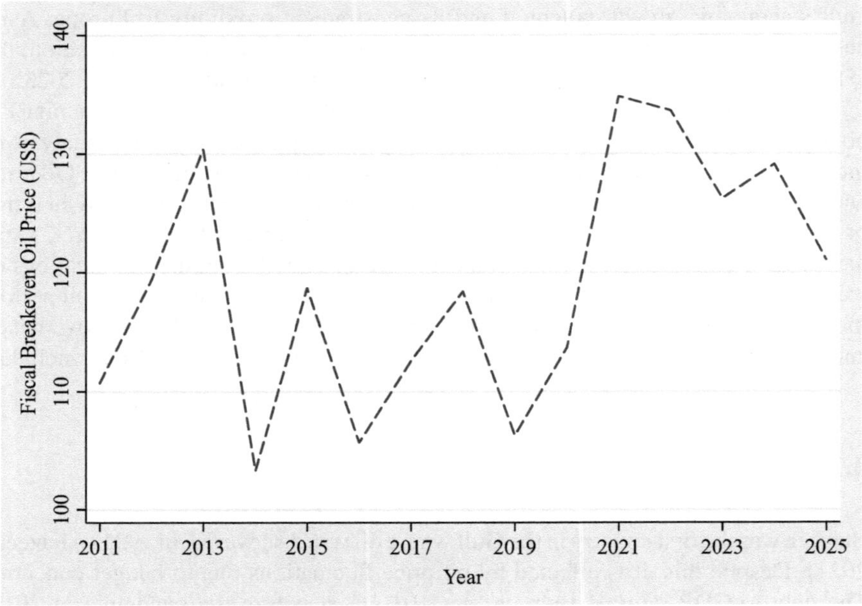

Fig. 4.2 Fiscal breakeven oil price (per barrel). Source: 2011–2019 data is from Oil and the Political Economy of the Middle East (Beck & Thomas, 2023, p. 5); 2020–2024 data is from the IMF and 2025 data is from Fitch

budget deficit, which also directly impacted energy investments. Due to the external shocks created by the pandemic, some companies faced the jeopardy of bankruptcy and needed government intervention. The Bahraini government provided loan relief and interest rate discounts to those companies. In addition, stimulus checks, bond issuance, and utility assistance were delivered (Young, 2023a, p. 102).

In terms of institutional quality and democratization, Fig. 4.3 below illustrates the Corruption Perception Index (CPI) and participatory democracy scores of Bahrain. In CPI data, which is between 0 and 100, where the higher score represents less corruption, Bahrain did not perform well during the post-Arab-Spring period; however, there has been a slight upward trend since the pandemic, albeit a recent decline. Moreover, the participatory democracy[1] score has been increasing since 2018, even before the pandemic, which is promising for Bahraini democracy. Corruption and participatory democracy scores underline the improvement in institutional quality,

[1] V-Dem has electoral, liberal, participatory, and deliberative democracy scores. To the best of my knowledge, utilizing the participatory democracy variable is more meaningful for the purpose of this study. However, the analysis does not alter with other variables. Also, corruption scores are indexed in World Governance Indicators (WGI), International Country Risk Guide (ICRG), and Varieties of Democracy (V-Dem), but there is a strong positive correlation between these indices and CPI (Kırşanlı, 2023a, p. 52) and the results do not change with alternative measures (Kırşanlı, 2023b).

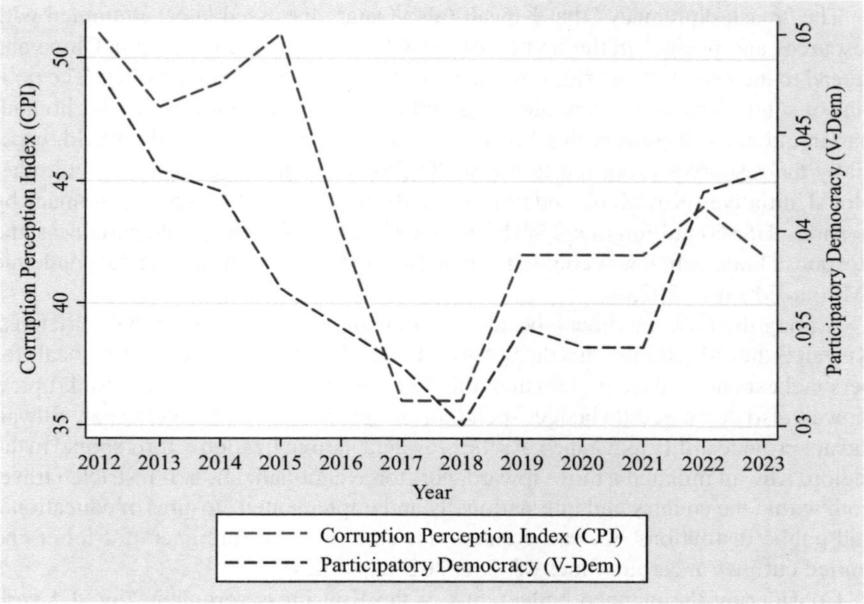

Fig. 4.3 Corruption Perception Index (CPI) and participatory democracy performance. Source: Transparency International and V-Dem

which helps the country to be more resilient to future shocks; nevertheless, that needs further emphasis to reach globally accepted standards. On the other hand, Kuwait has made massive progress in institutional quality and accomplished the highest level of democracy and freedom of expression in the region, whose case is examined next.

4.3 Kuwait

Kuwait has a 2035 Vision, and the country wants to benefit from the Chinese Belt and Road Initiative (CBRI) and from Iraq and Iran because of its proximity and geostrategic location (Hoetjes, 2021). As a small state, Kuwait attempted to be resilient in the past and put more weight on its capacity to be more impactful in the diplomatic process of conflict resolution and humanitarian aid. The strategic move of leveraging its status in diplomatic relations to promote multilateralism has been significant in Kuwait's foreign policy, yet due to the failures of the world's advanced nations and elite institutions during the pandemic, this meaningful cornerstone policy was reduced in diplomatic and economic arenas. Malfunctioning in global governance and partnership lowered the significance and capability of small states that may aspire to punch above their weight (Milton-Edwards, 2023a, 2023b).

The "mask diplomacy" that Kuwait (also Qatar, discussed later) attempted with resources and pledges in the first phases of COVID-19, which started in China and spread to the rest of the world, were important for Kuwait's foreign policy. The policies of sending resources and pledging millions of dollars of donations for humanitarian aid are soft powers that boost the country's importance in the world, especially the resources promised to the WHO (World Health Organization) and other global initiatives. Kuwait proved its moral and ethical status through its diplomacy by donating US$60 million to the WHO for the COVID-19 Strategic Preparedness and Response Plan, which was counted one of the largest donations during the pandemic (Milton-Edwards, 2023a).

Among the GCC members, besides the donations made to the global initiatives, Kuwait is the only country that did not issue bonds. It sent stimulus checks to locals for personal expenditures (e.g., fees and rents) and to SMEs to prevent their bankruptcy. Kuwait also decreased its budget spending. Together with Oman and Qatar, Kuwait did not provide utility assistance. As the prevalent nationalization efforts spread in the region, Kuwait initiated a move toward workforce nationalization. It restricted travel both within the country and internationally and implemented closures of educational and public institutions. Kuwait granted visa extensions to foreigners for relief and started curfews at certain times (Young, 2023a, p. 102).

Considering the planned budget cuts of the Kuwaiti government, Fig. 4.4 verifies the necessity of budget discipline in spending. The fiscal breakeven oil prices exceeded US$85 in 2021, and the IMF foresees around US$80 per barrel of oil for a balanced budget in 2024 and 2025. Albeit lower than Bahrain, the requirement of US$80 of oil per barrel for a prolonged period is affected by global geopolitics and volatilities. In addition to the projections of low demand for oil from 2035 and beyond, Kuwait needs a transformational shift from oil and to diversify its economy.

Kuwait's main advantage in transforming and diversifying its economy is its freer parliament and democracy. As Fig. 4.5 below demonstrates, Kuwait's corruption and participatory democracy scores have also improved since 2020, giving an incentive that through institutional resilience, economic diversification and lower oil dependency can be accomplished. To do so, institutional quality (i.e., corruption) needs to be improved locally by initiating policies for participatory democracy.

With a GDP of over US $40,000 per capita, Kuwait is approaching being among the high-income countries, which helps transform the economy with structural reforms outside of oil that advance development and welfare. When Kuwait manages to be more resilient toward external shocks with higher institutional quality and democracy scores, then it can set an example for the rest of the Gulf region and attain the power it desires through its "mask diplomacy". Another member of the GCC that is resilient to internal and external shocks through its peace diplomacy is Oman.

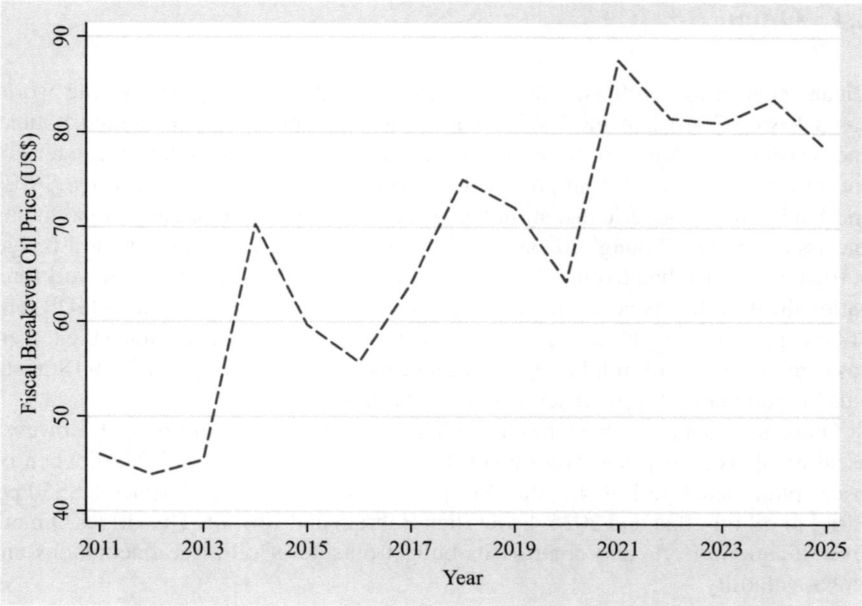

Fig. 4.4 Fiscal breakeven oil price (per barrel). Source: FRED Economic Data—St. Louis FED

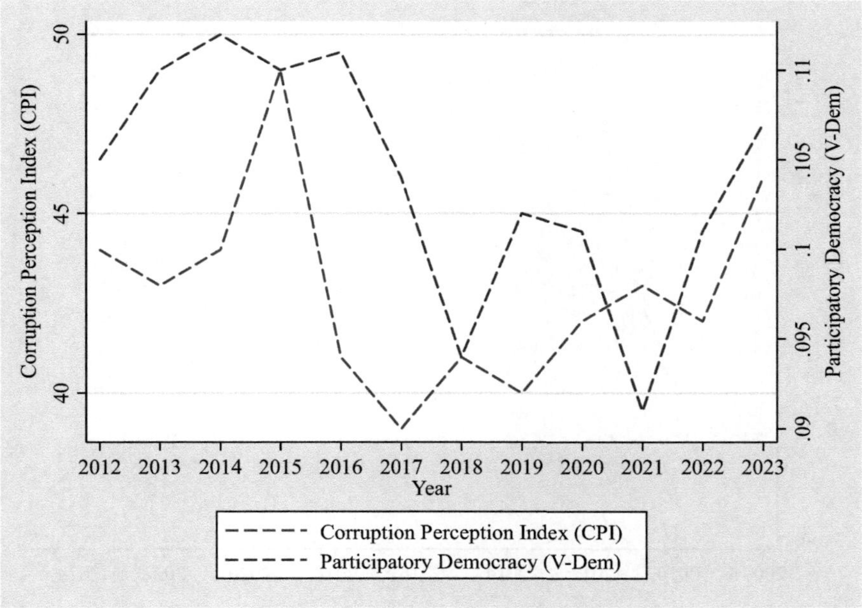

Fig. 4.5 Corruption Perception Index (CPI) and participatory democracy performance. Source: Transparency International and V-Dem

4.4 Oman

Oman, neutral to conflicts, wars, and turmoil in the Gulf region and the world, has a Vision 2040 like the UAE. During the pandemic, it implemented policies and restrictions similar to those of other Council members. Besides visa extension and utility assistance, Oman provided all necessary support to its citizens, SMEs, and banks. It provided loans, stimulus checks, and interest rate cuts on payments, and issued bonds (Young, 2023a, p. 102). These, expectedly, came with a budget burden that Oman had to cut. Oman implemented closures, curfews, and workforce nationalization for its economy. Contrary to Bahrain, the public-debt-to-GDP ratio of Oman jumped to 68% during the first year of the pandemic, and since then, there has been a downward trend in the debt ratio, which will be 36.5% in 2029 (Statista, n.d.-b); good news for the macroeconomic performance.

Oman's advantage is that it has lower breakeven oil prices for its budget. However, fiscal breakeven oil prices were much higher from late 2014 until 2016, when oil prices plummeted. In Fig. 4.6, the IMF projects a requirement of nearly US$50 per barrel of oil for 2024 and 2025, lower than Bahrain and Kuwait. This drives Oman's diversification efforts and controls its budget outside of oil price fluctuations and global volatility.

Among the diversification programs, Oman introduced a new policy for immigrants that encourages investors and protects skilled long-term residents. Oman also

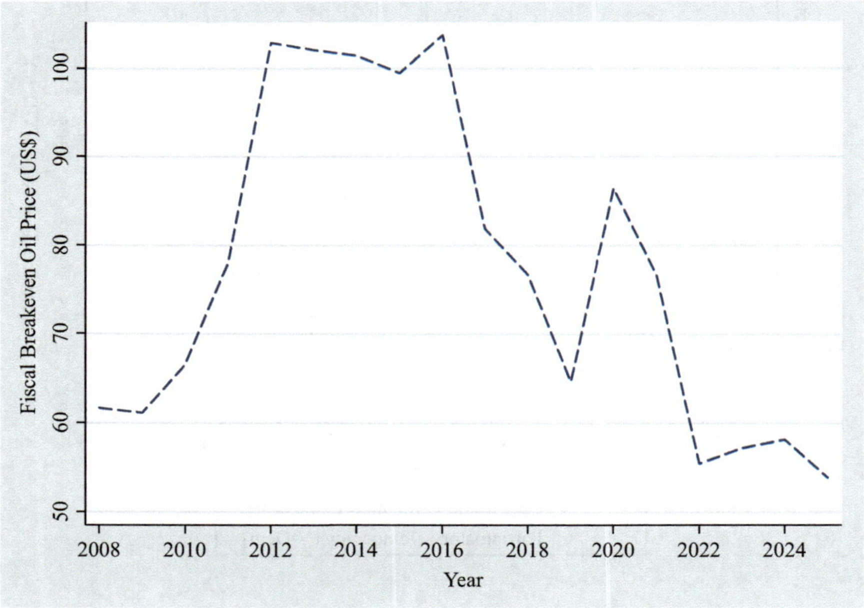

Fig. 4.6 Fiscal breakeven oil price (per barrel). Source: FRED Economic Data—St. Louis FED

encouraged domestic demand for the hospitality and retail sectors (Reuters, 2023). Environmental sustainability plays an important role in diversification policies. The *Tanfeedh* (Implementation) program aims to achieve three environmental goals: environmental protection; crisis risk management; and innovation in technology and science (Al-Sarihi et al., 2023).

An advantage that all countries in the GCC benefit from, including Oman, is the prevalence of Islamic banking, which COVID-19 did not have much effect on. Islamic banks were less exposed to the negative impacts of the pandemic, and Islamic financial institutions in Oman were unaffected by the COVID-19 outbreak. Nevertheless, the Islamic financial institutions were less resilient to the pandemic than to the 2008 financial crisis, which implies that they require further improvements in human capital, technological reforms, and R&D investments (Akkaş & Hazem, 2022). Islamic banks are also like their conventional counterparts in that both are equally affected by the external shock of the pandemic. It may even be said that social distancing harms Islamic banks more than conventional banks due to their higher cost structure (Ashraf et al., 2022).

Besides its resilience toward the pandemic, Oman is in dire need of improving institutional quality and democracy scores (Fig. 4.7). The negative trend continued from 2019 to 2022. Although the scores have been slightly up for participatory democracy since 2022, the corruption score has sustained its decline. Even a relatively higher democracy score cannot carry Oman to the world standards for achieving sustainable development. Thus, Oman's prerequisite is the progress in its institutional quality needed to catch up with other GCC members, particularly Qatar, which has natural gas (LNG), putting diversification efforts ahead of other Council members due to the inherently less environmental degradation that causes, which is discussed next.

4.5 Qatar

Qatar has the lowest youth unemployment rate in the Gulf and the entire MENA region (Arab News, 2023). Qatar has an advantage in that its oil dependence is much lower than that of other GCC members, and it has a large amount of proven natural gas (LNG) resources. This advantage can be used to diversify its economy (Wright, 2023) and leveraged to improve its institutional quality and participatory democracy. Among other GCC members, Qatar targets environmental development in air quality, fishery, climate change, sanitation, drinking water, and waste management. Vision 2030 aims for economic development and environmental protection simultaneously, without sacrificing one for the other. Qatar, like other GCC members except Oman and Kuwait, has made its Nationally Determined Contributions (NDC), which is part of the Paris Agreement, in parallel with economic diversification attempts (Al-Sarihi et al., 2023).

Like Oman, Qatar has introduced new immigration policies to encourage property investment, boost domestic demand for the hospitality sector, and protect skilled

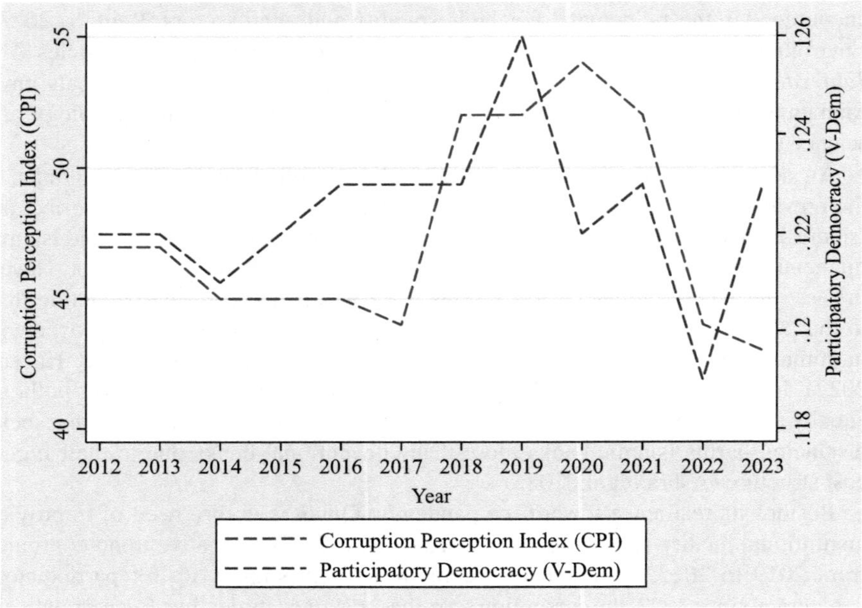

Fig. 4.7 Corruption Perception Index (CPI) and participatory democracy performance. Source: Transparency International and V-Dem

foreign long-term residents (Foxman, 2020). Like Kuwait, Qatar has used "mask diplomacy" and attempted to be weighed more heavily on the international scale due to its resilience in diplomatic arenas. Qatar tried multilateralism as part of its foreign policy for conflict resolution and institutionally strategic maneuver capability (Milton-Edwards, 2023a, 2023b). However, the failure of global leadership during the pandemic weakened Qatar's diplomatic attempts and even threatened its sovereignty. Qatar pledged millions of dollars to global health programs and initiatives, which helped to further its power and dominance (Milton-Edwards, 2023b). However, the blockade of Qatar by Saudi Arabia, the UAE, and Bahrain and the decision to host the FIFA World Cup in 2022 before the pandemic changed Qatar's foreign policy and increased the pressures on the *kafāla* system.

Kafāla works with two-year short-term contracts for blue-collar employees, which promotes a constant turnover in the workforce. *Kafāla* provided resilience to Qatar during the pandemic because it restricted the state's commitments to immigrant workers' long-term social and economic welfare and gives employers external control mechanisms (i.e., businesspeople obtain a permit from the government) that control worker rights and responsibilities. Qatari citizens want to keep the *kafāla* system, but rights and labor laws should be improved. The attitude of Qatari citizens shows the resilience and sustainability of the system itself. That is, the *kafāla* system helped Qatar to adapt to the unexpected external shock of COVID-19 due to its importance in the Qatari economy. By adjusting to the demand to adapt, the *kafāla* system has become more flexible and more effective in response to crises (Ewers et al., 2023).

As with other GCC members, COVID-19 adversely affected Qatar's budget, international trade, inflation, growth, and housing market. The challenge faced by Qatar was greater during the pandemic due to the blockade of other countries in the region and Qatar's previous commitment to host the FIFA World Cup 2022. The importance of Islamic finance has been at the forefront of Qatar's resilience. Besides the *kafāla* system, Islamic financial alternatives such as *zakat* (obligatory tax), *infaq* (voluntary donation), *qard al-hassan* (finance without interest), and *waqf* (Islamic institutions) were resilient to the COVID-19 pandemic, which helped to regain the stability that derailed during the first phases of the pandemic.

Besides Islamic finance, Qatar has implemented policies such as providing 75 billion Qatari riyals to the private sector, postponing loan repayments through mechanisms created by the Qatar Central Bank (QCB), increasing the size of the stock market and liquidity for banks, introducing tax exemption on food and medical goods and rents for SMEs, and supporting the hospitality and tourism sectors and commercial complexes. The Qatari government's economic and social resilience through its stimulus packages, exemptions, and new mechanisms has helped the country to be more stable and to regain stability faster than other countries in the region.

In addition, as Fig. 4.8 shows, Qatar needs a little over US$40 per barrel of oil to keep its budget balanced in 2024 and 2025. This also encourages Qatar to become a gas-rich nation, which will lower its oil dependency and, therefore, increase the pace of its economic diversification.

However, Qatar's institutional quality and democratization processes need to be reinvigorated. The downward trend in CPI and participatory democracy scores are apparent, and this substantial negative trend has accelerated since 2012, even before the pandemic, as Fig. 4.9 clearly illustrates. There was a slight increase right after the pandemic, but that did not go much further, and this indicates that Qatar should work toward institutional stability and quality along with its democracy. The recent attempts to organize the International Anti-Corruption Excellence Award program (ACE) can undergird institutional improvement efforts. Like Qatar, the Kingdom of Saudi Arabia also wants to improve its institutional quality, especially the pace of diversification for Vision 2030, and the Saudi case will be examined next.

4.6 Saudi Arabia

Saudi Arabia has the largest economy and population in the GCC. Like Bahrain, Kuwait, Oman, and Qatar, Saudi Arabia's Vision 2030 focuses on numerous economic diversifications. The government is increasing tax revenue (e.g., Saudi Arabia increased VAT from 5% to 15% during the pandemic's immediate impact) and decreasing subsidies. Saudization policies are in place, and that makes expatriates more costly to employ and causes them to depart the country. Innovative SMEs are invited and encouraged to enter domestic markets. However, the criticism is that

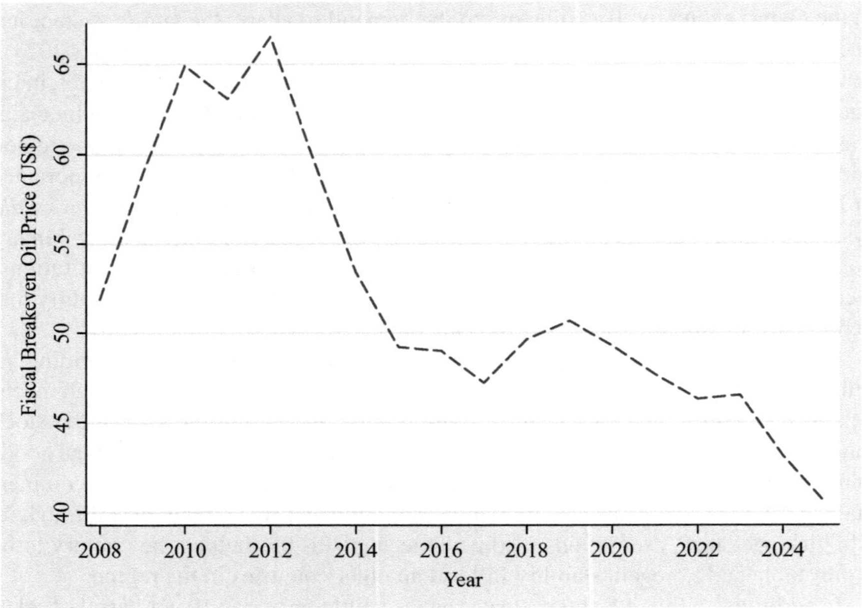

Fig. 4.8 Fiscal breakeven oil price (per barrel). Source: FRED Economic Data—St. Louis FED

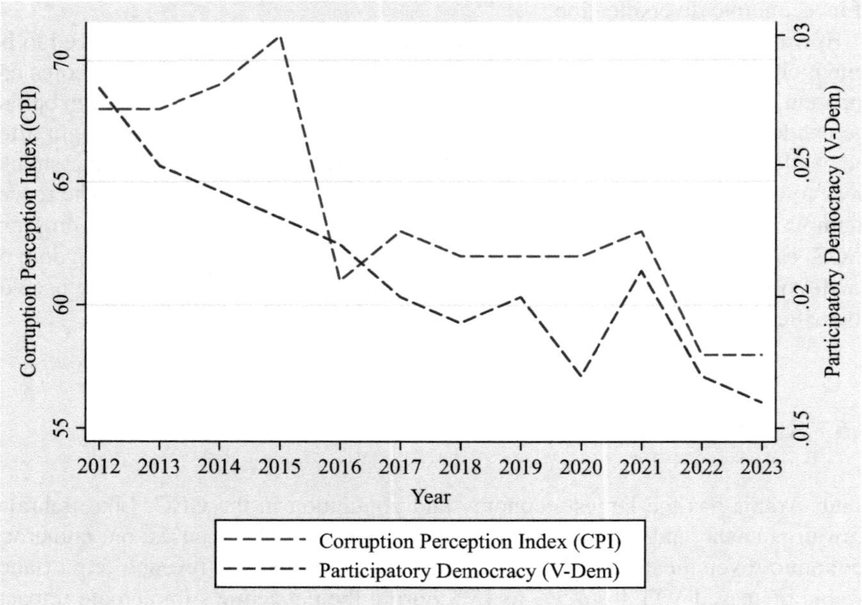

Fig. 4.9 Corruption Perception Index (CPI) and participatory democracy performance. Source: Transparency International and V-Dem

Vision 2030 is overly focused on diversification and needs to create more jobs for Saudis. It also requires more credibility to achieve its targets as it emphasizes job creation, Saudization, and private sector development.

The Saudi DRRS (Disaster Risk Reduction System) worked at three levels during the pandemic. The strategic level worked as a planning division that monitored the responses to crises at the operational and tactical levels. The outcome was related to good governance at the local level and flexibility at the regional level (Jaziri & Mohammad, 2021). The Kingdom's National Renewable Energy Program (NREP) aspires to take 35 projects nationwide. Among the goals of this program, Saudi Arabia targets relying on an equal combination of renewable resources and gas by 2030, thus accomplishing a balanced energy mix. In addition, Saudis aim to make a sustainable transition to utilizing natural gas for the electricity sector and renewable energy by 2030 (Abi-Aad & Nikolay, 2023).

Saudi Arabia is located in one of the most strategic locations, ranking sixth in the world in the utilization of solar and wind energy. The country has immense potential for solar energy and ranks in 13th place for land-based wind production. In order to control carbon emissions, the Kingdom proposed a Circular Carbon Economy Initiative (CCE) during its G-20 presidency, which was accepted unanimously, seeking to implement a comprehensive and pragmatic approach. CCE considers national contexts and offers policies and options from a diverse package. It primarily focuses on lowering greenhouse gas emissions, system efficiency, natural resources, and social, political, economic, and environmental development panoramas (Al-Sarihi et al., 2023).

Along with environmental developments, Saudi Arabia is projecting to increase the number of private schools by 2030, to reach 25% of total K-12 schools, from the current level of 17% (Globe Newswire, 2020). Early years schooling is also expected to increase from the current level of kindergarten enrollment of about 15% (Hoteit et al., n.d.). While the government targeted expanding private schools, the pandemic impeded this project. Approximately 45% of current enrollment in private schools has transitioned to public schools due to the high costs of private schooling and its impracticality (Arab News, 2021).

Besides positive developments and engagements in environmental protection for economic diversification and educational transformation, the pandemic has exacerbated some concerns. The plunge in oil prices due to the pandemic has financially restricted the Saudi government and reduced oil revenues by around 45% in the first half of 2020, which lowered tax revenue by 75% (Al-Rajhi, 2020). As mentioned, Saudi Arabia has the highest population in the GCC, and that sizeable population has significant implications for oil and gas revenue, because their budget calculation is on a per capita basis. The small populations of Kuwait, Qatar, and the UAE do not matter, but this implication of oil revenue distribution is important in Saudi Arabia. The public sector wage bill is proof of this, and it will reach $200 billion, which is 60% of total net income from gas and oil, in 2030 (Al-Saffar, 2021).

Ambitious projects (e.g., NEOM) may cause delays due to their unpragmatic structure (Diwan, 2021) and the public debt climbing to 213 billion from 26 billion between 2014 and mid-2020 (Al-Rajhi, 2020). However, overall, environmental changes (e.g., solar and wind energy, CCE comprehensive package), privatization of education, increasing taxes, lowering subsidies, providing incentives for SMEs, and loans for individuals are important policies to be undertaken during the pandemic.

Regarding oil prices, for a balanced budget, the breakeven oil prices for the Kingdom in 2024 and 2025 need to be around US $90. Figure 4.10 shows that fiscal breakeven levels have mostly been above US$80 per barrel since 2012, which puts extra pressure on the Saudi economy. Oil dependence will make the Kingdom's economy more fragile and vulnerable unless it diversifies its economy. Furthermore, Saudi Arabia struggles with lower institutional quality and democratic participation.

Figure 4.11 illustrates that the institutional quality score on corruption has a positive trend that has improved since 2016; after the pandemic, it decreased but is still higher than before. However, the participatory democracy score is going down, although there was a little upward trend in the 2021–2022 period. Considering the Kingdom's population, its expatriate labor force, and other macroeconomic indicators, the figure above clearly sends a signal that to have a stable economy and resilient response mechanisms, democratic improvements need to level up, and the corruption score needs to be improved as well.

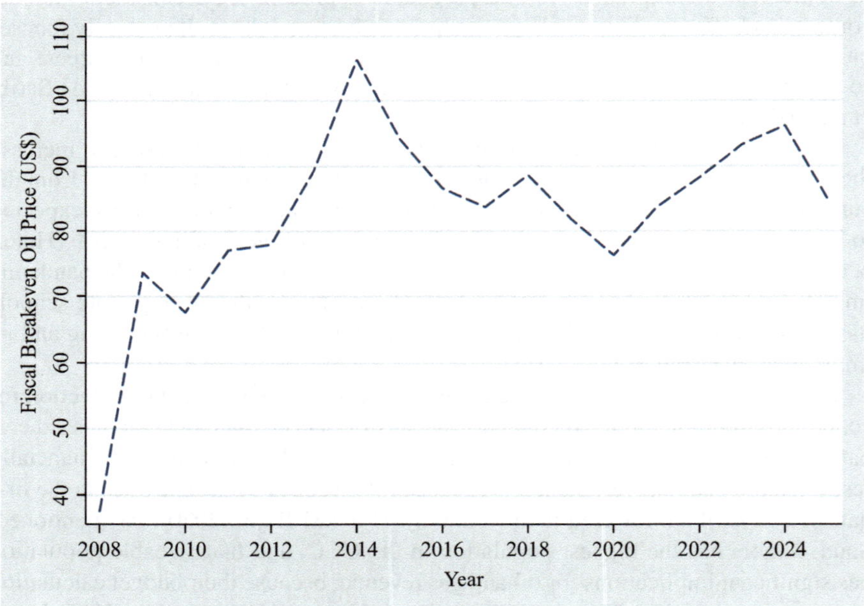

Fig. 4.10 Fiscal breakeven oil price (per barrel). Source: FRED Economic Data—St. Louis FED

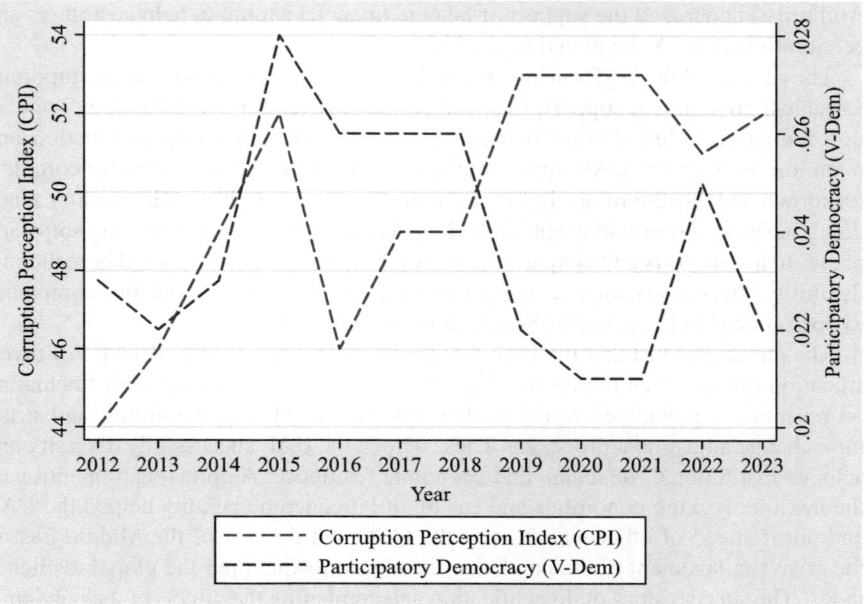

Fig. 4.11 Corruption Perception Index (CPI) and participatory democracy performance. Source: Transparency International and V-Dem

4.7 The United Arab Emirates

The UAE is the region's least oil-dependent country, with the highest level of diversification. The UAE also has the highest score on business environment measures (Mehra, 2024). During the pandemic, most countries around the globe, including advanced nations, struggled to alleviate adverse health and economic effects; however, the Emirati government ensured there was no supply shortage and that the services sector functioned well. This pace of implementation to make sure that there were no supply issues is the consequence of strong support and stimulus by the government to supply chain and logistics companies (Atayah et al., 2022). Considering that 90% of commodities are imported from overseas in the UAE, all these products go through the supply chain industries' operations, which depend on the resilience of financial institutions.

Atayah et al. (2022) find that the reason why the world struggled with supply chain bottlenecks but the Emirati government did not was the solid financial support of the government during the pandemic. This minimal impact of the external shock shows that the UAE is resilient and has strong institutions that could spread the financial packages throughout the country. The UAE has resilient institutions and a robust financial system that help to impact specific sectors quickly and make the necessary changes at full speed. In addition to conventional banking, the Higher Shari'ah

Authority encouraged the support of Islamic financial capital to help customers and economic sectors (Mohammed et al., 2020).

The success of the UAE during the pandemic is due to its focus on seven important variables: government support, financial resources availability, technology application, innovative value addition, service-delivery efficiency, customer satisfaction, and retention. Further, the UAE implemented specific restriction measures for complete lockdown and closure of air flights in the first 6 months of 2020. The country made doing business easier and extended its logistics webs by quickly allowing ships and planes to use its ports and airspace for transit. Due to its capabilities and bureaucratic flexibility, this permission encouraged major global trading and logistics companies to open offices in the country (Sundarakani & Okey, 2021).

The paradigm shift that the UAE has managed to achieve after decades of diversification projects was the result of years of experience in dealing with fluctuating oil revenues, a globalized world market, environmental responsibilities, and structural change adjustment processes. These helped the UAE successfully diversify and achieve institutional, financial, and economic resilience. As previously mentioned, the flexible working conditions and easing of bureaucratic rigidity helped the UAE and put it ahead of other members of the GCC and the rest of the Middle East in the crisis management of the pandemic, making it resilient per the global resilience index. The success story of diversification and combating the effects of the pandemic is the outcome of the UAE's efforts over a long time. However, the connectedness between diversifying an economy, making it resilient, and not sacrificing growth and prosperity is essential. The UAE, with its partnership of private and public sector companies along with academics and NGOs, is at the core of incorporating sustainable development goals (SDGs) into the Gulf economy, for growth, green growth, prosperity, and development in the long run.

When the diversification efforts of the country are undergirded by creating more jobs, supporting green transition further, alleviating climate change repercussions, eliminating cyber security risks, separating public finance from oil revenues, incentivizing high-value chain services, entrepreneurial support for eco-friendly growth, credit availability for SMEs and start-ups, and encouraging sustainable finance, then the UAE will be much more resilient to subsequent external shocks (El et al., 2023). The UAE has this capability with a 0.866 human development index score, the highest in the region, and also has the highest vaccination rate in the region of 72.9% (Sadiki & Layla, 2023).

As diversification processes continue at full pace, along with creating jobs, facilitating the businesses, and being a global shipping hub the UAE government has managed to lower the fiscal breakeven oil price to almost below US$50 in 2024 and 2025, as shown in Fig. 4.12, which is the second lowest in the GCC, following Qatar. Nevertheless, Qatar's proven LNG reserves automatically made oil dependency secondary for the Qatari government. In terms of the initiative to lower oil dependency, the Emirati government has shown a substantial accomplishment. Besides reducing the need for oil, the UAE has been improving its fight against corruption, yet there is still room for democratization.

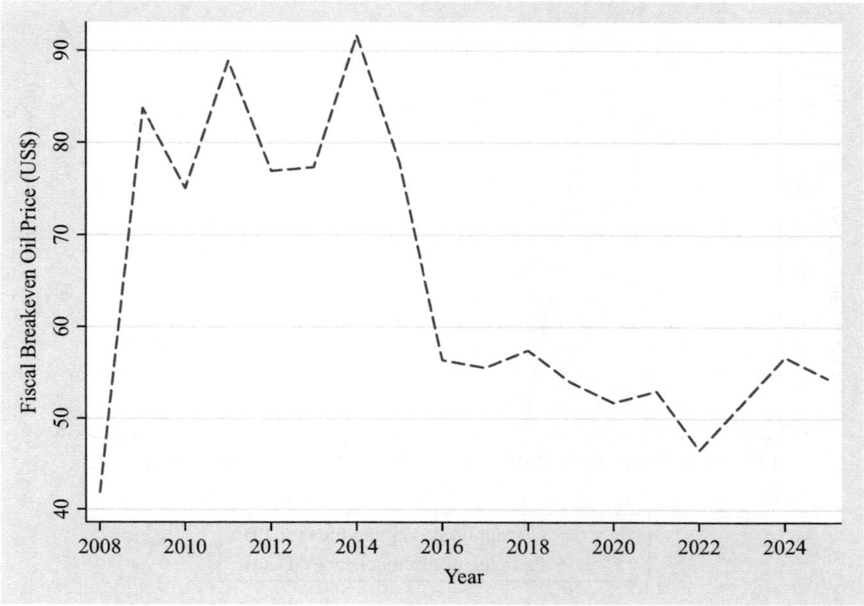

Fig. 4.12 Fiscal breakeven oil price (per barrel). Source: FRED Economic Data—St. Louis FED

The corruption and participatory democracy scores in Fig. 4.13 show that since 2016 there is an upward trend in the UAE's performance, which ceases during the pandemic, until 2022, and then increases again. However, the CPI score is the highest in the UAE, meaning it is the cleanest country in the GCC and MENA region. The UAE is focused on eliminating corruption in the private and public spheres. However, the UAE needs to bolster this cleanliness score by performing better in democratization, by encouraging people to participate in local and national politics. There is a blanket rule in the GCC that when taxation is non-existent, there is insufficient representation in the public and democratic arenas. Considering the policies undertaken during the pandemic (i.e., increasing VAT or introducing new taxes), this situation may be ruled out, due to the requests of people to know where the collected taxes are spent, and that accountability may increase institutional resilience to future external shocks.

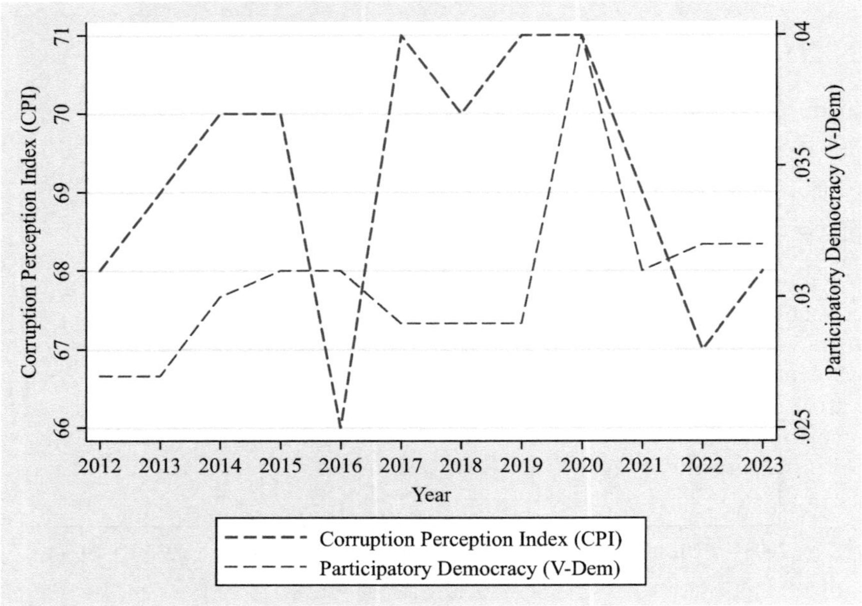

Fig. 4.13 Corruption Perception Index (CPI) and participatory democracy performance. Source: Transparency International and V-Dem

4.8 Conclusion

According to the projections, the total global production loss due to the pandemic will be a massive $35 trillion between 2020 and 2025 (McKibbin & David, 2020). GCC members, especially those with solid financial institutions, with the capacity to support conglomerates, banks, SMEs, and households, that adopted comprehensive disaster management programs, and have sufficient vaccines for their nations, have not been heavily affected by the total global output loss. These countries have placed a distance between themselves and the rest of the countries in the MENA region.

The pandemic threatened the GCC members because they had a large number of foreign workers in their lands—about 90% in the UAE and Qatar, 70% in Kuwait, 60% in Bahrain, and around 40% in Saudi Arabia and Oman are foreign workforce (Akkaş, 2022, p. 237). The GCC countries nevertheless granted comprehensive support packages. The studies show that their measures were sufficient, timely, and restricted enough (Chen et al., 2024). Along with providing vaccines, the GCC members have protected foreigners' rights and supported them financially by improving their access to health services and ensuring their essential living conditions. Studies verify that GCC members have had comprehensive packages for both nationals and expatriates and have taken necessary steps to be more resilient in potential future pandemics (Chen et al., 2024). Nevertheless, their corruption and participatory democracy scores lack comprehensiveness and need improvement for better governance.

Bahrain, where the breakeven oil prices are fairly high, has balanced budget pressures and institutional quality predicaments; thus, the pace of diversification and improving institutional quality must be enhanced. Participatory democracy helped Kuwait to have more holistic and comprehensive packages, but institutional quality still needs improvement. Oman also needs to diversify its economy and improve institutional quality. Qatar is in an advantageous position because it has proven gas reserves, which may make its green energy transition easier than the rest of the Gulf. However, Qatar has one of the lowest participatory democracy scores.

With the largest economy and population, Saudi Arabia has ambitious Vision 2030 plans to nationalize the workforce. However, it must create more local jobs, and fluctuating oil prices pressure its breakeven fiscal levels. Yet, it is trying to diversify and financialize its economy at full speed. Lastly, the UAE has the most diversified and resilient economy and is the cleanest of corruption, but it lacks democracy. Flexible bureaucratic conditions, ease of doing business, and being the hub of global logistics companies may help the UAE democratize, since these structural transformations provide paths toward institutional and financial resilience.

This chapter analyzed the political economy of Gulf members by examining their economic resilience to the pandemic and their institutional governance. As country-specific investigations highlight, GCC members were resilient to the pandemic and its adverse effects on their economies. Due to their oil revenues, GCC members had sufficient financial resources to support businesses and households. They restricted the tourism, hospitality, transportation, education, and health sectors. While they were resilient to the repercussions of the pandemic, their governance to combat corruption and their democratic transitioning need to be improved overall. Without improving institutional quality and transitioning to strong local and national politics, diversification efforts, Visions, and resilience will be short-term success stories. Therefore, the correct decision-making on diversification and economic resilience should be undergirded by robust, transparent, and justice-oriented institutions.

References

Abi-Aad, N., & Nikolay, K. (2023). Challenges facing the Gulf petroleum industry: Covid-19, new technologies, energy transition. In N. Kozhanov, K. Young, & J. Qanas (Eds.), *GCC hydrocarbon economies and COVID* (1st ed., pp. 13–29). Springer. https://doi.org/10.1007/978-981-19-5462-7_2

Akkaş, E. (2022). Körfez Sermayesinin Ekonomi Politiği: Ekonomik Reform ve Yapısal Sorunlar. In T. Eğri & M. H. Mercan (Eds.), *Orta Doğu Ekonomi Politiği* (1st ed., pp. 229–256). Marmara Üniversitesi.

Akkaş, E., & Hazem, A. S. (2022). Are Islamic financial institutions more resilient against the COVID-19 pandemic in the GCC countries? *International Journal of Islamic and Middle Eastern Finance and Management, 15*(2), 331–358. https://doi.org/10.1108/IMEFM-07-2020-0378

Aljazeeri, S. (2021). Upgrading towards neoclassical rentier governance: Bahrain's Post-2014 oil price decline adjustment. In M. Beck & T. Richter (Eds.), *Oil and the political economy in the*

Middle East: Post-2014 adjustment policies of the Arab gulf and beyond (1st ed., pp. 36–57). Manchester University Press.

Al-Rajhi, B. (2020). *Monthly economic report*. Retrieved from https://www.alrajhi-capital.com/-/media/Feature/AlRajhiCapital/ResearchListing/Economic-Research/Saudi_Economy_Report_August_2020_English.pdf

Al-Saffar, A. (2021). Outlook for producer economies. In G. Luciani & T. Moerenhout (Eds.), *The political economy of the middle east* (1st ed., pp. 55–71). Springer. https://doi.org/10.1007/978-981-15-5728-6_4

Al-Sarihi, A., Noura, M., & Sarah, A. O. (2023). GCC renewable energy development amid COVID-19 pandemic. In N. Kozhanov, K. Young, & J. Qanas (Eds.), *GCC hydrocarbon economies and COVID* (1st ed., pp. 57–89). Springer. https://doi.org/10.1007/978-981-19-5462-7_4

Arab News. (2021). *45% of students in Saudi Arabia shifted to government schools amid pandemic*. Arab News. Retrieved from https://arab.news/9ad56

Arab News. (2023). *Qatar boasts lowest unemployment rate in the world: Report*. Arab News. Retrieved from https://arab.news/27h6f

Ashraf, B. N., Mosab, I. T., & Mohammad, K. H. (2022). Are Islamic banks more resilient to the crises Vis-à-Vis conventional banks? Evidence from the COVID-19 shock using stock market data. *Pacific-Basin Finance Journal, 73*(June), 101774. https://doi.org/10.1016/j.pacfin.2022.101774

Atayah, O. F., Mohamed, M. D., Khakan, N., & Guilherme, F. F. (2022). Impact of COVID-19 on financial performance of logistics firms: Evidence from G-20 countries. *Journal of Global Operations and Strategic Sourcing, 15*(2), 172–196. https://doi.org/10.1108/JGOSS-03-2021-0028

Barclays. (2021). *GCC contrasting fiscal performance*, London.

Beck, M., & Thomas, R. (2023). *Oil and the political economy in the Middle East: Post-2014 adjustment policies of the Arab gulf and beyond*. Manchester University Press.

Beutel, J. (2021). Economic diversification and sustainable development of GCC countries. In G. Luciani & T. Moerenhout (Eds.), *When can oil economies be deemed sustainable?* (1st ed., pp. 99–151). Springer. https://doi.org/10.1007/978-981-15-5728-6_6

Chen, S., Lei, G., Yewei, X., Di, D., Rana, S., Mohammed, A., Adwa, A., et al. (2024). Government responses to the COVID-19 pandemic of the Gulf cooperation council countries: Good practices and lessons for future preparedness. *Global Health Research and Policy, 9*, 10. https://doi.org/10.1186/s41256-024-00349-y

Dale, S., & Bassam, F. (2018). *Peak oil demand and long-run oil prices*. Oxford Institute for Energy Studies. https://www.oxfordenergy.org/publications/peak-oil-demand-long-run-oil-prices/

Diwan, I. (2021). Fiscal sustainability, the labor market, and growth in Saudi Arabia. In G. Luciani & T. Moerenhout (Eds.), *When can oil economies be deemed sustainable?* (1st ed., pp. 31–53). Springer. https://doi.org/10.1007/978-981-15-5728-6_3

El, A., Amany, A., & Usman, K. (2023). From diversification resistance to sustainable diversification: Lessons from the UAE's public policy shift. *Management & Sustainability: An Arab Review, 2*(1), 47–66. https://doi.org/10.1108/MSAR-06-2022-0025

Ewers, M., Abdoulaye, D., Kien, T. L., & Lina, B. (2023). Resilience and sustainability in the Gulf migration regimes: Kafāla in the era of Covid-19. *Journal of Immigrant & Refugee Studies, 21*(1), 28–44. https://doi.org/10.1080/15562948.2022.2128496

Fattouh, B., & Anupama, S. (2021). Economic diversification in Arab oil-exporting countries in the context of peak oil and the energy transition. In G. Luciani & T. Moerenhout (Eds.), *When can oil economies be deemed sustainable?* (1st ed., pp. 73–97). Springer. https://doi.org/10.1007/978-981-15-5728-6_5

Fitch Solutions. (2023). *GCC will drive MENA growth*. Retrieved from https://www.fitchsolutions.com/bmi/country-risk/gcc-will-drive-mena-growth-2024-15-09-2023

Foxman, S. (2020). Al Jazeera. Qatar loosens restrictions on foreign property ownership. https://www.aljazeera.com/economy/2020/10/6/bbqatar-loosens-restrictions-on-foreign-property-ownership

Globe Newswire. (2020). *KSA private K12 education market: Growth, trends, and forecast (2020-2025)*. Retrieved from https://www.globenewswire.com/news-release/2020/05/14/2033151/0/en/KSA-PRIVATE-K12-EDUCATION-MARKET-GROWTH-TRENDS-AND-FORECAST-2020-2025.html

Hanieh, A., & Rafeef, Z. (2022). Pandemic effects: COVID-19 and the crisis of development in the Middle East. *Development and Change, 53*(6), 1308–1334. https://doi.org/10.1111/dech.12727

Hoetjes, G. (2021). Stalled reform: The resilience of Rentierism in Kuwait. In M. Beck (Ed.), *Oil and the political economy in the middle east: Post-2014 adjustment policies of the Arab gulf and beyond* (1st ed., pp. 58–78). Manchester University Press.

Hoteit, L., Maya, E. H., Patrick, E., & Serene, F. (n.d.). *Where to invest now in GCC private education*. Boston Consulting Group.

International Monetary Fund. (2022). *Countering the cost-of-living crisis. World economic outlook*. International Monetary Fund.

Jaziri, R., & Mohammad, S. M. (2021). The impact of crisis and disasters risk management in COVID-19 times: Insights and lessons learned from Saudi Arabia. *Ethics, Medicine and Public Health, 18*(September), 100705. https://doi.org/10.1016/j.jemep.2021.100705

Kırşanlı, F. (2023a). Corruption and economic growth nexus: What has the Arab spring changed? *Current Research in Social Sciences, 9*(1), 41–57.

Kırşanlı, F. (2023b). Unemployment–corruption relationship in OECD countries: System GMM approach. *Business and Economics Research Journal, 14*(1), 1–17.

McKibbin, W., & David, V. (2020). Global macroeconomic cooperation in response to the COVID-19 pandemic: A roadmap for the G20 and the IMF. *Oxford Review of Economic Policy, 36*(Supplement_1), S297–S337. https://doi.org/10.1093/oxrep/graa032

Mehra, D. (2024). *The 5 pillars of the UAE economy in 2024*. Retrieved from https://www.cbd.ae/docs/default-source/default-document-library/economyandmarketwatch-february_2024.pdf?Status=Temp&sfvrsn=db58536b_2

Milton-Edwards, B. (2023a). Small states of the Arab gulf diplomatic dimensions of a pandemic. In L. Sadiki & L. Saleh (Eds.), *COVID-19 and risk society across the MENA region: Assessing governance, democracy, and inequality* (1st ed., pp. 31–45). I.B. Tauris.

Milton-Edwards, B. (2023b). The survival of small states in the Gulf region: COVID-19, pandemic politics and the future of niche diplomacy. *International Politics, 60*(2), 390–405. https://doi.org/10.1057/s41311-021-00343-y

Moerenhout, T. (2021). Fuel and electricity reform for economic sustainability in the Gulf. In G. Luciani & T. Moerenhout (Eds.), *When can oil economies be deemed sustainable?* (1st ed., pp. 191–213). Springer. https://doi.org/10.1007/978-981-15-5728-6_8

Mohammed, M. M., Muhammed, E. T., & Syed, N. A. (2020). The potential of Islamic finance in reinforcing and regaining economic stability in Qatar. *Journal of Economic Cooperation and Development, 41*(4), 191–217.

Reuters. (2023). *Oman introduces long-term residencies for foreign investors*. Retrieved from https://www.reuters.com/world/middle-east/oman-offers-up-10-year-residency-foreign-investors-trade-ministry-2021-06-23/

Sadiki, L., & Layla, S. (2023). MENA risk society and the 'pandemic condition'. In L. Sadiki & L. Saleh (Eds.), *COVID-19 and risk society across the MENA region assessing governance, democracy, and inequality* (1st ed., pp. 1–29). I.B. Tauris.

Statista. (n.d.-a). *Bahrain - National Debt in Relation to Gross Domestic Product (GDP) 2019–2029*. Retrieved from https://www.statista.com/statistics/525466/national-debt-of-bahrain-in-relation-to-gross-domestic-product-gdp/

Statista. (n.d.-b). *Oman - National Debt in Relation to Gross Domestic Product (GDP) 2019–2029*. Retrieved from https://www.statista.com/statistics/457221/national-debt-of-oman-in-relation-to-gross-domestic-product-gdp/

Sundarakani, B., & Okey, P. O. (2021). Fast, furious and focused approach to Covid-19 response: An examination of the financial and business resilience of the UAE logistics industry. *Journal of Financial Services Marketing, 26*(4), 237–258. https://doi.org/10.1057/s41264-021-00118-9

SWFI. (n.d.). *Top 100 largest sovereign wealth fund rankings by total assets*. Retrieved from https://www.swfinstitute.org/fund-rankings/sovereign-wealth-fund

The Atlas of Economic Complexity. (n.d.). Retrieved from https://atlas.cid.harvard.edu/

Wright, S. (2023). Qatar's energy sector in the post-COVID era. In N. Kozhanov, K. Young, & J. Qanas (Eds.), *GCC hydrocarbon economies and COVID* (1st ed., pp. 31–56). Springer. https://doi.org/10.1007/978-981-19-5462-7_3

Young, K. E. (2023a). The Covid response in the GCC: Making a durable economic policy agenda. In N. Kozhanov, K. E. Young, & J. Qanas (Eds.), *GCC hydrocarbon economies and COVID* (1st ed., pp. 91–109). Springer. https://doi.org/10.1007/978-981-19-5462-7_5

Young, K. E. (2023b). *The economic statecraft of the Gulf Arab states: Deploying aid, investment and development across the MENA. Middle East Institute Policy Series*. I.B. Tauris.

Chapter 5
Examining the Dynamic Nexus Among Stock Markets, Environmental Quality, and Energy Consumption Across the GCC Countries

Muhammad Zubair Mumtaz and Anis Khayati

Abstract Over time, technological advancement and industrial development have caused severe threats to our environment. Almost all countries are affected by a surge in environmental issues. The GCC countries were also influenced by increased production over a number of years, contributing to increased carbon emissions. In line with this argument, this study aims to examine the dynamic nexus among stock markets, environmental quality, and energy consumption across the GCC countries, using the sample from 1990 to 2020. We apply both ARDL and nonlinear ARDL techniques to examine the effect of stock market development on environmental quality. Regarding the linear relationship, this study reports that energy consumption and population density are the statistically significant factors of environmental quality in the short run. In contrast, energy consumption, stock traded, economic growth, and population density are robust predictors of environmental quality in the long run. However, our analysis finds no nonlinear relationship between stock market development and CO_2 emissions. Results of the causality test show that bidirectional causality flows between energy consumption, economic growth, population density, and environmental quality.

Keywords Stock markets · Energy consumption · Carbon emissions · GCC countries

M. Z. Mumtaz (✉) · A. Khayati
College of Business Administration, University of Bahrain, Sakhir, Bahrain
e-mail: mmumtaz@uob.edu.bh

A. Khayati
e-mail: aelkhayati@uob.edu.bh

H. Miniaoui and M. Bellalah (eds.), *Green Growth Opportunities and Sustainable Finance in the Gulf Cooperation Council Region*, Gulf Studies 24,
https://doi.org/10.1007/978-981-96-8287-4_5

5.1 Introduction

Owing to industrial and technological advancements, countries have produced goods on a large scale, which has resulted in greenhouse gas (GHG) emissions on the one hand and deteriorated environmental quality on the other. Over time, mature firms are listed on stock markets to raise funds for expanding and sustaining their business activities; however, a question arises about how they behave toward change in environmental quality. Regarding financial development, stock markets act as a catalyst to determine equity market activities and boost investors' confidence (Altan et al., 2021).

Generally, stock market capitalization offers an opportunity to obtain funds by issuing shares and investing in green technologies. This shows that firms may shift from traditional manufacturing methods to new technologies and utilize renewable energy resources.

Those firms that raise funds tend to increase their production capacity, thus contributing to economic growth but also worsening carbon emissions. Firms are less reluctant to acquire green technology due to loosened financial constraints; however, stock market capitalization acts as a catalyst in reducing likely carbon emissions by investing in green resources. Thus, firms can invest in green technologies and reduce the magnitude of carbon emissions.

Earlier studies (Akinlo & Apanisile, 2023; Azeem et al., 2023) report mixed findings regarding the linkages between stock markets and CO_2 emissions. Some researchers argue that stock market parameters decrease the likelihood of carbon emissions (Ouyang & Li, 2018; Paramati et al., 2018) while others emphasize that stock market parameters increase the quantum of carbon emissions (Azeem et al., 2023; Raza et al., 2020). Theoretically, the impact of stock market capitalization on carbon emissions is subject to diversity, mainly based on the efficiency of the financial environment. We found no prior evidence relating to stock market parameters and CO_2 in GCC countries. To explore this relationship, the researchers investigate whether stock market development affects CO_2 in GCC countries, using the sample from 1999 to 2020. We use stock market capitalization and total value of stock traded as proxies of stock market development.

This chapter explores the linear and nonlinear nexus between stock market development and CO_2 using auto-regressive distributed lags (ARDL) and nonlinear ARDL techniques. Other explanatory variables in our analysis include energy consumption, GDP, and population density. Regarding the linear (symmetric) association, this study reports that energy consumption and population density are robust predictors influencing CO_2 emissions in the short run. Regarding longer-term effects, energy consumption, stock traded, economic growth, and population density are the significant factors that predict CO_2 emissions. Thus, our findings report a linear relationship between the stock market and CO_2. Moreover, the nonlinear (asymmetric) relationship documents that stock market parameters do not affect CO_2 in the short or long run. Regarding the causal relationship, we find that bidirectional causality flows between energy consumption, economic growth, population density, and environmental quality, but that there is no causal relationship between stock market parameters and CO_2.

5.2 Review of Earlier Studies

Stock market development is crucial in stimulating various advantages for firms that contribute to the standard of living. Stock market activities assist in gathering funds by reducing the financing cost, by making available a mix of optimum debt and equity, helping participation in large projects, sharing risk, and providing modernized technologies. Another potential benefit is participating in advanced low CO_2 emission technologies to improve environmental quality. Thus, firms in developed and developing countries use modernized technologies that may increase renewable resources and provide an opportunity to rely on environmentally friendly production to overcome carbon emissions (Pineiro Chousa et al., 2009).

Earlier studies (Alam et al., 2021; Ouyang & Li, 2018; Paramati et al., 2018; Sharma et al., 2021) have extensively documented that stock market capitalization is a function of environmental quality, which describes a linear association between carbon emissions and stock market development; however, owing to annual exponential growth, this relationship is nonlinear. Kuvshinov and Zimmermann (2022) considered a sample of 17 advanced economies from 1870 to 2018 to examine nonlinearity evidence. They determined new evidence, such as a "hockey stick" graph, illustrating a variation regarding the monotonic association between stock market development and CO_2.

Azeem et al. (2023) analyzed the behavior of stock markets in relation to CO_2 in 40 countries from 1996 to 2018 and found an inverted U-shape linkage between stock markets and CO_2, which illustrates that an increase in stock market capitalization leads to a likely increase in CO_2; however, after reaching a certain point, carbon emissions increase at a diminishing rate. In their analysis, other factors influencing carbon emissions include urbanization, industrialization, and energy intensity.

Al-Mulali et al. (2019) analyzed the nonlinear behavior in Malaysia using a sample period between 1980 and 2017 to investigate the nexus between the stock market and air pollution. They reported that a surge in stock market activity leads to increased CO_2 in the short run. However, long-run elasticities reveal that a rise in the development of stock markets has the probability of creating CO_2 and vice versa (Paramati et al., 2017).

Zafar et al. (2019) studied the dynamics of the banking sector and stock market development in relation to CO_2 during 1990–2016. They used various measures to construct banking and stock market development indicators and validate them using principal component analysis. Their cointegration results identified a long-run relationship among all the variables. Regarding the dynamics of the long run, they determined that renewable energy enhances environmental quality by reducing CO_2 emission among the sample countries. They found mixed results in terms of stock market and banking development parameters. As per their findings, banking development reduced CO_2, and stock market development inflated CO_2 in G-7 countries and vice versa in N-11 countries.

Akinlo and Apanisile (2023) employed the sample from 1981 to 2020 to investigate the nexus between stock markets and CO_2 in Nigeria. They found that the nonlinear ARDL bounds technique confirms the evidence of the long-run association between CO_2, GDP per capita, energy intensity, and population density. Their analysis determined that positive shocks in stock trades do not influence carbon emissions; however, adverse shocks in stock market development significantly reduce them. According to their findings, policymakers must ensure strict adherence to environmental regulations to monitor carbon emissions, which may increase due to a manufacturing activity surge by enhancing stock market development.

Mahmood (2020) examined the effect of financial market development on CO_2 in GCC countries using the sample from 1980 to 2018. He reported that financial market development and carbon emissions have an insignificant relationship in short- and long-run analyses. Bekhet et al. (2017) analyzed the association between CO_2, financial development, energy consumption, and GDP in GCC countries from 1980 to 2011. Using the ARDL technique, they found a long-term relationship between CO_2 and economic growth, energy consumption, and financial development in all GCC countries except the UAE.

In another study, Salahuddin et al. (2015) studied the long-run association between GDP, CO_2 emissions, electricity utilization, and financial development in GCC countries from 1998 to 2012. They used various methodologies to test their hypotheses and reported that long-run relationships exist in their analysis. They found that GDP and electricity consumption both positively influence CO_2. In contrast, they identified a negative relationship between CO_2 and financial development. At the same time, the results of the Granger causality test showed that a bidirectional relationship persists between GDP and CO_2, whereas a unidirectional relationship flows between electricity consumption and CO_2. Interestingly, they reported no causal association between CO_2 and financial development.

In the literature, we found no evidence of researchers examining the effect of stock market development on CO_2 in GCC countries. It is a general phenomenon that when firms produce more, there is a probability of higher carbon emissions. However, if the firms are listed on stock markets and involved in manufacturing activities, a question arises as to how they react to these carbon emissions. Earlier studies have documented the mixed results regarding the nexus between CO_2 and stock markets. Hence, this study explores the evidence of linear and nonlinear relationships between stock markets and CO_2 in GCC countries.

5.3 Data and Methodology

5.3.1 Data

This study employs the dataset of the six GCC countries, Bahrain, Kuwait, Oman, Qatar, Saudi Arabia, and the United Arab Emirates, from 1990 to 2020. In this study, we use carbon emissions as a dependent variable—a proxy of environmental quality. Independent variables include energy consumption, stock market development, GDP per capita, and population density. The data was gathered from WDI.

Table 5.1 reports the descriptive statistics. Among the sample countries, Qatar produced the highest per capita carbon emissions (38.693 metric tons), followed by the UAE (24.788 metric tons). The average carbon emission per capita is 22.637 metric tons, with a standard deviation of 9.398. The dispersion between maximum and minimum values shows the highest variation of carbon emissions produced across the GCC countries. Similarly, Qatar used the most energy over the sample period, followed by Bahrain. The result shows that, on average, per capita energy consumption is 9627 kg of oil equivalent, with maximum and minimum values of 21,421 kg and 2009 kg per capita, respectively.

Regarding the proxy for stock market activity, two variables are used, stock market capitalization and stock traded. The stock market capitalization ranges between 31% and 83% of GDP. SMC is the highest in Kuwait and the lowest in Oman. The mean value of SMC is 65.463%, and the mean value of ST is 22.468%. On average, GDP per capita is US$ 26,377, with a standard deviation of 98,041. Population density is the highest in Bahrain (1309.17) and the lowest in Oman (9.55). On average, the value of PD is 279.27, with a standard deviation of 492.90. Table 5.2 presents the results of correlation between variables.

Table 5.1 This table presents the descriptive statistics of variables used in this study

Country	CO_2	EC	SMC	ST	GDP	PD
Bahrain	22.103	10,842.207	79.167	3.081	14,248.072	1309.174
Kuwait	23.727	9506.614	83.344	29.034	28,665.998	149.538
Oman	13.070	4739.190	31.116	6.017	13,761.119	9.552
Qatar	38.693	18,028.073	81.312	18.595	48,118.099	115.815
Saudi Arabia	13.439	5417.649	71.874	66.490	14,500.784	12.036
United Arab Emirates	24.788	9466.639	44.965	11.593	35,967.119	79.124
Consolidated						
Observations	180	180	180	180	180	180
Mean	22.637	9624.977	65.463	22.468	26,376.865	279.270
Maximum	47.657	21,420.629	330.818	372.260	18,659.895	1915.626
Minimum	5.841	2009.408	10.308	0.543	5984.610	6.123
Std. dev.	9.398	4504.246	37.282	43.870	98,041.362	492.889

Table 5.2 Correlation matrix

	CO_2	EC	SMC	ST	GDP	PD
CO_2	1.000					
EC	0.620	1.000				
SMC	0.474	0.606	1.000			
ST	0.128	0.027	0.312	1.000		
GDP	0.638	0.685	0.409	0.200	1.000	
PD	0.393	0.620	0.528	−0.348	0.359	1.000

Table 5.2 demonstrates the findings of pairwise correlation, which is employed to capture the multicollinearity problem between the variables. The probability of having a multicollinearity issue arises when the coefficient value between the variables is more than 0.7 (Mumtaz, 2023; Sheraz et al., 2022). The coefficient between stock market capitalization and CO_2 is 0.474, which confirms there is no evidence of a multicollinearity issue. The value of other coefficients is less than 0.7, which suggests that our data is accessible from multicollinearity.

Figure 5.1 demonstrates the graphics of CO_2 emissions across GCC countries and describes their behavior over time. Figure 5.2 depicts the position of stock market capitalization as a percentage of GDP, which varies across the GCC countries.

Various panel unit root tests are employed to determine the stationarity of the data, and the results are reported in Table 5.3. The findings of all tests show that a few variables are stationary at level; however, all the variables are stationary at first difference. To resolve the stationarity problem, this study uses ARDL methodology to estimate our results.

5.3.2 Model Specification

This study investigates the linear and nonlinear association between stock market parameters, energy use, and CO_2. Thus, we hypothesize that stock market activity and higher energy usage may lead to an increase in GHG emissions, which, as a result, cause the environmental quality to deteriorate. To test our hypotheses, we propose the following model:

$$\ln CO_{2\,\mathrm{it}} = \beta_0 + \beta_1 \ln EC_{\mathrm{it}} + \beta_2 \ln SMD_{\mathrm{it}} + \beta_3 \ln GDP_{\mathrm{it}} + \beta_4 \ln PD_{\mathrm{it}} + \varepsilon_{it} \quad (5.1)$$

where $CO_{2\mathrm{it}}$ shows carbon emissions/capita in metric tons in country *i* at time *t*, which is a proxy of environmental quality, SMD_{it} represents stock market development, which is a proxy of stock market capitalization as a % of GDP and total value of stock traded as a % of GDP, GDP_{it} is a per capita gross domestic product, PD_{it} is a population density/capita, and β_i shows the long-run coefficient covariates. The ensuing paragraphs discuss linear and nonlinear ARDL models.

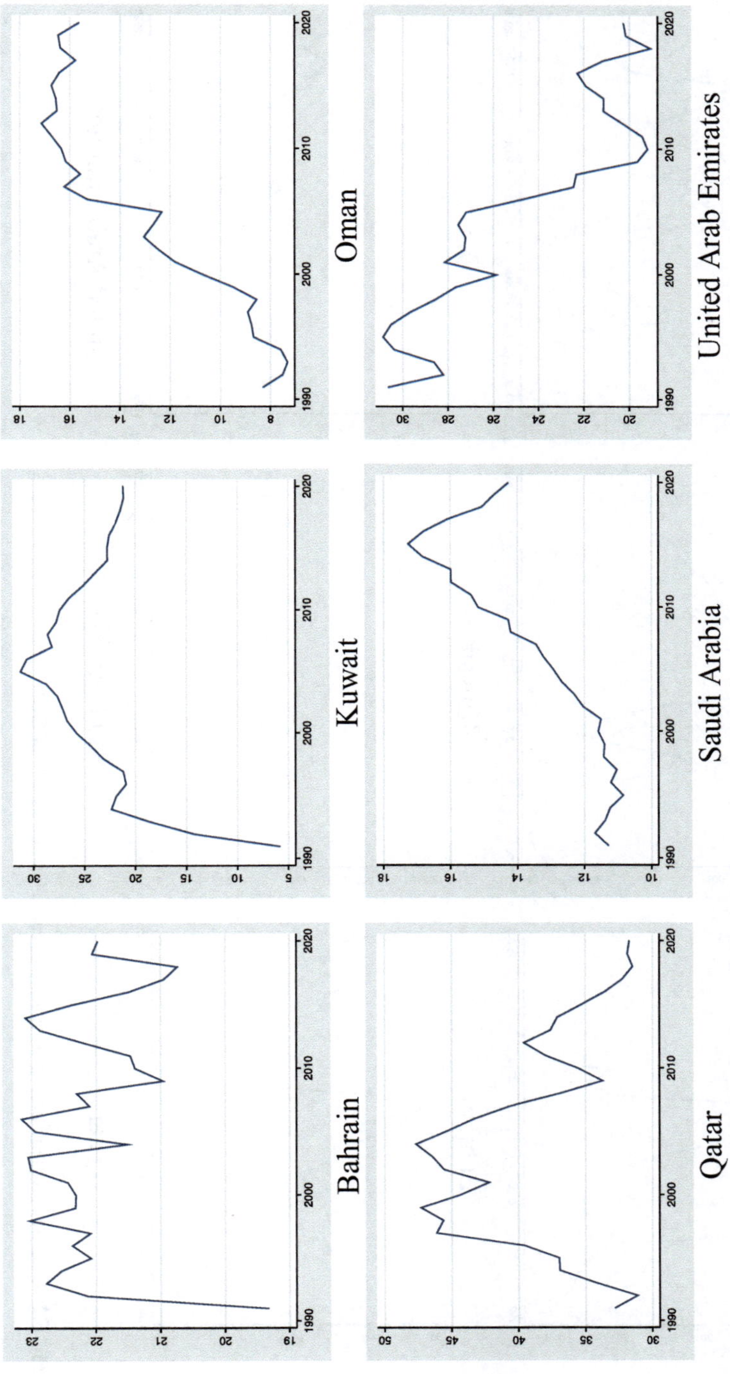

Fig. 5.1 Graphic representation of carbon emissions per capita in metric tons

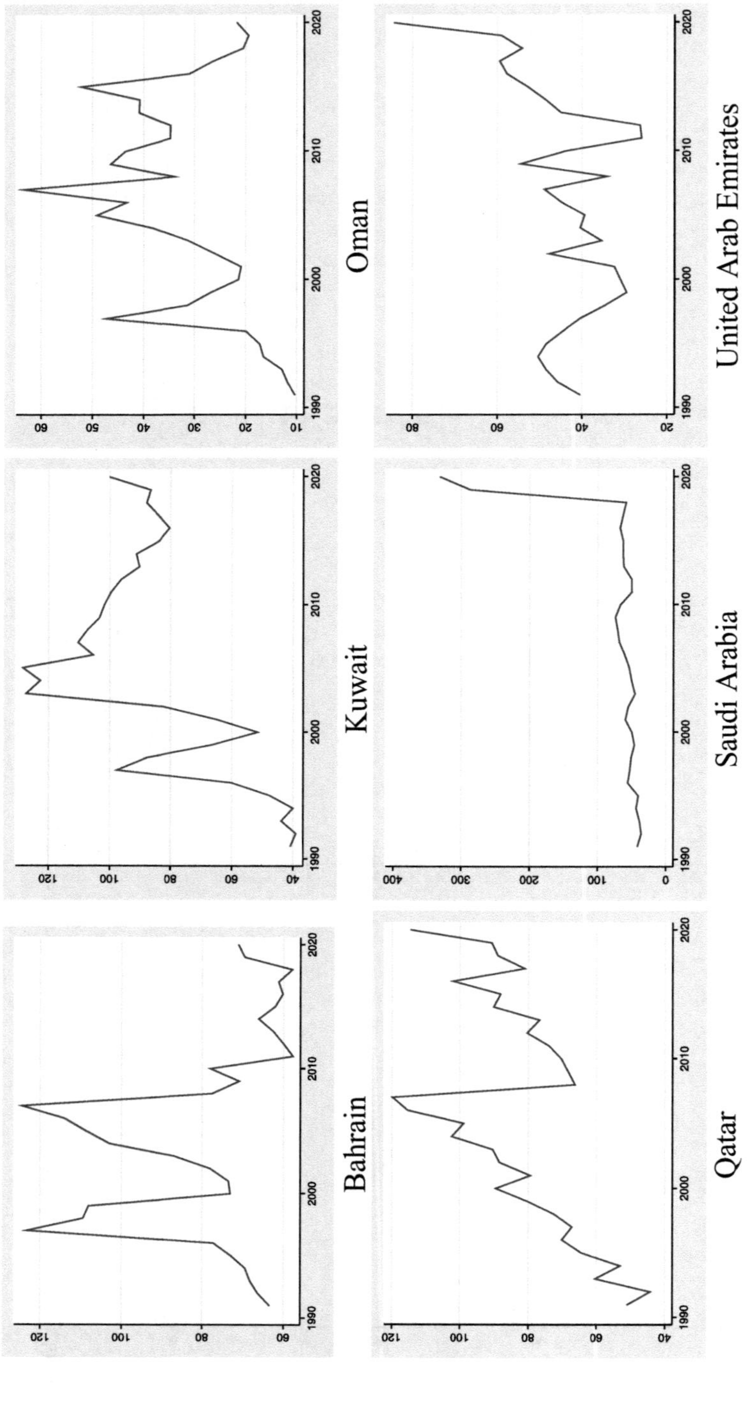

Fig. 5.2 Graphic representation of stock market capitalization as a % of GDP

Table 5.3 This table reports the results of the panel unit root testing

	Levin-Lin-Chu		Im-Pesaran-Shin		ADF		PP		Harris-Tzavalis	
	Level	1st diff.	Level	1st diff.	Level	1st diff.	Level	1st diff.	Level	1st diff.
$\ln CO_2$	−0.520	-3.462^{***}	−0.454	-5.361^{***}	13.748	53.692^{***}	32.132^{***}	98.239^{***}	0.854	0.315^{***}
lnEC	1.071	-3.383^{***}	−0.249	-7.567^{***}	15.159	75.321^{***}	50.277^{***}	128.669^{***}	0.767^{***}	0.201^{***}
lnSMC	-1.302^{*}	-5.537^{***}	−0.987	-5.687^{***}	16.846	5.485^{***}	15.869	113.923^{***}	0.795^{***}	-0.086^{***}
lnST	-3.994^{***}	-5.516^{***}	-4.152^{***}	-6.322^{***}	40.351^{***}	61.881^{***}	24.252^{**}	104.147^{***}	0.723^{***}	0.149^{***}
lnGDP	-1.566^{*}	-5.627^{***}	0.446	-5.780^{***}	6.824	55.302^{***}	8.506	80.928^{***}	0.927	0.124^{***}
lnPD	-2.581^{***}	-3.546^{***}	0.995	-2.345^{***}	7.516	25.383^{**}	3.933	28.451^{***}	0.991	0.738^{***}

***, **, and * show significance at 1%, 5%, and 10% levels, respectively

5.3.2.1 The Linear Model

First, we examine the symmetric association between stock market development and CO_2. We employ the ARDL model suggested by Pesaran and Shin (1998) to test our hypothesis. This method permits the researchers to evaluate the cointegration association among CO_2 emissions and its predictors. The ARDL model is written as:

$$\Delta y_t = \sum_{i=1}^{n} \beta_i \Delta y_{t-1} + \sum_{i=0}^{n} \vartheta_i \Delta x_{t+i} + \omega_1 \Delta y_{t-1} + \omega_2 \Delta x_{t-1} + \mu_t \quad (5.2)$$

To examine the relationship between CO_2 and independent variables, Eq. (5.2) is transformed as follows:

$$\begin{aligned}\Delta \ln CO_2 = & \beta_0 + \alpha_1 \ln CO_{2\,t-1} + \alpha_2 \ln EC_{t-1} + \alpha_3 \ln SMD_{t-1} + \alpha_4 \ln GDP_{t-1} \\ & + \alpha_5 \ln PD_{t-1} + \sum_{j=0}^{p} \beta_{1i} \Delta \ln EC_{it-j} + \sum_{j=0}^{p} \beta_{2i} \Delta \ln SMD_{it-j} \\ & + \sum_{j=0}^{p} \beta_{3i} \Delta \ln GDP_{it-j} + \sum_{j=0}^{p} \beta_{4i} \Delta \ln PD_{it-j} + \varepsilon_{it} \end{aligned} \quad (5.3)$$

where $\sum_{j=0}^{p} \beta_{2i}$ shows the short-run impact of stock market development on CO_2 while α_3 indicates the long-run influence of stock market activity on CO_2. The error correction term of Eq. (5.3) is shown as:

$$\begin{aligned}\Delta \ln CO_2 = & \beta_0 + \sum_{j=0}^{p} \beta_{1i} \Delta \ln EC_{it-j} + \sum_{j=0}^{p} \beta_{2i} \Delta \ln SMD_{it-j} \\ & + \sum_{j=0}^{p} \beta_{3i} \Delta \ln GDP_{it-j} + \sum_{j=0}^{p} \beta_{4i} \Delta \ln PD_{it-j} + \theta_i ECT_{(t-1)} + \varepsilon_{it} \end{aligned} \quad (5.4)$$

where θ_i represents the long-run equilibrium speed of adjustment after the shock in the short run.

5.3.2.2 The Nonlinear Model

Researchers (see Mhadhbi et al., 2021; Sharma et al., 2021) have documented that asymmetries exist regarding stock market parameters on various dependent variables. This evidence translates to positive or negative shocks influencing the stock market parameters, which tend to create asymmetric impacts on other variables. To test the above hypothesis, we employ a nonlinear ARDL model to investigate short- and long-run asymmetric effects of CO_2 emissions on direct and indirect shocks on the parameters of stock markets. Following Stickles and Horrace (2014), the nonlinear ARDL model is expressed as:

$$\begin{aligned}\Delta \ln CO_2 =& \vartheta_0 + \gamma_1 \ln CO_{2\,t-1} + \gamma_2 \ln EC_{t-1} + \gamma_3 \ln SMD^{+}_{t-1} \\ &+ \gamma_4 \ln SMD^{-}_{t-1} + \gamma_5 \ln GDP_{t-1} + \gamma_6 \ln PD_{t-1} \\ &+ \sum_{j=0}^{p} \xi_{1i} \Delta \ln EC_{it-j} + \sum_{j=0}^{p} \xi_{2i} \Delta \ln SMD^{+}_{it-j} + \sum_{j=0}^{p} \xi_{3i} \Delta \ln SMD^{-}_{it-j} \\ &+ \sum_{j=0}^{p} \xi_{4i} \Delta \ln GDP_{it-j} + \sum_{j=0}^{p} \xi_{5i} \Delta \ln PD_{it-j} + \varepsilon_{it} \end{aligned} \tag{5.5}$$

where $\sum_{j=0}^{p} \xi_{2i}$ and $\sum_{j=0}^{p} \xi_{3i}$ measure the direct and indirect influence of stock market indicators on CO_2 in the short run while γ_3 and γ_4 measure the long-run influence of stock market indicators on CO_2. Finally, the error correction model can be reflected as follows:

$$\begin{aligned}\Delta \ln CO_2 =& \vartheta_0 + \sum_{j=0}^{p} \xi_{1i} \Delta \ln EC_{it-j} + \sum_{j=0}^{p} \xi_{2i} \Delta \ln SMD^{+}_{it-j} \\ &+ \sum_{j=0}^{p} \xi_{3i} \Delta \ln SMD^{-}_{it-j} + \sum_{j=0}^{p} \xi_{4i} \Delta \ln GDP_{it-j} \\ &+ \sum_{j=0}^{p} \xi_{5i} \Delta \ln PD_{it-j} + \tau_i ECT_{(t-1)} + \varepsilon_{it} \end{aligned} \tag{5.6}$$

5.4 Results

5.4.1 Analyses of Linear Relationship

First, we examine the linear relationship between stock market parameters and energy consumption across the GCC region, using the ARDL model. The results of short- and long-run relationships are reported in Table 5.4. We took the natural logarithms of all variables to explain the coefficients as elasticities. Two models are estimated to test our hypotheses. Model 1 uses stock market capitalization, and Model 2 employs total value of stock traded to determine their impact on carbon emissions.

Model 1 indicates that the covariate of energy usage is directly associated with CO_2 in the short run, which illustrates that a 1% rise in energy usage leads to an increase of 0.87% in CO_2. Our finding corroborates with earlier studies (Mhadhbi et al., 2021; Mumtaz, 2023), which suggest that if countries utilize higher energy

Table 5.4 This table presents the findings of the linear model

	Model 1	Model 2
ECT	−0.263*** (0.096)	−0.255*** (0.093)
Short-run relationship		
d.lnEC	0.870*** (0.177)	0.868*** (0.198)
d.lnSMC	−0.219 (0.311)	
d.lnST		−0.120 (0.126)
d.lnGDP	0.104 (0.888)	0.090 (0.858)
d.lnPD	−0.777** (0.317)	−0.958*** (0.304)
Long-run relationship		
lnEC	0.609*** (0.203)	0.497** (0.233)
lnSMC	0.201 (0.594)	
lnST		0.434* (0.231)
lnGDP	0.476*** (0.092)	0.540*** (0.104)
lnPD	−0.768*** (0.115)	−0.827*** (0.125)
Constant	−0.103*** (0.024)	−0.827*** (0.192)
Observations	174	174

*** $p < 0.01$, ** $0.01 < p < 0.05$, and * $0.05 < p < 0.10$

consumption, the probability of increased carbon emissions would be higher. Model 2 also endorsed the significant linkage between energy use and CO_2. Regarding regressing stock market parameters, we find an insignificant effect on CO_2 over the short horizon. We find that GDP per capita and CO_2 have no impact over the short horizon. The coefficient of population density is inverse and significant in both models, which means that a 1% increase in population density reduces the probability of CO_2 by 0.78% and 0.96% in Models 1 and 2, respectively.

Over the long-run dynamics, it is notable that energy consumption usage is lower in both models than in the short-run horizon. However, energy usage significantly affects CO_2 in the long run. In Model 1, the energy usage parameter is 0.609, which shows that a 1% increase in the usage of energy resources may generate an increase in CO_2 of 0.61%, which is more than in Model 2.

In Model 1, stock market capitalization does not affect carbon emissions. However, the long-run dynamic shows that a 1% rise in traded stocks enhances CO_2 by 0.434%. This evidence corroborates previous research (Mhadhbi et al., 2021; Mumtaz et al., 2017; Paramati et al., 2017; Zafar et al., 2019). This evidence shows that listed firms whose stocks are traded contribute to inflating carbon emissions and negatively impact environmental quality. This linkage can be elaborated on through financial development, GDP, energy use, and CO_2 emissions. Financial development is usually supposed to enhance energy intensity and GDP and generate CO_2, as new financing extensions permit firms to expand their production activities, contributing to energy consumption, such as fossil fuel energy (Sharma et al., 2021). Considering the negative effect of stock market parameters on carbon emissions, we suggest that GCC countries should devise and execute environmentally friendly policies to reduce CO_2 without affecting their production and energy consumption levels.

The findings of long-run elasticities show that the coefficient of GDP is positive and significantly impacts carbon emissions in both models. As mentioned above, we can argue that higher production activities of firms cause higher emissions and contribute to economic growth. Finally, we find that population density has an inverse relationship with CO_2 in the long run, which illustrates that an increase in population density by 1% decreases CO_2 emissions by 0.77% and 0.83% in Models 1 and 2 respectively.

5.4.2 Analyses of Nonlinear Relationship

This section determines the nonlinear linkages between stock market parameters and CO_2. Hence, we employ the Nonlinear ARDL (NARDL) technique to investigate the relationship. The findings of nonlinear relationships are demonstrated in Table 5.5. Elasticities in short- and long-run horizons suggest that energy consumption and carbon emissions have direct and significant linkages, indicating that a 1% increase in energy use enhances carbon emissions by 0.9%. Short-run analysis suggests that economic growth does not influence carbon emissions; however, the results illustrate that a 1% increase in GDP tends to increase CO_2 by 0.5% in the long-run dynamics.

Population density negatively and significantly influences CO_2 over short- and long-run horizons. This finding implies that a 1% increase in population density decreases the probability of CO_2 roughly by 0.8%. We find that stock market capitalization and total value of stock traded are insignificant determinants of CO_2 in the short and long run across GCC countries. However, previous researchers (Mhadhbi et al., 2021; Sheraz et al., 2022) reported different results, where researchers identified that parameters of stock market development have a distinct behavior with CO_2, which illustrates that parameters of the stock market have an asymmetric effect on CO_2. In their analysis, they identified that the development of stock markets affects CO_2 in the short- and long-run, implying that the liquidity level in the stock market has a nonlinear influence on CO_2.

Regarding our findings on stock market development, we can infer that stock market capitalization and stock traded have insignificant asymmetric effects on environmental degradation in the short and long run. However, in the case of a nonlinear relationship, it can be argued that when stock market activity increases, firms can more easily raise funds for green projects and acquire new resources, which, as a result, assist in reducing consumption of fossil fuels.

Table 5.5 This table presents the results of the nonlinear ARDL model

	Model 1	Model 2
ECT	-0.249^{***} (0.088)	-0.250^{***} (0.096)
Short-run relationship		
d.lnEC	0.913^{***} (0.171)	0.850^{***} (0.213)
d.lnSMC^{+}	−0.145 (0.480)	
d.lnSMC^{-}	−0.131 (0.495)	
d.lnST^{+}		−0.086 (0.110)
d.lnST^{-}		−0.022 (0.114)
d.lnGDP	0.136 (0.930)	0.045 (0.815)
d.lnPD	-0.748^{**} (0.348)	-0.915^{***} (0.317)
Long-run relationship		
lnEC	0.552^{**} (0.226)	0.625^{***} (0.205)
lnSMC^{+}	0.451 (0.660)	
lnSMC^{-}	0.401 (0.667)	
lnST^{+}		0.400 (0.246)
lnST^{-}		0.165 (0.306)
lnGDP	0.479^{***} (0.100)	0.496^{***} (0.101)
lnPD	-0.775^{***} (0.124)	-0.779^{***} (0.124)
Constant	-0.876^{***} (0.201)	-0.103^{***} (0.024)
Observations	174	174

*** $p < 0.01$, ** $0.01 < p < 0.05$, and * $0.05 < p < 0.10$

5.4.3 Panel Granger Causality Results

We use Dumitrescu & Hurlin, 2012 causality test to identify a causal relationship between CO_2 and its determinants (Table 5.6), which indicates that bidirectional causality flows between energy consumption, GDP, population density, and carbon emissions. These findings illustrate that energy consumption causes carbon emissions; eventually, higher carbon emissions mean firms utilize more energy for their manufacturing concerns. Similarly, economic growth contributes to CO_2, and as a result, higher CO_2 leads to a higher likelihood of economic growth. The results of the panel causality test endorsed the results of the ARDL model that stock market capitalization and stock traded do not affect carbon emissions and have neither bidirectional nor unidirectional relationships. This evidence indicates that an asymmetric association between stock market parameters and CO_2 does not exist.

Table 5.6 This table exhibits the results of panel Granger causality test (Dumitrescu and Hurlin (2012)

Null hypothesis	W-stat.	p-value	Causality flow
EC → CO_2	4.372***	0.000	EC ↔ CO_2
CO_2 → EC	5.773***	0.000	
SMC → CO_2	1.779	0.296	SMC ≠ CO_2
CO_2 → SMC	0.957	0.851	
ST → CO_2	1.640	0.403	ST ≠ CO_2
CO_2 → ST	0.854	0.731	
GDP → CO_2	4.359***	0.000	GDP ↔ CO_2
CO_2 → GDP	2.480**	0.036	
PD → CO_2	5.320***	0.000	PD ↔ CO_2
CO_2 → PD	3.847***	0.000	
SMC^+ → CO_2	1.248	0.804	SMC^+ ≠ CO_2
SMC^- → CO_2	1.426	0.606	SMC^- ≠ CO_2
ST^+ → CO_2	1.447	0.584	ST^+ ≠ CO_2
ST^- → CO_2	0.529	0.405	ST^- ≠ CO_2

*** $p < 0.01$, ** $0.01 < p < 0.05$, and * $0.05 < p < 0.10$

5.5 Conclusions and Policy Implications

This chapter examines the dynamic nexus among stock markets, environmental quality, and energy consumption in GCC countries using a dataset from 1999 to 2020. We use two proxies for stock market development: (a) stock market capitalization and (b) total value of stock traded. Applying various panel unit root testing methods, we find that a few are stationary at level; however, at first difference, all variables are stationary. We apply ARDL and NARDL techniques to test our hypotheses. In terms of linear relationship, we find that in the short run, energy consumption and population density are the robust determinants; however, in long-run analysis, energy consumption, stock traded, economic growth, and population density are the significant predictors of CO_2 emissions. We can infer from our results that stock market development and CO_2 have a linear association in GCC countries. We also investigate the nonlinear nexus between stock market development and carbon emissions. The results report no asymmetric effect between GCC countries' stock market parameters and CO_2 emissions. The causality test reports that bidirectional causality flows between energy consumption, economic growth, population density, and environmental quality. However, we did not find any causal relationship between stock market parameters and CO_2 emissions, which illustrates that an asymmetric relationship does not exist in our analysis.

This study is helpful for policymakers and government agencies, as they may consider necessary measures to help firms shift to renewable energy resources to reduce carbon emissions. To move toward renewable energy, governments must provide incentives so that firms may obtain funds to support their manufacturing activities. In this regard, financial institutions in GCC countries may chalk out a comprehensive plan for providing green finance to make the environment healthier. On the other hand, governments may restrict firms from using plants that are older than a specified age (say 15 years), which will compel them to invest in green technology and achieve sustainable development. To reduce the magnitude of carbon emissions, governments may impose carbon taxes to incentivize a shift away from fossil fuels. Governments must provide funding for research and development in renewable energy technologies, enhancing efficiency and making renewables more reliable and cost-effective, in order to achieve sustainability. For future research, the results for GCC countries may be compared with those in other regions to find the differences in whether stock market development affects carbon emissions and explore the reasons for any differences in findings.

References

Akinlo, A. E., & Apanisile, O. (2023). Stock markets and carbon emissions nexus in Nigeria: Evidence from nonlinear ARDL and causality approaches. *Energy and Climate Change, 4*, 100091.

Alam, M. S., Pergis, N., Paramati, S. R., & Fang, J. (2021). The impacts of R&D investment and stock markets on clean-energy consumption and CO_2 emissions in OECD economies. *International Journal of Finance & Economics, 26*(4), 4979–4992.

Al-mulali, U., Solarin, S. A., & Ozturk, I. (2019). Examining the asymmetric effects of stock markets on Malaysia's air pollution: a nonlinear ARDL approach. *Environmental Science and Pollution Research, 26*, 34977–34982.

Altan, I. M., Aslan, L., Aydin, U., Berk, C., Bilik, M., et al. (2021). *Financial systems, central banking, and monetary policy during COVID-19 pandemic and after*. Rowman & Littlefield.

Azeem, A., Naseem, M. A., Ul-Hassan, N., Butt, I., Aslam, M. T., Ali, S., & Jadoon, A. K. (2023). A novel lens of stock market capitalization and environmental degradation. *Environmental Science and Pollution Research, 30*, 11431–11442.

Bekhet, H. A., Matar, A., & Yasmin, T. (2017). CO_2 emissions, energy consumption, economic growth, and financial development in GCC countries: Dynamic simultaneous equation models. *Renewable & Sustainable Energy Reviews, 70*, 117–132.

Dumitrescu, E. I., & Hurlin, C. (2012). Testing for granger non-causality in heterogenous panels. *Economic Modelling, 29*(4), 1450–1460.

Kuvshinov, D., & Zimmermann, K. (2022). The big bang: Stock market capitalization in the long run. *Journal of Financial Economics, 145*(2), 527–552.

Mahmood, H. (2020). Impact of financial market development on the CO_2 emissions in GCC countries. In *MPRA Paper No. 109134*. MPRA. Retrieved from https://mpra.ub.uni-muenchen.de/109134/1/MPRA_paper_109134.pdf

Mhadhbi, M., Gallali, M. I., Goutte, S., & Guesmi, K. (2021). On the symmetric relationship between stock market development, energy efficiency, and environmental quality: A nonlinear analysis. *International Review of Financial Analysis, 77*, 101840.

Mumtaz, M. Z. (2023). What factors cause ocean CO_2 emissions? A panel data analysis. *Environmental Science and Pollution Research, 30*, 123111–123125.

Mumtaz, M. Z., Smith, Z. A., & Khan, A. H. (2017). An analysis of Chinese outward foreign direct investment in emerging and developing countries: Implications for Pakistan under the CPEC. *Journal of Social, Political, and Economic Studies, 42*(3/4), 327–367.

Ouyang, Y., & Li, P. (2018). On the nexus of financial development, economic growth, and energy consumption in China: New perspective from a GMM panel VAR approach. *Energy Economics, 71*, 238–252.

Paramati, S. R., Alam, M. S., & Apergis, N. (2018). The role of stock markets on environmental degradation: A comparative study of development and emerging economies across the globe. *Emerging Markets Review, 35*, 19–30.

Paramati, S. R., Mo, D., & Gupta, R. (2017). The effects of stock market growth and renewable energy use on CO_2 emissions: Evidence from G20 countries. *Energy Economics, 66*, 360–371.

Pesaran, H. H., & Shin, Y. (1998). Generalized impulse response analysis in linear multivariate models. *Economic Letters, 58*(1), 17–29.

Pineiro Chousa, J., Tamazian, A., & Vadlamannati, K. C. (2009). Does higher economic and financial development lead to environmental degradation: Evidence from BRIC countries. *Energy Policy, 37*(1), 246–253.

Raza, S. A., Shah, N., Qureshi, M. A., Qaiser, S., Ali, R., & Ahmed, F. (2020). Nonlinear threshold effect of financial development on renewable energy consumption: Evidence from panel smooth transition regression approach. *Environmental Science and Pollution Research, 27*(25), 32034–32047.

Salahuddin, M., Gow, J., & Ozturk, I. (2015). Is the long-run relationship between economic growth, electricity consumption, carbon dioxide emissions and financial development in Gulf Cooperation Council Countries robust? *Renewable and Sustainable Energy Review, 51*, 317–326.

Sharma, G. D., Tiwari, A. K., Erkut, B., & Mundi, H. S. (2021). Exploring the nexus between non-renewable and renewable energy consumptions and economic development: Evidence from panel estimations. *Renewable and Sustainable Energy Reviews, 146*, 111152.

Sheraz, M., Deyi, X., Mumtaz, M. Z., & Ullah, A. (2022). Exploring the dynamic relationship between financial development, renewable energy, and carbon emissions: A new evidence from belt and road countries. *Environmental Science and Pollution Research, 29*, 14930–14947.

Stickles, R. C., & Horrace, W. C. (2014). *Modelling asymmetric cointegration and dynamic multiplier in a nonlinear ARDL framework. Festschrift in Honor of Peter Schmidt* (pp. 281–314). Springer.

Zafar, M. W., Zaidi, S. A. H., Sinha, A., Gedikli, A., & Hou, F. (2019). The role of stock market and banking sector development, and renewable energy consumption in carbon emissions: Insights from G-7 and N-11 countries. *Resources Policy, 62*, 427–436.

Chapter 6
Central Bank Transparency in the GCC Countries: Enhancing Financial Resilience

Mohamed Zitouni Bechri and Héla Miniaoui

Abstract The role and mandates of central banks have become broader and more complex since the 2008 global financial crisis. More transparency is required to maintain public support, safeguard independence, and enhance policy effectiveness.

This chapter examines the transparency performance in the Gulf Cooperation Council (GCC) countries. Compared to benchmarks such as the Hong Kong Monetary Authority, the GCC central banks scored very low in economic, procedural, and policy transparency, due to lack of data, policy models, analysis of the outcomes of unexpected macroeconomic shocks, and explanation of the way monetary policy decisions are taken.

Central banks' transparency should be at the top of the reform agenda in the Gulf Arab States, since it will require other stakeholders to improve their statistics collection, processing, and publication, with wider implications for investor confidence, and the economy at large. Gulf central banks can bolster transparency by adopting credible reform agendas focused on the publication of accurate statistics, disclosure of analysis and forecasting methodology, and forward-looking explanations of policy decisions. This will be a powerful tool in building more resilient financial sectors that can weather global shocks more effectively.

Keywords Central Banks · Transparency · GCC · Financial resilience

---

Present Address:
M. Zitouni Bechri (✉)
The Economist Group, London, UK
e-mail: mohamed.bechri2020@gmail.com

H. Miniaoui
Lusail University, Doha, Qatar
e-mail: hminiaoui@lu.edu.qa

H. Miniaoui and M. Bellalah (eds.), *Green Growth Opportunities and Sustainable Finance in the Gulf Cooperation Council Region*, Gulf Studies 24,
https://doi.org/10.1007/978-981-96-8287-4_6

6.1 Introduction

Central bank transparency focuses on the flow and accessibility of information so the public and markets are informed in an accurate, timely manner about the decisions made to achieve a central bank's goals and the broader economic and financial developments that shaped such decision-making. Transparency matters because it makes for more accurate expectations by well-informed economic agents, thereby making policy announcements more credible and policy decisions better informed, more predictable, and more effective.

The role and mandates of central banks have become broader and more complex since the 2008 global financial crisis. The unconventional nature and growing scale of interventions (as seen again during the COVID-19 pandemic) have attracted much closer scrutiny. More transparency and accountability are therefore required to maintain public support, safeguard independence, and enhance policy effectiveness (Adrian et al., 2020). In this context, transparency is expected to enhance financial sector resilience, i.e., improve the ability of financial institutions to withstand unexpected shocks.

The International Monetary Fund (IMF) developed a Central Bank Transparency Code (CBTC) in April 2019, to allow central banks and their stakeholders to map national transparency practices onto international best practices. The purpose of the CBTC is to enhance the transparency and accountability of the central bank and contribute to policy effectiveness.

The currencies of the Gulf Arab states are pegged to the dollar, thereby limiting in significant ways the room for maneuver in central banks' decision-making, as interest rates need to follow the decisions of the U.S. Federal Reserve. Transparency remains important nonetheless to these banks' broader governance and accountability frameworks, as well as to their independence from political interference. And conversely, lack of transparency can lead to uncertainty about the ability of Gulf monetary authorities to continue defending the peg, which may encourage speculation and eventually lead to the collapse of the fixed exchange rate regime.

This chapter examines the performance of central bank transparency in the Gulf Cooperation Council (GCC) countries and explains its importance as a powerful tool in building more resilient financial systems that can weather global shocks more effectively and pave the path to sustainable finance.

The remainder of this chapter is as follows. Section 6.2 discusses the nexus between transparency and financial sector resilience. Section 6.3 examines the performance of central bank transparency in the GCC countries. Section 6.4 presents the priority reforms in the GCC. Section 6.5 concludes the chapter by focusing on the role of transparency in achieving a flexible and resilient financial sector.

6.2 Transparency for Financial Sector Resilience in the GCC Countries

Several studies have found that greater central bank transparency leads to more predictable policy decisions, which can reduce market volatility and uncertainty (Blinder et al., 2008). By openly communicating its policy framework, models, and forecasts, a transparent central bank allows markets to better understand the central bank's reaction function. This improved predictability may facilitate more efficient pricing of risks by financial markets (Kahn et al., 2019).

Central bank transparency has been extensively studied in the context of monetary policy effectiveness and financial stability (Crowe & Meade, 2008; Geraats, 2002). However, the linkages between transparency and financial sector resilience have received relatively little attention in academic literature. In fact, transparency plays a crucial role in enhancing the resilience of financial sectors, i.e., the ability of financial institutions, particularly banks, to withstand external shocks, owing to high capital adequacy, sufficient liquidity, and few and sufficiently provisioned non-performing loans, in addition to helping stabilize inflation and output variability.

In particular, transparency in central bank operations can mitigate financial stress and promote stability by reducing information asymmetry and fostering confidence. The GCC countries, with their emerging financial markets, can benefit significantly from increased transparency, which aligns with global trends observed since the late 1990s.

Central bank transparency reduces also financial stress, as it allows for better-informed decision-making by market participants. By providing clear and timely information, central banks can prevent the kind of opaque practices that contribute to past financial crises, such as the 2008 global financial crisis. Meanwhile, a transparent monetary policy is crucial for financial stability, as it helps align market expectations with policy objectives (Horvath & Vaško, 2013). The GCC countries, with their unique economic structures, can leverage transparency to enhance financial stability, especially during periods of economic uncertainty.

While central bank transparency is generally beneficial, its impact can vary depending on the specific economic and political context of a country. In the GCC, where financial markets are still developing, the effects of transparency might be less pronounced compared to more developed economies. However, as these countries continue to evolve their financial systems, the strategic implementation of transparency can play a pivotal role in fostering long-term economic stability and growth.

Studies of the GCC banking sector have generally concluded that it is relatively resilient, due mostly to the fact that it is well capitalized with a strong deposit base and highly profitable. The IMF Financial Sector Assessment of the Saudi banks (IMF, 2024), as an example, concluded that banks "appear resilient to macroeconomic shocks," including to interest rate shocks decided by the Federal Reserve in the U.S. And this is especially the case for domestically systemic banks.

Implementing transparency faces challenges nonetheless, including resistance from financial institutions and technological hurdles. These risks can be aggravated by bank exposure to large infrastructure projects, the vulnerable Small and Medium Enterprises sector, and the volatile real estate sector, as examples.

Enhanced central bank transparency can play a crucial role with regards to data gaps by increasing their availability and timeliness, and the credibility of policy decisions.

Transparency about the objectives, criteria, and impact of these initiatives can help mobilize private sector participation and promote the development of green financial markets. This will be a powerful tool in building more sustainable financial systems that can weather global shocks more effectively.

6.3 Central Bank Transparency in the Gulf States

For years, the International Journal of Central Banking has published the "Central Bank Transparency and Independence Index," which evaluates central bank transparency with scores between 0 and 15. The most recent version, published in Dincer et al. (2022), focused on monetary policy transparency, with updated yearly country scores during the period 1998–2019.[1]

The Gulf states' overall transparency indexes were very low in 1998, ranging from 1.5 in Kuwait to 2.5 in the UAE, 3.5 in Oman, Qatar, and Saudi Arabia, and 4.5 in Bahrain. One reason may be that the Gulf central banks do not need much transparency, as their main target is to continuously defend the fixed exchange rate of their currencies with the dollar, which makes them unique even among oil exporters. (Norway, as an example, adopts a free-floating exchange rate regime.) The value of the Kuwaiti dinar has been determined since 20 May 2007, by virtue of the Decree No. 147/2007, by reference to an undisclosed basket of international currencies. It is generally thought nonetheless that the U.S. dollar remains determinant in this regard, and therefore, the transparency requirements for Kuwait do not differ much from the other GCC member countries.

The authors of the transparency index argue that "exchange rate targeters feel less urgency about improving their information disclosure, perhaps because their monetary policy is automatically adjusted to maintain the exchange rate target, the achievement of which is easily observable." But this is not a solid argument because it overlooks the importance of transparency for policy credibility and effectiveness. Moreover, some exchange rate targeters had much higher transparency scores in 1998 than the Gulf Arab states, such as Hong Kong (5.5 out of 15), which points to transparency issues in Gulf central banks that cannot be fully explained by their currencies being pegged to the U.S. dollar.

Moreover, the Gulf countries' overall transparency scores improved little between 1998 and 2019. In the latter year, Saudi Arabia achieved the highest score of 5.5,

[1] https://eml.berkeley.edu/~eichengr/data.shtml.

Table 6.1 Transparency index (2019)

	T1	T2	T3	T4	T5	Overall score Transparency index 2019
Bahrain	3	0	1	0	1	5
Kuwait	1.5	0	0	1.5	0	3
Oman	2.5	0	0	0	1.5	4
Qatar	1.5	0	1	0	1	3.5
Saudi Arabia	2.5	0	0	1	2	5.5
UAE	2	0.5	1	0	1.5	5
Benchmark: HKMA	2.5	2	1.5	1	1.5	8.5

due to reforms the Kingdom implemented under pressure as a member of the G20. However, Saudi Arabia's score remained well below the scores of the Hong Kong Monetary Authority (8.5) that year (Table 6.1).

Overall transparency ratings are calculated by evaluating performance on five issues with scores between 0 and 3. These are: political transparency, economic transparency, procedural transparency, policy transparency, and operational transparency. For each component, the scores out of 1 are given to 3 main questions (a, b, and c), based on information available in the English-language section of the monetary authority's website.

6.3.1 Political Transparency

Political transparency measures the openness about the policy objective(s), namely: (a) the existence of a formal statement about the monetary policy objective(s) and prioritization in case of multiple objectives, (b) the existence of well quantified objective(s) such as a 2% inflation target, and (c) the existence of an explicit contract/ arrangement with the government.

For the latter, in the case of the U.K, as an example, the remit for the Monetary Policy Committee (MPC) requires an exchange of open letters between the Governor of the Bank of England and the Chancellor of the Exchequer if inflation moves away from the target by more than 1% point in either direction. The open letter should be published alongside the minutes of the following Monetary Policy Committee meeting, referring as necessary to the Bank's latest Monetary Policy Report and forecasts. The reason for publishing the letter at that time is to allow the MPC time to form and communicate its strategy regarding returning inflation to the target.

Enhanced political transparency impacts financial sector resiliency by clearly stating the commitment to maintain the peg arrangement to the U.S., dollar, thereby limiting exchange rate uncertainty and risks for financial institutions and others, such as corporates and securities markets.

The scores (out of 1) for each sub-component are summarized for GCC countries and the Hong Kong Monetary Authority (HKMA), as a benchmark, in Table 6.2.

Bahrain gets a full score, followed by Oman. The relatively high political transparency scores of Gulf central banks related to policy objectives could be because their main policy objective is clearly stated (maintaining the fixed exchange rate of the local currency to the U.S. dollar) and well quantified (e.g., 3.75 Saudi riyals per U.S. dollar).

This exceptional performance was due to the fact that the Central Bank of Bahrain (CBB) has a formal statement of the objective(s) of monetary policy. In fact, the CBB clearly states that it is responsible for setting and implementing monetary policy in the Kingdom of Bahrain. The Kingdom maintains an exchange rate peg at 0.376 Bahraini dinars to the U.S. dollar, which provides an anchor for monetary policy, which contributes to controlling inflation and protecting the external value of the currency. For the quantification of the primary objective, the CBB offers a set of deposit and lending standing facilities in Bahraini dinars to all retail banks. The interest rates on the standing facilities are the CBB policy interest rates. The CBB standing facilities rates are decided by a Monetary Policy Committee at the CBB. They are updated on a daily basis on the CBB website and the CBB's Reuters and Bloomberg pages "CBOB." The policy interest rates are updated at 10 a.m. each business day.

In contrast, the UAE, as an example, gets half the score in the first two sub-components due to the fact that there is no comprehensive document explaining the rationale for the currency peg arrangement, the exchange rate target, and what to do in case there is pressure in the FOREX market for the exchange rate target to diverge from the target of 3.6725 Dirhams per U.S. dollar. Similarly, other central banks could increase their overall scores for political transparency by being more explicit about their central banks' independence—for example, by clearly writing it out in their central bank laws—and the type of institutional arrangement with the government to achieve the stated objectives.

Table 6.2 Political transparency scores (2019)

	a-Formal statement	b-Quantified objective (s)	c-contract with government	Overall score
Bahrain	1	1	1	3
Kuwait	0.5	0.5	0.5	1.5
Oman	1	1	0.5	2.5
Qatar	0.5	0.5	0.5	1.5
Saudi Arabia	1	1	0.5	2.5
UAE	0.5	0.5	1	2
Benchmark: HKMA	1	1	0.5	2.5

Table 6.3 Economic transparency scores (2019)

	a-Public availability of economic data	b-Macroeconomic policy model (s)	c-Publication of macroeconomic forecasts	Overall score
Bahrain	0	0	0	0
Kuwait	0	0	0	0
Oman	0	0	0	0
Qatar	0	0	0	0
Saudi Arabia	0	0	0	0
UAE	0	0	0.5	0.5
Benchmark: HKMA	0.5	1	0.5	2

6.3.2 *Economic Transparency*

Economic Transparency measures the extent of economic information used to develop monetary policy. This is a particularly important aspect for the GCC countries as they generally suffer from lack of statistics, especially data that require a well-resourced and empowered national statistics center and the collaboration of several stakeholders (finance and economics ministries, customs, chambers of commerce…).

For economic transparency, all Gulf states scored 0 except for the UAE (0.5) compared to 2 for Hong Kong (Table 6.3). First, the five required economic variables are either published with long lags (money supply, inflation, and GDP) or they may be missing (quarterly unemployment and capacity utilization rates). Second, the macroeconomic models used for policy analysis are not posted on the websites; and they may be missing altogether. Third, forecasts are rarely made public, if they ever exist. The UAE was a notable exception with half the possible score due to the publication of some forecasts in its Quarterly Economic Review and Annual Report.

6.3.3 *Procedural Transparency*

For procedural transparency, which concerns decision-making, including openness about the monetary policy frameworks and policy deliberations, they impact financial sector resilience by providing banks and other financial institutions with the needed information. This transparency significantly enhances financial sector resilience by reducing uncertainty, aligning expectations, and enabling informed decision-making among financial institutions.

Bahrain, Saudi Arabia, and the UAE all scored 1, while the other Gulf states scored 0 (Table 6.4). The low scores are due to the central banks' failure to publish

Table 6.4 Procedural transparency scores (2019)

	a-Explicit monetary policy strategy	b-Policy deliberations	c-Voting balance/ records (within 3/8 weeks)	Overall score
Bahrain	1	0	0	1
Kuwait	0	0	0	0
Oman	0	0	0	0
Qatar	1	0	0	1
Saudi Arabia	0	0	0	0
UAE	1	0	0	1
Benchmark: HKMA	1	0.5	0	1.5

accounts of their deliberations. Press releases following the board of directors' meetings, for example, are limited to the announcements of the decisions without further explanations, and no minutes are published after that.

6.3.4 Policy Transparency

Policy Transparency relates to the prompt disclosure of policy decisions, with clear explanations of the rationale and an indication of likely future policy actions. Kuwait scored 1.5 and Saudi Arabia scored 1, while the other Gulf states scored 0, compared to 1 for Hong Kong (Table 6.5).

First, for the prompt announcement of the main operating instrument or target, Kuwait and Saudi Arabia got the full score. These central banks, CBK and SAMA, announce decisions about adjustments to the main operating instrument or target on

Table 6.5 Policy transparency scores (2019)

	a-Prompt announcement of adjustments to policy instruments/targets	b-Announcing dates of future policy decisions	c-Announcement of future policy orientation (forward guidance)	Overall score
Bahrain	0	0	0	0
Kuwait	1	0.5	0	1.5
Oman	0	0	0	0
Qatar	0	0	0	0
Saudi Arabia	1	0	0	1
UAE	0	0	0	0
Benchmark: HKMA	1	0	0	1

the day of implementation through their websites. Second, for explanation of policy decisions, the only exception was Kuwait, which got half score as the CBK provides an explanation when policy decisions change, but only superficially. Third, for the future inclination of policy, or forward guidance, all GCC countries got zero (which was also the case for Hong Kong). This may be due to the fact that this task falls on the authority of the anchor currency. But this is no excuse, as monetary authorities should give their own assessment of the future policy orientation, backed by updated data. Future actions should also hint at the possible use of macroprudential tools which may differ from the Fed's.

6.3.5 Operational Transparency

Operational Transparency evaluates openness about challenges in policy implementation. Their impact is similar to the previously stated one. The impact may be higher, though, as it relates to implementation. Unlike transparency in policy formulation (e.g., disclosing decision-making processes), it focuses on the real-world application of policies where theoretical plans meet practical realities.

Saudi Arabia ranked first among Gulf states in operational transparency with a score of 2, followed by Oman and the UAE with 1.5, Bahrain and Qatar with 1, and Kuwait with 0, compared to 1.5 for Hong Kong (Table 6.6).

First, with regards to the regular evaluation of achieving the main policy operating target, i.e., achieving the target fixed exchange rate of the local currency against the U.S. dollar, all countries got the full score except Kuwait, which fails to announce the basket of currencies it is using, thereby failing to disclose the basis for such an evaluation. Kuwait needs to catch up with the publication of the basket of international currencies and their respective weights.

Table 6.6 Operational transparency scores (2019)

	a-Evaluation of achieving the main policy operating targets	b-The information provided on unexpected macro disturbances	c-Evaluation of monetary policy outcomes	Overall Score
Bahrain	1	0	0	1
Kuwait	0	0	0	0
Oman	1	0	0.5	1.5
Qatar	1	0	0	1
Saudi Arabia	1	0.5	0.5	2
UAE	1	0.5	0	1.5
Benchmark: HKMA	1	0	0.5	1.5

Second, GCC countries are very weak at providing information on (unanticipated) macroeconomic disturbances that affect the transmission process of monetary policy (zero for all except for Saudi Arabia and the UAE, which got 0.5 each). This is due to the fact that the latter provide some analysis in this regard in their quarterly and annual reports.

Third, only Saudi Arabia and Oman got half the score on evaluating the outcomes of unexpected macroeconomic shocks. As on the previous sub-component, Saudi Arabia got half the full score thanks to the persuasive explanations provided in its regular reports on issues such as inflation, economic developments, and financial stability. Oman got half the score since the Central Bank of Oman (CBO) provides superficially an evaluation of the policy outcome in light of its macroeconomic objectives, mainly the inflationary pressures, through its financial stability and macroeconomic stability reports.

The GCC central banks scored very low in economic, procedural, and policy transparency, due to various factors such as lack of analysis of the outcomes of unexpected macroeconomic shocks, the way monetary policy decisions are taken, and the delays in disclosing their policy decisions, together with an explanation of the decision.

6.4 Priority Reforms

The importance of transparency increased further after the global financial crisis. As a result, the International Monetary Fund started asking authorities to provide more data in a timely manner. This translated into legal requirements, with countries improving the quality of statistics and monetary authorities adding financial stability to their core objectives in central bank laws, which reinforced data and analysis requirements. Therefore, if Gulf central banks aim to close the gap with their peers, they should prioritize the timely and accurate provision of key missing quarterly statistics, such as data on job markets, the economic output gap, the equilibrium exchange rate, and fiscal outlook.

This will require capacity building and coordination with stakeholders, such as national statistics authorities and ministries of labor. Moreover, data should be calculated using the appropriate methodology (e.g., International Monetary Fund manuals), thoroughly checked for accuracy, and published in a timely manner.

Gulf central banks will also have to explain technical issues to increase their transparency. Bahrain, for example, got a full score in political transparency thanks to its publication of its monetary policy framework, the rationale for its currency peg regime, and the expected stability outcome of the fixed exchange rate arrangement of the Bahraini dinar to the dollar. Other Gulf central banks can look to Bahrain as a model and provide a clear, timely picture of these issues based on macroeconomic models of policy simulations and forecasting and developed with technical assistance from the AMF and the IMF. Similarly, they can improve transparency by quickly announcing policy decisions and providing information on related deliberations, the time frame, and expected outcomes. Minutes of board of directors and other standing committee meetings should be published within 3 weeks to achieve maximum transparency scores.

Gulf central banks can bolster transparency by informing the public and markets about the future orientation of monetary and exchange rate policies. Even though such forward guidance is generally associated with inflation targeting, exchange rate peggers, such as the Gulf states, still need to regularly renew their commitment to the peg arrangement and, therefore, the policy orientation of the anchor currency (the U.S. Federal Reserve), while disclosing their foreign currency position on a regular basis. The "foreign currency cover"—a required ratio of official foreign currency reserves to the monetary base, or money in circulation—should be clearly stated in Gulf states' central bank laws with a clear definition of the numerator and denominator, and related data should be published monthly to support the credibility of their commitment to the peg.

By implementing these reforms, Gulf central banks can close the gaps with global peers, mitigate risks, and enhance financial resilience through greater trust, predictability, and data-driven policymaking.

6.5 Putting the Pieces Together: Toward More Resilient Financial Sectors

Central banks' transparency should be at the top of the reform agenda in the Gulf Arab states. It will require other stakeholders to improve their statistics collection, processing, and publication, with wider implications for investor confidence and the economy at large.

To enhance financial resilience, Gulf central banks can increase transparency by adopting credible reform agendas focused on the publication of accurate statistics, disclosure of analysis and forecasting methodology, and forward-looking explanations of policy decisions (Bechri, 2024). Moreover, they can adopt international standards such as the IMF's Code of Practices on Monetary Policy Transparency, ensuring clarity in decision-making and accountability. Publishing detailed monetary policy reports, including rationales for decisions, economic forecasts, and risk assessments, would also reduce market uncertainty and build investor confidence. Regular, proactive communication through press conferences, digital dashboards, and accessible disclosures of stress test results can demystify operations and demonstrate robust oversight of financial institutions. Strengthening governance via independent audits and stakeholder engagement forums would foster trust, while leveraging technology to provide real-time data access ensures widespread dissemination of critical information.

Implementation can be enhanced with technical assistance from regional and international institutions and successful peers, especially exchange rate targeters, such as the Hong Kong Monetary Authority, which share similar concerns and constraints. Related issues, such as central bank governance and independence, can be part of the same reform strategy, which should be launched within a short time frame to limit delays and the risk of reform reversal. Moreover, central banks can benefit from IMF reviews of transparency practices to identify and address gaps.

These reforms will spill over onto stakeholders, such as the finance authorities and the national statistics bureaus, and will improve their data coverage and disclosure, which will strengthen investor confidence, further attract FDI, and enhance Gulf countries' credit ratings and their access to international capital markets. This will constitute a milestone in the GCC countries' long-term visions of diversification away from hydrocarbons and the buildup of a knowledge-based and sustainable economy.

By aligning with global best practices and prioritizing open dialogue, Gulf central banks can mitigate systemic risks, enhance crisis preparedness, and solidify their role as pillars of economic stability in a volatile global landscape.

References

Adrian, T., Shabsigh, G., & Khan, A. 2020 *Transparency makes central banks more effective and trusted*. IMF blog. Retrieved from https://www.imf.org/en/Blogs/Articles/2020/07/30/blog-transparency-makes-central-banks-more-effective-trusted

Bechri, Z. M. (2024). *The urgent need for transparency reform in Gulf Central Banks*. AGSIW. Retrieved from https://agsiw.org/the-urgent-need-for-transparency-reform-in-gulf-central-banks/

Blinder, A. S., Ehrmann, M., Fratzscher, M., De Haan, J., & Jansen, D. J. (2008). Central bank communication and monetary policy: A survey of theory and evidence. *Journal of Economic Literature, 46*(4), 910–945.

Crowe, C., & Meade, E. E. (2008). Central bank independence and transparency: Evolution and effectiveness. *European Journal of Political Economy, 24*(4), 763–777.

Dincer, N., Eichengreen, B., & Geraats, P. (2022). Trends in monetary policy transparency: Further updates. *International Journal of Central Banking, 18*(1), 331–348.

Geraats, P. M. (2002). Central bank transparency. *The Economic Journal, 112*(483), F532–F565.

Horvath, R., & Vaško, D. (2013). *Central bank transparency and financial stability: Measurement, determinants and effects*. Research Papers in Economics. Retrieved from http://www.fiw.ac.at/fileadmin/Documents/Publikationen/Working_Paper/N_113-HorvathVasko.pdf

IMF. (2024). *Saudi Arabia: Financial sector assessment program-financial system stability assessment*. Retrieved from https://www.imf.org/en/Publications/CR/Issues/2024/09/03/Saudi-Arabia-Financial-Sector-Assessment-Program-Financial-System-Stability-Assessment-554533

Kahn, M. E., Mohaddes, K., Ng, R. N., Pesaran, M. H., Raissi, M., & Yang, J. C. (2019). *Long-term macroeconomic effects of climate change: A cross-country analysis* (IMF Working Paper No. 19/215).

Chapter 7
Smart Cities in GCC Countries

Samir Brahim Belhaouari, Ashhadul Islam, Sami Miniaoui, and Siredj Eddine Benaichouche

Abstract This chapter explores the concept and implementation of smart cities in the Gulf Cooperation Council (GCC) countries, examining the unique challenges and opportunities presented by the region's economic, social, and technological landscape. It analyzes the motivations behind smart city initiatives in the GCC, including urbanization, economic diversification, and sustainability goals. The role of advanced technologies in shaping urban environments is discussed, alongside comparing the GCC's approach to smart city development with global practices. Key challenges and opportunities are identified, focusing on governance, infrastructure, and cultural factors. The chapter provides insights into the future outlook for smart cities in the GCC, focusing on strategies for long-term success and sustainable urban development. This comprehensive analysis contributes to the broader understanding of smart city dynamics in rapidly developing regions, offering valuable perspectives on the intersection of technology, urban planning, and socio-economic development in the context of the GCC countries.

Keywords Smart cities · Gulf Cooperation Council (GCC) · Technological integration · Sustainable urban development

---

S. B. Belhaouari (✉) · A. Islam
College of Science and Engineering, Hamad Bin Khalifa University, Education City, Qatar Foundation, Doha, Qatar
e-mail: sbelhaouari@hbku.edu.qa

A. Islam
e-mail: asislam@hbku.edu.qa

S. Miniaoui
College of Engineering and IT, University of Dubai, Dubai, UAE
e-mail: sminiaoui@ud.ac.ae

S. E. Benaichouche
Computer Science Department, University of Science and Technology Oran - Mohamed Boudiaf, Oran, Algeria
e-mail: siredjeddine.benaichouche@univ-usto.dz

H. Miniaoui and M. Bellalah (eds.), *Green Growth Opportunities and Sustainable Finance in the Gulf Cooperation Council Region*, Gulf Studies 24,
https://doi.org/10.1007/978-981-96-8287-4_7

7.1 Introduction

The rapid growth of urbanization and technological advancements in the twenty-first century have led to the emergence of smart cities, necessitating a new approach to urban development in the Gulf Cooperation Council (GCC) countries. With the global urban population expected to expand, more than two-thirds of the world's people are projected to live in cities by 2050 (Mathew & Bangwal, 2024). The GCC region faces unique challenges and opportunities in preparing for this urbanized future.

The concept of smart cities has evolved from a technology-centric focus to a broader framework encompassing social well-being, environmental sustainability, and economic development objectives. This evolution is particularly relevant in the GCC, a region characterized by high levels of urbanization and rapid economic growth (Elian & Kisswani, 2023). Several factors drive this focus on smart urbanism, including the need for economic diversification beyond reliance on oil and gas (Al-Saidi & Zaidan, 2024), the imperative to improve urban sustainability in resource-intensive cities, and the ambition to position GCC cities as global hubs of innovation and high-quality living standards. The urgency of transitioning to smart cities is underscored by unique environmental challenges in the region, such as water scarcity, extreme heat, and the need for sustainable energy solutions. Smart city initiatives in the GCC are notable for their scale and ambition, encompassing the construction of new cities like Masdar City in Abu Dhabi and NEOM in Saudi Arabia (Mutambik et al., 2023), as well as extensive digital upgrades in existing urban areas, such as Smart Dubai (Breslow, 2021). Cutting-edge technologies, including the Internet of Things (IoT), Artificial Intelligence (AI), and Big Data analytics, are being leveraged to improve urban efficiency, sustainability, and overall quality of life (Talari et al., 2017).

However, the adoption of smart city concepts in the GCC faces several challenges. Traditional governance models often rely on top-down decision-making and may conflict with emerging bottom-up approaches emphasizing community involvement. This highlights the need for new governance frameworks that balance centralized planning with participatory engagement (Al-Saidi & Zaidan, 2024). Moreover, smart city strategies must account for the region's unique social and cultural contexts. Balancing technological advancements with preserving cultural values requires a thoughtful approach that integrates local nuances into urban design, mobility innovations, and digital service implementation. Given the region's harsh climate and limited natural resources, environmental sustainability is another critical dimension of smart cities in the GCC. Innovations such as smart grid systems, building automation technologies, and AI-driven water management solutions are instrumental in addressing these challenges. For example, smart grid systems can significantly reduce energy consumption, while predictive maintenance systems can optimize water resource management.

The GCC nations also recognize the importance of fostering innovation ecosystems to drive technological progress and entrepreneurial activity. Creating environments that attract talent, support research and development, and encourage start-ups aligns with broader economic diversification goals. These efforts position GCC cities as leaders in areas such as clean energy, AI applications, and smart mobility solutions (Mathew & Bangwal, 2024). Collaboration and knowledge sharing are integral to the advancement of smart cities in the GCC. Initiatives like the Smart Cities Global Network facilitate the exchange of best practices and experience among experts, accelerating the adoption of innovative solutions.

This chapter examines smart city projects in the GCC, exploring their evolution, technological underpinnings, and future directions. It investigates how these projects promote sustainable urban growth, economic diversification, and social inclusion while addressing critical data privacy and cybersecurity issues. By analyzing real-life examples and offering in-depth insights, this chapter contributes to the discourse on smart urban development in the GCC and other rapidly urbanizing regions facing similar challenges and opportunities.

The Technological Integration and Sustainability in Smart City Development (see Sect. 7.2) highlights the pivotal role of technology in achieving sustainability within smart cities while addressing global challenges and their local implications. A discussion on The Evolution and Multifaceted Nature of Smart Cities (see Sect. 7.3) provides a comprehensive review of diverse definitions and conceptual frameworks, tracing their development and emphasizing the complexities of integrating various dimensions into unified models. The Role of Technology in Smart City Development (see Sect. 7.4) explores transformative technologies shaping urban planning, infrastructure, and service delivery. Prominent projects in the Gulf Cooperation Council (GCC) region are examined in Smart City Initiatives in GCC Countries (see Sect. 7.5), showcasing successful implementations and their impacts on local communities. Drivers and Motivations for Smart City Development in the GCC (see Sect. 7.6) delves into the economic, environmental, and social incentives driving this regional transformation. Challenges and Opportunities in Implementing Smart Cities in the GCC (see Sect. 7.7) addresses obstacles to smart city projects and opportunities arising from overcoming these challenges. A comparative analysis in Distinctive Approaches to Smart City Development in the GCC and Global Contexts (see Sect. 7.8) examines unique strategies in the GCC compared to other regions, focusing on economic, social, and technological dynamics. The chapter concludes with a Future Outlook (see Sect. 7.9), offering insights into emerging trends and pathways for innovation and sustainable growth in GCC smart cities.

7.2 Technological Integration and Sustainability in Smart City Development: Challenges and Global Perspectives

The rising trend of urbanization worldwide and the emergence of smart cities signify profound changes in population patterns and technological advancements, alongside a growing emphasis on sustainable urban growth. Urban areas, now driving approximately 80% of the global GDP, are projected to host 70% of the world's population by 2050 (Hayat, 2016). This rapid urban expansion, driven by demographic shifts and economic opportunities, places immense pressure on resources, infrastructure, and the environment (Hayat, 2016). Currently, urban areas consume 75% of the planet's energy and account for roughly 70% of global greenhouse gas emissions, underscoring the urgency of implementing environmentally sustainable urban strategies (Hayat, 2016). In addressing these challenges, the integration of advanced technologies has proven crucial. Over the past two decades, significant advancements in communication, computing, and data analytics have enabled city planners to design intelligent urban spaces that prioritize resilience and the well-being of residents. Key technological developments include the Internet of Things (IoT), where sensors and actuators monitor and manage urban infrastructure, and adopting cloud and edge computing for data storage and real-time processing. Big Data analytics further empowers cities to derive actionable insights from extensive datasets (Lea, 2017). For instance, IoT sensors embedded in energy grids can reduce costs, enhance the reliability of energy distribution systems, and provide transparent information to suppliers and consumers, thereby improving efficiency and accountability (Neirotti et al., 2014).

Smart city projects vary significantly in scope and design, reflecting the diverse socio-economic contexts and priorities of individual cities rather than adhering to a universal model. These initiatives can be categorized broadly into two domains: the "hard" domain, focusing on physical infrastructure such as energy systems, waste management, transportation, and public safety, and the "soft" domain, which encompasses education, culture, social integration, and governance services (Neirotti et al., 2014). For example, hard domain initiatives include Singapore's implementation of smart energy systems, with smart grids integrated into 30% of households, and San Francisco's commitment to renewable energy, aiming for zero waste by 2020 (Hayat, 2016). Conversely, soft domain initiatives emphasize human and social dimensions, such as Barcelona's open government platforms, which enhance citizen participation in governance and urban planning, and Rio de Janeiro's Rio Datamine platform and digital literacy programs aimed at fostering social inclusion and learning opportunities in low-income communities (Hayat, 2016). In Europe, many cities prioritize investments in human capital and creativity, often surpassing physical infrastructure in importance. Wealthier cities with robust governance systems typically lead these investments (Neirotti et al., 2014).

Cities worldwide are adopting renewable energy sources, green building practices, and smart mobility solutions to mitigate environmental impacts (Neirotti et al., 2014). For instance, Barcelona promotes sustainable building projects, while Singapore's

Green Mark certification encourages energy efficiency in urban design (Hayat, 2016). Modern transportation systems increasingly utilize information and communication technology (ICT) to enhance public transit and alleviate traffic congestion, as seen in Rio de Janeiro's ongoing efforts to improve its transport infrastructure (Hayat, 2016). Additionally, urban planners are incorporating disaster risk reduction measures into smart city designs to enhance resilience against climate change and natural disasters such as floods and cyclones (Hayat, 2016). Despite these advancements, significant challenges persist. Financial constraints limit the capacity for extensive smart city initiatives, particularly in less developed regions. Social inclusion also poses a critical challenge, as technological advancements risk exacerbating existing inequalities, particularly in access to digital infrastructure and services (Neirotti et al., 2014). Ensuring the equitable distribution of benefits, especially to marginalized groups such as low-income populations, the elderly, and persons with disabilities, remains a priority. Furthermore, an over-reliance on technology without adequate consideration of societal and ethical dimensions may lead to unintended issues, including privacy concerns and cybersecurity risks (Neirotti et al., 2014). Cities must address these challenges to ensure that smart city projects are inclusive, secure, and capable of fostering holistic urban development.

7.3 The Evolution and Multifaceted Nature of Smart Cities: A Comprehensive Review of Definitions and Conceptual Frameworks

The smart city concept has evolved significantly over the years, transitioning from a purely technology-driven perspective to a comprehensive approach that integrates human needs, social dynamics, and infrastructure development. This evolution underscores the aim of creating sustainable, vibrant urban spaces that prioritize residents' well-being. Mahizhnan (1999) highlighted the dual importance of leveraging technology while simultaneously enhancing citizens' daily lives. This perspective laid the foundation for subsequent studies, such as the study conducted by Giffinger et al. (2007), which characterized smart cities through six performance-based dimensions: economy, mobility, environment, people, living, and governance. These dimensions emphasize the role of informed and proactive residents in the effective use of urban resources and actions. The notion of smart cities as an "urban labeling" phenomenon was explored by Hollands (2008), who highlighted the emphasis on learning, innovation, and creativity among residents and institutions, supported by digital communication networks. This perspective illustrates the interplay between human capital and technological infrastructure in fostering smart urban environments. Chen (2010) provided a more practical perspective on smart cities, stating that these urban centers will utilize communication and sensor technologies embedded within city infrastructures to enhance electricity usage and transportation systems, as well as other logistical functions that facilitate daily life activities, resulting in an enhancement of overall quality of life for all inhabitants.

The concept presented by Harrison et al. (2010) presents a perspective on smart cities by defining them as urban areas that link together physical, IT, social, and business infrastructures to harness collective intelligence within cities. This interpretation highlights the importance of integrating different aspects of urban systems to enhance the effectiveness and adaptability of the urban environment. Thite in his research on smart cities emphasized how human capital plays a crucial role in their development by stating that "The nexus of competitive advantage has shifted to those regions that can generate, retain, and attract the best talent" (Thite, 2011). Thuzar in his analysis of smart cities discussed the importance of urban development policies that cater to all residents, with a focus on uplifting disadvantaged communities as well (Thuzar, 2011). He mentioned that smart cities prioritize enhancing residents' quality of life while investing in human and social capital alongside modern communication infrastructure for sustainable economic growth. Moreover, Thuzar emphasized implementing policies to manage natural resources effectively in line with social and environmental concerns (Thuzar, 2011). This narrative underscores the significance of balancing economic growth with social and environmental considerations. Winters offered a simple, education-focused definition, considering "smart cities" to be metropolitan areas where a large share of the adult population has a college degree (Winters, 2011). According to Caragliu & Nijkamp's research, a city is considered smart when it invests in human and social capital and balances transportation with modern (ICT) communication infrastructure to promote sustainable economic growth and improve the quality of life through effective management of natural resources via participatory governance (Caragliu et al., 2011). Cretu in his research on smart cities pointed out two areas of interest: one concerning fresh perspectives on governance and the economy and another concentrating on sensor networks and integration of smart devices, real-time data, and ICT across various aspects of human existence (Cretu, 2012). Kourtit and Nijkamp defined what constitutes smart cities by describing them as the result of knowledge-intensive and creative strategies focused on improving various aspects of cities, such as the socio-economic, ecological, logistic, and competitive performance aspects (Kourtit & Nijkamp, 2012). Smart cities are said to rely upon a blend of assets such as human capital (e.g., skilled labor force), infrastructural capital (e.g., high-tech communication facilities), social capital (e.g., intense and open network linkages), and entrepreneurial capital (e.g., creative and risk-taking business activities) (Kourtit & Nijkamp, 2012). The writers also mentioned that the success of smart cities is underpinned by trustworthy governance structures and open-minded, innovative individuals who can boost economic output locally and provide fresh sustainable solutions (Kourtit & Nijkamp, 2012). According to Kourtit and Nijkamp in another study, smart cities are known for their efficiency due to well-educated individuals and knowledge-based professions with a focus on innovative endeavors and sustainability efforts (Kourtit et al., 2012). They emphasize how human capital, economic activities, and environmental consciousness are interconnected in modern urban environments (Kourtit et al., 2012). Lazaroiu took the approach of describing a smart city as a community of average technology size that is interconnected and emphasizes safety, in addition to technology usage for comfort and attractiveness (Lazaroiu & Roscia, 2012). Adding to this perspective in the same year, Lombardi et al. stressed the importance of information and communications technology (ICT), focusing on human capital development and education, social and relational capital, and environmental issues (Lombardi et al., 2012). This description highlights how technology plays a role in improving facets of urban life such as the social and environmental dimensions (Lombardi et al., 2012). According to Barrionuevo et al., a smart city utilizes all advancements and available technology intelligently and cohesively to create interconnected urban areas while also being livable and sustainable (Barrionuevo et al., 2012). This explanation underscores how technology and resources are strategically employed to accomplish multiple urban development objectives concurrently.

Zygiaris defined smart cities as places that blend socio-technical and socio-economic aspects for development purposes, integrating green infrastructure to protect the environment, that foster innovation, reduce CO_2 emissions, have a well-connected digital economy, and can generate valuable information through real-time data analysis in an intelligent manner (Zygiaris, 2013). Smart cities were defined by the British Standards Institute as "the effective integration of physical, digital and human systems in the built environment to deliver a sustainable, prosperous and inclusive future for its citizens" (City Framework–Guide, 2014). This definition emphasizes the integration of various systems and the ultimate goal of improving citizens' lives. According to Moura and de Abreu e Silva's analytical standpoint, a smart city is defined as a region where electronic data collection sensors (IoT) are integrated into infrastructure such as buildings, vehicles, and institutions, providing real-time data on city operations (Moura & de Abreu Silva, 2019). The information collected is then integrated into ICT platforms to enable city managers and decision-makers to enhance the effectiveness and durability of activities and services. Their definition also emphasizes the opportunity to promote good governance practices, enhance cooperation among economic actors, encourage innovative business models, inspire approaches to entrepreneurship, and support environmentally friendly urban growth.

Marsal-Llacuna et al. delved into the real-world implementations of smart city projects by emphasizing the utilization of data, information, and information technologies (IT) to optimize service delivery for residents, monitoring, and infrastructure while fostering collaboration among various economic actors to promote innovative business models across public and private domains (Marsal-Llacuna et al., 2015). Smart cities are defined by the ISO/IEC JTC1, as cited in the paper of Cheu et al., which highlights the use of technology-centric definition, as the application of new generation information technologies, like the Internet of Things (IoT), cloud computing, big data, and integration of space/geographical information to improve urban planning, development management, and delivery of smart services in cities (Cheu et al., 2015). A study by Barth et al. found that the majority of definitions of a smart city emphasize the connection between two areas: the city's control system (which includes infrastructures, economics, politics, governance spaces, and geographical factors) and how information is used by the city's stakeholders (such as residents, companies, administrations, tourists, etc.) (Barth et al., 2017). This perspective underscores the relationship between urban systems and human behavior within smart city environments. As shown in Table 7.1, various authors have contributed key concepts to the understanding of smart cities.

In conclusion, the smart city concept has expanded to cover a set of factors such as quality of life, citizen engagement, and promoting sustainable development. Smart cities aim to be resilient and interconnected, and to prioritize the overall welfare of society, balancing technological advancement with human and social needs. This comprehensive strategy guarantees that smart cities are not only technologically advanced but also socially and environmentally responsible, by integrating various systems to establish adaptable and inclusive urban environments. In addition to that, they combine technology with government and society to facilitate features like a smart economy, smart mobility, smart environment, smart people, smart living, and smart governance. However, as pointed out by Monzon, an agreed-upon definition for a Smart City has not been universally recognized yet, which indicates the changes and intricacies of the concept as it adjusts to emerging technologies, urban challenges, and societal requirements (Monzon, 2015).

7.4 The Role of Technology in Smart City Development

The significance of technology in the progress of smart cities is crucial, as innovations in the Internet of Things (IoT), Big Data analytics, and Artificial Intelligence (AI) are fundamental in enriching urban environments and enhancing the quality of life for residents. The IoT acts as the core infrastructure of smart cities by linking diverse devices and sensors that interact instantly to facilitate the continuous gathering and sharing of information across different urban systems. This connectivity supports a wide range of applications ranging from traffic management to environmental monitoring. This includes the deployment of smart sensors across the city that gather real-time data on air quality levels, traffic conditions, and energy usage patterns. This data is valuable for planners and officials as it assists them in making informed decisions to enhance city operations and services effectively. These sensors play a role in enhancing municipal services by easing traffic congestion using advanced traffic light systems and enhancing energy conservation through smart grid technologies (Talari et al., 2017). Moreover, cloud computing integration enables the storage and analysis of large volumes of data produced by Internet of Things (IoT) devices, providing effective solutions to managing the growing requirements for data processing. Utilizing this data promptly enables cities to address concerns like sudden increases in pollution levels or infrastructure failures, and thereby emergency responses are enhanced, leading to improved overall effectiveness and durability of urban systems.

Big Data analytics enhances smart city capabilities by enabling the integration and analysis of complex datasets, like social media interactions and IoT-generated information, which are crucial for comprehending the intricacies of urban life

Table 7.1 Key concepts in smart city research

Author(s)	Year	Key concept
Mahizhnan	1999	Importance of balancing information technologies with improving quality of life for citizens
Giffender et al.	2007	Six characteristics of smart cities: economy, mobility, environment, people, living, and governance
Hollands	2008	Smart cities as territories with high capacity for learning and innovation through human capital and technological infrastructure
Chen	2010	Use of communications and sensor capabilities to optimize city operations and improve quality of life
Harrison et al.	2010	Integration of physical, IT, social, and business infrastructures to leverage collective city intelligence
Thite	2011	Focus on human capital as the key to smart city competitiveness
Thuzar	2011	Emphasis on sustainable development benefiting all residents, including the poor
Winters	2011	Smart cities are metropolitan areas with a large share of college-educated adults
Caragliu et al.	2011	Smart cities invest in human/social capital, infrastructure, and participatory governance
Cretu	2012	Two research streams: governance and economy, and ICT integration in all aspects of human life
Kourtit and Nijkamp	2012	Smart cities result from knowledge-intensive strategies enhancing socio-economic, ecological, and competitive performance
Kourtit, Nijkamp, and Arribas	2012	High productivity through educated citizens, creative activities, and sustainability
Lazaroiu and Roscia	2012	Smart cities are interconnected, sustainable, and secure communities
Lombardi et al.	2012	Application of ICT focused on human capital, social/relational capital, and environmental concerns
Barrionuevo et al.	2012	Strategic use of technology and resources for integrated, habitable, and sustainable urban development
Zygiaris	2013	Focus on socio-technical and socio-economic growth, green infrastructure, and real-time data use
British Standards Institute	2014	Effective integration of physical, digital, and human systems for a sustainable, inclusive future
Moura and de Abreu e Silva	N/A	Smart cities use IoT to optimize city operations and promote participatory governance and sustainable urban development
Marsal-Llacuna et al.	2015	Practical applications of smart city initiatives using data and IT to improve services and infrastructure
ISO/IEC JTC1	2015	Smart Cities apply technologies like IoT, cloud computing, and Big Data for planning and services
Barth et al.	2017	Smart cities operate at the interface of city control systems and stakeholders' information behavior

(Talari et al., 2017). Through the examination and interpretation of these extensive datasets, cities can obtain insights to enhance public services, such as transportation systems improvement and more efficient waste management practices, in addition to enhancing emergency response capabilities. For example, Big Data analytics can help examine traffic flow in specific areas in order to predict traffic jams and prevent them by taking actions like tweaking traffic light timings or redirecting vehicles instantly. This ability also applies to fields like energy consumption. For instance, smart grids leverage Big Data to manage supply and demand equilibrium for an efficient and smarter energy distribution network. Moreover, the use of predictive analysis allows municipal authorities to anticipate forthcoming developments, like population increases or changes in energy demands, enabling strategic long-term planning and more effective resource distribution (Talari et al., 2017). In addition, Big Data is essential in enhancing safety by identifying patterns of activity and areas of heightened criminal activity, thus enabling law enforcement to deploy resources efficiently and strengthen security measures overall.

Artificial Intelligence (AI) is commonly referred to as the "brain" behind smart city innovations, playing a role in interpreting and responding to the information gathered by IoT devices and analyzed through Big Data analytics systems. The impact of AI on urban management is diverse and far-reaching, ranging from improving transportation systems to enhancing healthcare provisions. Within the transportation sector, AI-enabled smart traffic control systems utilize real-time data to ease congestion, enhance traffic flow efficiency, and facilitate the incorporation of self-driving cars. AI technology also helps improve transportation systems by optimizing routes and schedules using real-time data to shorten wait times for commuters and promote the use of public transport. This ultimately aids in decreasing the city's carbon footprint (Allam & Dhunny, 2019). Additionally, in the energy industry, AI plays a role in managing grids by analyzing information from IoT devices and Big Data platforms to predict energy consumption accurately and facilitate the incorporation of renewable energy sources. This helps to make sure that energy is distributed sustainably as part of the larger sustainability objectives of smart cities. In addition to that point, AI is quite crucial in the healthcare sector of smart cities, where it helps diagnose illnesses, plan treatments, and keep track of patients. AI-driven technology can examine information instantly, enabling timely medical treatments. This proves advantageous for handling chronic ailments and providing assistance to the elderly. Artificial-intelligence-based remote monitoring tools can notify professionals about any anomalies promptly, ensuring effective healthcare services (Allam & Dhunny, 2019). Advanced technologies like AI play a role in enhancing safety by implementing sophisticated surveillance and security measures that leverage facial recognition and predictive analytics to detect potential risks and handle emergencies with greater efficiency. These initiatives not only boost the safety of individuals but also contribute to a heightened sense of security in urban environments, which ultimately improves the quality of city living. In terms of governance, AI helps city officials make decisions based on real-time data analysis that facilitates both data-driven and transparent policy-making. This technology enhances public services by allowing citizens to engage through feedback and participation platforms provided by AI (Allam &

Dhunny, 2019). This interaction fosters a partnership between residents and city officials, resulting in efficient public services. In addition to that, AI contributes to sustainability efforts by enhancing the efficiency of resource utilization and waste management protocols using data-driven insights.

IoT (Internet of Things), Big Data analytics, and AI (Artificial Intelligence) are the components of smart city technology that help cities operate more smartly and sustainably while meeting the demands of their residents effectively when combined in an integrated system approach for managing city infrastructure, with data-driven insights to inform decision-making processes and improve public services for sustainable growth. In the Gulf Cooperation Council (GCC) countries, where rapid urbanization and unsustainable development are the concerns, the utilization of these technologies holds great importance in addressing the challenges brought by population growth and the need for environmental sustainability.

7.5 Smart City Initiatives in GCC Countries

The Gulf Cooperation Council (GCC) countries have embraced smart city initiatives as part of broader strategies to enhance urban living through digital transformation. Bahrain's Vision 2030 emphasizes cloud-based e-government services to optimize infrastructure and public services (Elian & Kisswani, 2023). Oman's Digital Oman Vision 2030 plan focuses on IT development and smart city projects such as the Duqm Special Economic Zone, with particular attention to utilities and waste management (Elian & Kisswani, 2023). Qatar's Lusail Smart City integrates cutting-edge technology and sustainable infrastructure to align with its National Vision 2030 objectives (Elian & Kisswani, 2023). Saudi Arabia's Vision 2030 includes initiatives like NEOM City that target urban environments with advanced technology (Elian & Kisswani, 2023). In the Middle East region, the UAE stands out for programs such as Dubai's Smart Dubai and Masdar City in Abu Dhabi which showcase progress in digital connectivity (Elian & Kisswani, 2023). These endeavors demonstrate an effort within the region to tackle issues using cutting-edge technologies; however, the effectiveness and influence of these projects may differ.

Looking at specific examples of these initiatives offers an understanding of how each GCC country implements its smart city strategies.

In the UAE, efforts to embrace smart cities are essential in light of the rapid urban growth witnessed in recent years. According to data from 2014, 85% of the population resides in urban areas. This number is projected to soar to 91% by the year 2050. The smart city initiatives in the UAE are primarily geared toward incorporating digital innovations into various infrastructure sectors to enhance operational efficiency and reduce energy consumption. Noteworthy projects include the implementation of advanced metering systems in Abu Dhabi to regulate electricity demand and the introduction of smart meter technology in Dubai to encourage energy conservation and support electricity generation initiatives. These measures have resulted in reductions in energy and water usage, along with decreased carbon emissions (Riadh, 2022).

Saudi Arabia's Vision 2030 prioritizes economic growth and enhanced urban living through ambitious smart city initiatives. The $500 billion NEOM mega-city project stands out as an endeavor under this vision by striving to redefine living standards through eco-friendly practices such as zero carbon emissions and reliance on renewable energy sources, while incorporating intelligent infrastructure across various sectors, such as energy generation and distribution systems. Furthermore, the Yanbu Industrial Smart City venture leverages technologies to boost operational efficiency and sustainability in its development approach. Urban centers such as Riyadh and Jeddah are implementing smart initiatives to enhance traffic control and public amenities. Meanwhile, Makkah and Al-Madinah are utilizing smart technologies to enhance the pilgrimage journey (Belaıd et al., 2023).

The innovative ideas in Lusail City in Qatar are highlighted by the smart city concepts there, specifically, that centralize data from various smart systems known as the LCCC (Large Command and Control Center). This center coordinates information from systems to oversee transportation, using smart traffic lights and video surveillance alongside crash detection systems, all powered by cutting-edge fiber optics and 5G networks. The setup guarantees traffic flow management and immediate response to incidents as they happen in real-time. The city also gains from direction signs and a transport application that improves accessibility and navigation for residents (Shaaban & Adalbi, 2021).

Oman's efforts to build smart cities concentrate on addressing traffic congestion and enhancing public transportation. The public transport operator Mwasalat has received positive feedback on safety and cost-effectiveness, although further improvements are needed. Rising vehicle numbers continue to exacerbate traffic issues, emphasizing the need for comprehensive urban mobility solutions (Al Maqbali & Refeque, 2017).

Bahrain's plan to transform into a smart city focuses on making investments in technology, infrastructure, and a skilled workforce, as outlined in the Bahrain Vision 2030, blueprint for progress by the year 2030. The vision emphasizes the implementation of smart urban systems and renewable energy solutions while embracing digital innovation across industries such as e-government services and information and communication technology (ICT). Moreover, the country is actively enhancing its logistics operations and service offerings. Bahrain's reputation for logistics operations and a thriving business environment has positioned it as a leader in smart city initiatives among the Gulf Cooperation Council (GCC) nations (Mamlook et al., 2019).

Kuwait's Vision 2035 is focused on positioning the country as a leading trade center by implementing smart city initiatives that incorporate cutting-edge technologies and sustainable approaches. The initiatives involve smart traffic control systems, energy-efficient construction in Kuwait City, and the Kuwait Integrated Urban Development Project, which advocates for housing and public transportation options. These efforts highlight Kuwait's dedication to improving urban life through smart solutions and collaborations between the public and private sectors (Elian & Kisswani, 2023).

These various smart city projects in the GCC countries demonstrate their dedication to leveraging digital technologies to develop connected urban spaces efficiently. Each country's strategy mirrors its socio-economic objectives and infrastructure needs, the shared emphasis on innovation indicates a significant change in how urban issues are tackled in the region.

7.6 Drivers and Motivations for Smart City Development in the GCC

The rapid urbanization and population growth in the GCC region are major drivers for the implementation of smart city projects. Studies predict that by 2050 more than 90% of the population in GCC countries will reside in urban areas (Mutambik et al., 2023). This remarkable urban expansion places significant pressure on existing infrastructure and services, hence the need for innovative approaches to address these challenges effectively (Elian & Kisswani, 2023). Qatar and the UAE boast some of the highest urbanization rates, which further underscores the critical need for advanced urban management strategies (Asmyatullin et al., 2020).

Economic diversification plays a role in driving the development of smart cities in the GCC region, as these countries have traditionally relied heavily on oil and gas revenues, which makes their economies susceptible to changes in global oil prices. The push to decrease this dependence and establish resilient and diversified economies serves as an incentive behind the initiatives for smart city development (Al-Saidi & Zaidan, 2024). For example, Oman's Vision 2040 lays out objectives aimed at broadening income sources and enhancing competitiveness through the advancement of smart cities (Mathew & Bangwal, 2024). Saudi Arabia's Vision 2030 and the UAE's Vision 2021 include smart city initiatives, as elements of their strategies for economic development (Asmyatullin et al., 2020). The economic argument for smart city projects is boosted by the promise of a return on investment (ROI). For instance, Qatar's TASMU initiative, with a 6 billion QR investment, aims to yield 40 billion QR in returns over five years, a demonstration of stakeholders' confidence in the enduring financial advantages of such smart city initiatives (Shokeir & Yahia, 2020). GCC countries find such forecasts appealing as they aim to manage their resources during the shift away from reliance solely on oil-based economies.

Sustainability is another key driver, given the environmental and economic challenges faced by the GCC countries due to their climate and rapid growth. One approach to tackling these issues is through the implementation of smart city projects that aim to decrease emissions and manage water usage efficiently (Shokeir & Yahia, 2020). For instance, smart traffic management systems can enhance fuel efficiency and reduce emissions by optimizing traffic flow and guiding drivers on optimal routes (Shokeir & Yahia, 2020). Examples such as Masdar City in Abu Dhabi showcase a dedication to environmental sustainability by incorporating solar power and eco-friendly construction methods to establish an eco-friendly setting (Asmyatullin et al., 2020). Nevertheless, it is worth mentioning that certain ambitious endeavors, like Masdar City, have encountered obstacles to achieving their environmental sustainability objectives; this underscores the importance of planning and effective execution strategies (Mathew & Bangwal, 2024).

Enhancing residents' quality of life is a fundamental motivation behind the development of smart cities in the Gulf Cooperation Council (GCC) region. There is a growing understanding that smart cities need not only to focus on technological progress but also to prioritize human-centric designs by promoting citizen engagement and interaction (Al-Saidi & Zaidan, 2024). Smart technologies are anticipated to enhance public services' delivery efficiency, increase healthcare accessibility, optimize transportation systems, and foster urban spaces (Asmyatullin et al., 2020; Elian & Kisswani, 2023). Examples include the use of smart parking systems, intelligent lighting, and efficient waste management in cities to enhance safety and efficiency (Shokeir & Yahia, 2020). However, there is criticism that many current smart city projects do not prioritize a people-centric approach, and it is suggested that there should be an emphasis on citizen involvement to meet the needs and desires of local communities (Mathew & Bangwal, 2024). Technological advancements underpin the development of smart cities by leveraging the rapid progress of information and communication technologies (ICT). These include innovations like the Internet of Things (IoT), Big Data analysis tools, and Artificial Intelligence (Al-Saidi & Zaidan, 2024; Asmyatullin et al., 2020). These technologies empower cities to boost connectivity levels and enhance service delivery while promoting participation from citizens. Integrating these tools into urban services is widely viewed as indispensable for contemporary urban management practices that seek to optimize efficiency and service quality (Elian & Kisswani, 2023). Despite the emphasis on technological advancements in cities, there is sometimes a gap between these developments and the genuine requirements of urban dwellers' lives. A study conducted by Mathew and Bangwal discusses the importance of a governance approach that effectively links technology with the needs of communities to drive smart city projects (Mathew & Bangwal, 2024). Global competitiveness is also a driving force for smart city initiatives in the GCC as countries in the region aim to boost their appeal for investment and skilled workers in today's interconnected world (Asmyatullin et al., 2020). By developing smart cities and embracing innovation and technology, leadership roles are taken up by these nations, drawing attention from international businesses and skilled professionals (Asmyatullin et al., 2020). Efforts undertaken in cities such as Abu Dhabi and Dubai aim to attract international investment and skilled professionals through cutting-edge technology and innovative urban planning (Asmyatullin et al., 2020). The rivalry among GCC nations pushes cities to embrace technologies and creative urban planning approaches in order to stand out and draw in residents and businesses (Al-Saidi & Zaidan, 2024).

Government efforts and policies play a crucial role in shaping the progress of smart city initiatives. Countries like Bahrain with its Economic Vision 2030, Qatar with its National Vision 2030, and Kuwait with its Vision 2035, are examples of strategies that provide the structure and resources for smart city projects. These visions set objectives for smart city development and technology integration into planning (Asmyatullin et al., 2020; Elian & Kisswani, 2023). Additionally, these initiatives often involve e-government initiatives and specific smart city programs tailored to align with the objectives (Al-Saidi & Zaidan, 2024). Public–private partnerships (PPPs) are now widely acknowledged as crucial for the execution of smart city initiatives, according to a study by Mathew and Bangwal (2024). This research stresses the significance of adept stakeholder coordination and teamwork between government entities and private enterprises to achieve goals effectively, emphasizing the necessity of structured engagement strategies to boost stakeholder involvement and guarantee project alignment with citizen needs. The study conducted by Mutambik et al. (2023) emphasizes the importance of Open Government Data (OGD) in the progress of smart city projects around the world. While OGD platforms have been implemented in GCC nations to support city efforts initially, their efficacy has sometimes fallen short of anticipated standards, necessitating enhancements in usability and effectiveness for better outcomes. Utilizing data is perceived as crucial for promoting transparency within governance systems as well as encouraging citizen participation and facilitating well-informed decision-making processes. In the GCC regions, smart city initiatives are greatly influenced by cultural and social aspects as well as religious factors, as highlighted in the study by Mutambik et al. (2023). It is essential to consider these nuances when designing approaches for Open Government Data (OGD) and implementing smart city projects that truly connect with the local communities.

In conclusion, the reasons and driving forces behind the advancement of smart cities in the GCC nations are varied and interconnected, catering to the challenges and aspirations of the region. The swift urbanization and population expansion underscore the pressing demand for innovative urban management solutions. Economic variety emerges as a crucial factor, as GCC countries strive to diminish their dependency on oil earnings and foster resilient economies. Sustainability concerns also increase the importance of smart city endeavors, which aim to address issues while encouraging effective resource management. Improving the quality of life of residents is a goal that emphasizes the significance of human-centric design and engaging citizens in the process. Progress in technology plays a role in supporting these advancements by allowing cities to utilize solutions for better service provision and connectivity. Smart cities act as hubs to attract investment and skilled individuals as GCC countries aim for global competitiveness. Government programs and collaborations between public and private sectors are vital in guiding these initiatives to align with goals and meet community requirements. The effective implementation of smart city projects relies on overcoming obstacles by promoting citizen engagement and creating realistic, sustainable plans. The long-term success of smart city growth in the GCC lies in harmonizing technological advancements with cultural awareness and social integration while prioritizing sustainability goals.

7.7 Challenges and Opportunities in Implementing Smart Cities in the GCC

The implementation of smart cities in the Gulf Cooperation Council (GCC) countries showcases a mix of challenges and opportunities that mirror the economic, social, and technological contexts of the area. This intricate blend of elements requires a strategy for urban development that tackles various aspects at once.

One major obstacle that smart city projects in the GCC face is the traditionally centralized approach to urban planning in the region. This centralization can impede local-level efforts and community involvement, leading to a lack of flexibility in adapting to local needs and conditions (Al-Saidi & Zaidan, 2024). This challenge is compounded by the diverse demographics and varying community characteristics in parts of the GCC, making it tough to create smart city strategies that work universally and effectively (Al-Saidi & Zaidan, 2024). The varied nature of communities requires a method that considers the requirements of different communities while maintaining a unified plan for advancing smart city initiatives.

The variation in technological advancement between cities and communities within the GCC poses a notable challenge to progressing forward together effectively. Certain regions face difficulties due to a lack of infrastructure or expertise needed for the integration of advanced smart technologies. This situation leads to a gap that may worsen existing disparities (Al-Saidi & Zaidan, 2024). The issue of disparity goes beyond infrastructure; it also involves the skills and knowledge necessary to efficiently operate and sustain smart city systems. The research article by Shokeir and Yahia (2020) points out a lack of expertise in IoT systems and advanced technologies, with ongoing efforts to develop the required skills. This deficiency in skilled individuals could pose challenges to the execution and maintenance of smart city technologies, possibly resulting in less-than-optimal outcomes or even setbacks in smart city projects.

The challenge of regulatory frameworks is one of the challenges for smart city projects to tackle swiftly and effectively because current regulations may not be ideal for their rapid execution. Hence, the call for updated policies that encourage innovation and make it easier to incorporate new technologies (Al-Saidi & Zaidan, 2024). It is crucial to establish the laws and policies that support smart city initiatives, as without a supportive regulatory environment in place, the adoption of smart technologies could encounter significant obstacles (Elian & Kisswani, 2023). This regulatory hurdle becomes more complex when coupled with the need for robust governance practices to handle relationships with stakeholders and ensure positive project outcomes (Shokeir & Yahia, 2020).

Funding challenges are a concern in countries within the GCC region, where the success of smart city projects heavily hinges upon having enough financial support available (Shokeir & Yahia, 2020). Due to the reliance on oil and gas revenues for the economy in these regions, there is a hurdle in transitioning toward a sustainable smart city economy, which calls for carefully planned diversification strategies (Mathew & Bangwal, 2024). This shift in the economy is not just a matter of securing funds for smart city endeavors but also entails establishing new economic sectors and opportunities to sustain these projects in the long run.

Barriers rooted in culture and organization dynamics, like the "silo mentality" within companies, can hinder the progress of open government data (OGD) initiatives and smart city projects (Mutambik et al., 2023). In the Gulf Cooperation Council (GCC) political systems, often seen as authoritarian, may clash with the values of openness and transparency central to smart city concepts (Mutambik et al., 2023). This cultural resistance also influences perceptions; some individuals are cautious about sharing data because of privacy issues, which poses challenges in accessing data for smart services (Shokeir & Yahia, 2020).

Public involvement continues to pose a challenge, as there seems to be a lack of excitement among the public regarding technology-driven or data-focused services in smart city projects (Mutambik et al., 2023). This gap between technological progress and the genuine requirements of dwellers is emphasized as a noteworthy problem that hampers user engagement and contentment levels in smart city endeavors (Mathew & Bangwal, 2024). Raising awareness about the advantages and possibilities offered by smart city initiatives to the public is crucial, because informing individuals could enhance acceptance rates and mitigate reluctance toward new technologies (Shokeir & Yahia, 2020).

Securing systems and establishing the required infrastructure pose hurdles. This includes tackling concerns surrounding data privacy and cybersecurity while assessing the capability of the infrastructure to accommodate emerging technologies (Elian & Kisswani, 2023). Managing and leveraging data efficiently for the advancement of smart cities is a demanding task, due to the nature of the data ecosystem, involving various stakeholders and data channels (Mutambik et al., 2023).

Addressing issues like air pollution and the fight against global warming are key priorities that smart city projects aim to tackle (Aldegheishem, 2019). Dubai is often labeled as dependent on private cars, which sparks worries about economic sustainability and the use of resources (Gugler et al., 2021). This reliance presents a hurdle in developing a more sustainable urban environment, as it may result in higher levels of pollution and resource exhaustion.

Despite these challenges, the GCC region also presents significant opportunities for smart city development. The implementation of smart cities offers potential for economic advancement and diversity by lessening dependence on oil revenues and demonstrating capabilities globally (Shokeir & Yahia, 2020). Through investing in smart technologies and sustainable practices, GCC nations have the potential to establish new economic sectors and draw in foreign investments, thus promoting creativity across different domains (Mathew & Bangwal, 2024).

Smart city projects hold the promise of enhancing residents' daily lives through the use of technology to improve services and infrastructure while also addressing transportation and environmental concerns (Shokeir & Yahia, 2020). By prioritizing citizen involvement in decision-making and governance processes with a focus on community needs and well-being, improvement can be amplified in smart cities (Mathew & Bangwal, 2024).

Transitioning to smart cities presents chances to advance sustainability efforts and support global sustainability objectives while establishing GCC nations as pioneers in sustainable urban development (Shokeir & Yahia, 2020). By prioritizing eco-friendly strategies and innovations, GCC nations can attract investments and bolster their international reputation (Mathew & Bangwal, 2024).

The drive for smart cities promotes creativity and the integration of state-of-the-art technologies, which could nurture an environment attractive to tech start-ups and academic institutions (Shokeir & Yahia, 2020). The swift advancement of CT presents possibilities for the advancement of smart cities, allowing GCC countries to establish more effective, interconnected, and adaptive urban settings (Mathew & Bangwal, 2024).

Effective collaboration among stakeholders, such as government entities, private enterprises, and educational institutions, plays a role in advancing smart city projects by facilitating the exchange of information to address obstacles and expedite progress (Shokeir & Yahia, 2020). Public–private partnerships offer avenues to utilize resources, skills, and fresh ideas in the realm of smart city initiatives (Mathew & Bangwal, 2024).

The implementation of Open Government Data (OGD) platforms is considered a crucial move in advancing smart city projects. It aims to enhance transparency and promote citizen involvement (Mutambik et al., 2023). Through these efforts to foster transparency and engage citizens in government processes, using OGD platforms can bolster governance and decision-making procedures. This could result in transformations by encouraging greater citizen participation and empowerment (Mutambik et al., 2023).

Incorporating happiness metrics into government services presents a chance to customize offerings to better serve the community and local enterprises (Gugler et al., 2021). Prioritizing customer contentment can lead to enhanced services that are more agile and efficient in enhancing residents' overall well-being.

Connecting smart city plans with the United Nations' Sustainable Development Goals may enhance sustainability initiatives. This correlation could assist Gulf Cooperation Council nations in tackling local and worldwide difficulties while establishing themselves as pioneers in sustainable urban growth.

Smart mobility projects such as integrated transportation systems offer ways to improve accessibility and lessen the need for cars for an eco-friendly city environment (Gugler et al., 2021). These efforts aim to tackle the challenge of over-reliance on cars in cities like Dubai by encouraging effective transport alternatives. The economic influence of projects related to smart cities can be quite significant. For instance, the Smart Dubai Office estimates that open data initiatives could potentially contribute $2.8 billion to the economy by 2021 (Gugler et al., 2021). Smart city development is crucial not only for enhancing living conditions but also for fostering economic expansion and a variety of opportunities.

In conclusion, implementing smart cities in the GCC involves facing a journey with numerous obstacles and prospects to navigate carefully. The challenges stem from centralized urban planning systems in the region, existing technological gaps, and regulatory complexities, which are significant hurdles to overcome. Furthermore, cultural and organizational challenges add layers of difficulty to the task. Solving these issues demands a strategy focusing on policy changes, improving foundations, and nurturing specialized knowledge and capabilities. Encouraging community involvement through campaigns and promoting a feeling of responsibility among residents will play a role in overcoming reluctance to embrace change and boosting the integration of advanced technologies. Despite the challenges involved in this process, the Gulf Cooperation Council (GCC) stands to benefit from this shift. Smart city projects present an avenue for broadening the economy, lessening reliance on oil income, and establishing the region as a center for technological advancement and environmental sustainability. Investments in smart technologies have the potential to enhance city infrastructure and public services while opening up economic opportunities and supporting environmental sustainability goals to address climate change on a global scale. The effectiveness of smart city projects in the GCC region will depend on a rounded approach that combines technological progress with designs that prioritize human needs and cultural considerations alongside sustainable principles. Moreover, creating an environment involving government entities, businesses, educational institutions, and the public is crucial for developing innovative urban spaces. By seizing these chances, the Gulf Cooperation Council (GCC) could establish itself as a front-runner in smart urban growth, improving the quality of life of its inhabitants and serving as a model for upcoming cities globally.

7.8 Distinctive Approaches to Smart City Development in the GCC and Global Contexts: A Comparative Analysis of Economic, Social, and Technological Dynamics

Smart city initiatives in the Gulf Cooperation Council (GCC) countries and globally showcase unique approaches and methodologies that mirror economic, social, and technological environments. Global advancements in smart cities are often driven by shared challenges such as resource preservation and climate adaptation. The approaches and focuses in GCC countries frequently differ from those seen elsewhere in the world (Ringel, 2021).

In Gulf Cooperation Council (GCC) countries urban planning tends to take a centralized approach that impacts how smart city projects are designed and involve the community (Al-Saidi & Zaidan, 2024). This centralized system in the GCC requires a customized strategy for smart city development that can address specific local requirements while upholding an overall cohesive plan efficiently. In contrast, cities in regions such as Europe and North America often follow decentralized planning procedures that offer more flexibility when implementing smart city schemes (Neirotti et al., 2014). The technological landscape in the GCC is quite diverse across different cities and communities (Al-Saidi & Zaidan, 2024). There is an emphasis on IoT systems and high-tech solutions, as the region aims to lead the way in innovation

(Shokeir & Yahia, 2020). Skill development is a focus in smart city projects globally, but gains even more importance due to the swift technological advancements in some GCC nations.

The regulatory frameworks in the GCC are evolving to accommodate smart city initiatives, by making progress in establishing guidelines that promote creativity and enable the incorporation of new technologies (Al-Saidi & Zaidan, 2024). This changing regulatory scenario differs from the established frameworks in certain Western nations, where smart city policies may be better established but less flexible in their approach. A unique aspect of GCC smart city projects is their contribution to economic diversification plans. The region's reliance on income from oil and gas has prompted the consideration of smart city initiatives as drivers for shifting toward a broader range of knowledge-based economies (Mathew & Bangwal, 2024). This economic backdrop sets it apart from smart city projects worldwide, where funding sources are diverse and economic evolution may not be the primary focus. The development of smart cities in the GCC is heavily influenced by cultural and organizational factors unique to the region, shaping how data sharing and open government data (OGD) initiatives are approached compared to other parts of the world (Mutambik et al., 2023). These cultural nuances impact how smart city projects are carried out and the uptake of certain technologies in the region. Governance systems in the GCC countries are usually known for their centralized authority. This can impact the execution of smart city projects in a manner distinct from cities in the Western hemisphere (Breslow, 2021). The way this governance model works impacts parts of the development of smart cities, from how decisions are made to which technologies and services are given more importance. The involvement of the public in smart city projects within the GCC region is quite diverse and interesting; people show varying degrees of interest in technology-driven or data-focused services (Mutambik et al., 2023). Despite these variations, Gulf Cooperation Council (GCC) smart city projects align with global initiatives by emphasizing economic growth, innovation, and environmental and economic sustainability. The smart city ventures in the region stand out for their culturally distinctive nature and their role in demonstrating technological capabilities to the international community (Shokeir & Yahia, 2020). Through investments in technologies and sustainable strategies, GCC countries are creating new economic sectors and nurturing innovation ecosystems that can establish them as prominent technology hubs regionally and globally (Mathew & Bangwal, 2024). The adoption of Open Government Data (OGD) platforms in smart cities of the GCC region reflects the movement toward enhancing transparency and encouraging citizen participation (Mutambik et al., 2023). Nevertheless, the integration of these platforms in the GCC context could take paths shaped by concerns about data privacy and societal perspectives on sharing information.

In summary, although GCC nations and global smart city projects share overarching goals, their strategies and emphases reflect distinct contexts. The smart city frameworks in the GCC are distinguished by their emphasis on economic diversification, centralized governance systems, and rapid technology integration within culturally nuanced frameworks. These unique characteristics play a role in shaping smart city strategies that both draw from and contribute to the international discourse about urban growth in the digital era. The opportunity for shared learning among GCC smart city initiatives and those in other regions, like Europe, is providing chances for exchanging ideas and tactics in the continuous development of smart urban landscapes globally (Ringel, 2021).

7.9 Future Outlook: The Path Forward for Smart Cities in the GCC

The future for smart cities in the Gulf Cooperation Council (GCC) region represents an intricate road map built on strategic changes and technology adoption while focusing on sustainability and socio-economic transformation aspects. These countries prioritize transitioning to people-centric governance models that prioritize people's needs and preferences in their smart city plans. This strategy underscores the importance of engaging citizens in policy-making, to ensure that smart city projects align with the requirements and desires of the community. To ensure the success of smart city projects in the long run, it is crucial to involve citizens and stakeholders from the beginning, for increased transparency and building public trust, which will be essential for the long-term success of smart city projects (Mathew & Bangwal, 2024). Alongside governance changes, the GCC countries are focused on diversifying their economies beyond oil revenues by fostering industries such as technology, innovation, and services—aligning these sectors with sustainability goals and economic resilience in the face of a low-carbon future (Arif & Aldosary, 2023; Asmyatullin et al., 2020). The advancement of technology is vital in shaping the future of smart cities in the GCC region. Allocating resources to information and communication technologies (ICT) and establishing the digital framework required for supporting smart city initiatives will be crucial (Mathew & Bangwal, 2024). This involves improving services like transportation, health care, education, and governance by implementing automation and decision-making based on data analysis (Elian & Kisswani, 2023). Collaborations between governments and the private sector will play a role in reaching these objectives by leveraging expertise, enabling resource sharing, and fostering innovation (Mathew & Bangwal, 2024). Moreover, partnerships with international technology companies and research organizations can facilitate knowledge exchange, enabling GCC nations to adopt global best practices in smart city development (Asmyatullin et al., 2020; Elian & Kisswani, 2023). In the Gulf Cooperation Council countries, environmental sustainability plays a role in shaping the future of smart cities due to environmental issues like climate change and urbanization challenges that demand the integration of sustainable practices in urban planning and development strategies. Future smart city projects are anticipated to prioritize infrastructure development for improved sustainability, including energy efficiency, waste management, and reducing carbon strategies to meet global environmental targets (Arif & Aldosary, 2023). Creating spaces with low-carbon urban environments demands substantial funding for infrastructure enhancements that prioritize advanced technology and community involvement to guarantee social inclusivity and local relevance (Al-Saidi & Zaidan, 2024). Moreover, the countries in the GCC are urged to implement urban planning procedures that enable ongoing trial and adjustment processes. This strategy is crucial in handling the challenges of the region, including its climate conditions, the socio-political landscape, and rapid urban growth (Al-Saidi & Zaidan, 2024). Drawing insights from smart city projects worldwide can assist the GCC in developing approaches that encourage community engagement and technological assimilation (Mutambik et al., 2023). Studying the successes and challenges encountered by Southeast Asian cities could offer insights for developing and implementing smart city initiatives that cater to the specific requirements of Gulf cities (Arif & Aldosary, 2023). Engaging young people in the urban planning and governance of smart cities is a key focus for the GCC region as it seeks to tap into their creativity and align smart city strategies with the vision of future generations (Arif & Aldosary, 2023). This involvement not only fosters social inclusivity but also ensures that smart city projects remain adaptable to future needs. Comprehensive urban planning that integrates physical infrastructure with intangible elements like education, culture, and public services is vital for creating vibrant, livable cities (Arif & Aldosary, 2023). Recognizing the significance of having a policy framework to steer the development of smart cities is crucial for governments worldwide, as they aim to foster innovation and investment and incorporate new technologies into urban planning strategies effectively. The implementation of governance structures, such as the Gulf Urban Observatory, can play a role in monitoring development indicators and aiding data-informed decision-making processes (Arif & Aldosary, 2023). Additionally, ensuring the prosperous future of smart city initiatives in the GCC region will require overcoming challenges, such as adapting to climate change impacts, reducing youth unemployment rates, and managing urban settlements efficiently. Addressing these challenges may seem overwhelming; however, they also offer a chance to create smart city approaches that cater to the social, economic, and environmental requirements of the area (Arif & Aldosary, 2023; Asmyatullin et al., 2020).

In conclusion, the evolution of smart cities in the GCC will depend on a comprehensive strategy that integrates community-focused leadership, economic diversification, technological innovation, sustainability, and effective governance. By adopting global best practices, fostering public–private partnerships, and emphasizing citizen engagement—particularly among youth—the GCC has the opportunity to spearhead the advancement of smart cities that enhance residents' quality of life and support global initiatives for sustainable urban development

7.10 Conclusion

The development of smart cities in the Gulf Cooperation Council (GCC) countries represents a sophisticated paradigm of technological integration and strategic urban transformation. This multifaceted approach synthesizes technological innovation, sustainable infrastructure planning, and nuanced socio-economic diversification in response to complex global challenges. As the region navigates rapid urban growth and the imperative for economic adaptability, smart city initiatives have emerged as sophisticated mechanisms for addressing systemic vulnerabilities. These strategies have demonstrated remarkable resilience, particularly in confronting the COVID-19 pandemic, fluctuations in oil and gas prices, and the broader impacts of climate change. The architectural framework of these urban interventions prioritizes comprehensive sustainability, manifesting through strategic emphases on energy efficiency, renewable energy integration, and sophisticated resource optimization, aligning with global sustainability goals. Furthermore, smart cities in the GCC play a critical role in advancing sustainable finance frameworks by fostering investments in green technologies, infrastructure, and public–private partnerships, which are essential for long-term economic stability in a post-oil era. Crucially, these initiatives are anchored in a human-centric governance philosophy that balances technological advancement with social inclusivity. This approach addresses complex challenges surrounding data privacy, regulatory frameworks, and meaningful citizen engagement, thereby creating adaptive urban environments responsive to dynamic societal needs. The strategic deployment of smart city technologies in the GCC represents more than infrastructural innovation, it embodies a profound reimagining of urban potential, positioning the region as a potential global exemplar of sustainable, technologically sophisticated urban development.

This chapter contributes to the book's overarching themes by demonstrating how smart city development in the GCC aligns with green growth and sustainable finance while providing actionable insights into the region's capacity to respond to global shocks. Based on the GCC's experience, lessons can be learned that can be useful for policymakers, scholars, and stakeholders, and thus support the potential of smart cities to enable economic and environmental change in a world that is constantly developing.

References

Al Maqbali, N., & Refeque, M. (2017). Smart transportation for smart Oman: An enquiry into its potentials and implications. *Journal of Student Research*. https://doi.org/10.47611/jsr.vi.582

Aldegheishem, A. (2019). Success factors of smart cities: A systematic review of literature from 2000-2018. *TEMA Journal of Land Use, Mobility and Environment, 12*(1), 53–64.

Allam, Z., & Dhunny, Z. A. (2019). On big data, artificial intelligence and smart cities. *Cities, 89*, 80–91.

Al-Saidi, M., & Zaidan, E. (2024). Smart cities and communities in the GCC region: from top-down city development to more local approaches. *Frontiers in Built Environment, 10*, 1341694.

Arif, M., & Aldosary, A. S. (2023). Urban spatial strategies of the Gulf Cooperation Council: A comparative analysis and lessons learned. *Sustainability, 15*(18), 13344. https://doi.org/10.3390/su151813344

Asmyatullin, R. R., Tyrkba, K. V., & Ruzina, E. I. (2020). Smart cities in GCC: Comparative study of economic dimension. *IOP Conference Series: Earth and Environmental Science, 459*(6), 062045.

Barrionuevo, J. M., Berrone, P., & Ricart, J. E. (2012). Smart cities, sustainable progress. *IESE Insight, 14*(14), 50–57.
Barth, J., Fietkiewicz, K., Gremm, J., Hartmann, S., Ilhan, A., Mainka, A., & Stock, W. (2017). Informational urbanism. A conceptual framework of smart cities.
Belaıd, F., Amine, R., & Massie, C. (2023). Smart cities initiatives and perspectives in the MENA region and Saudi Arabia. In *Smart cities: Social and environmental challenges and opportunities for local authorities* (pp. 295–313). Springer International Publishing.
Breslow, H. (2021). The smart city and the containment of informality: The case of Dubai. *Urban Studies, 58*(3), 471–486. https://doi.org/10.1177/0042098020903233
Caragliu, A., Del Bo, C., & Nijkamp, P. (2011). Smart Cities in Europe. *Journal of Urban Technology, 18*(2), 65–82. https://doi.org/10.1080/10630732.2011.601117
Chen, T. M. (2010). Smart grids, smart cities need better networks. *IEEE Network, 24*(2), 2–3.
Cheu, R. L., Nazarian, S., Gates, A. Q., Villanueva-Rosales, N., Taboada Jimenez, H. A., Mondragon, O. A., & Bala, E. (2015). *Research challenges toward the implementation of smart cities in the United States (No. CAIT-UTC-060).* Rutgers University, Center for Advanced Infrastructure and Transportation.
City Framework–Guide. (2014). *Code of practice for Smart city framework–Guide to establishing strategies for smart cities and communities.*
Cretu, L. G. (2012). Smart cities design using event-driven paradigm and semantic web. *Informatica Economica, 16*(4), 57.
Elian, M. I., & Kisswani, K. M. (2023). Smart cities: GCC and Kuwait experience. In *Smart cities: Social and environmental challenges and opportunities for local authorities* (pp. 339–358). Springer International Publishing.
Giffinger, R., Fertner, C., Kramar, H., Kalasek, R., Pichler-Milanovic, N., & Meijers, E. J. (2007). Smart cities. Ranking of European medium-sized cities. Final Report.
Gugler, P., Alburai, M., & Stalder, L. (2021). *Smart city strategy of Dubai* (p. 27). Harvard Business School.
Harrison, C., Eckman, B., Hamilton, R., Hartswick, P., Kalagnanam, J., Paraszczak, J., & Williams, P. (2010). Foundations for smarter cities. *IBM Journal of Research and Development, 54*(4), 1–16.
Hayat, P. (2016). Smart cities: A global perspective. *India Quarterly, 72*(2), 177–191.
Hollands, R. G. (2008). Will the real smart city please stand up? Intelligent, progressive or entrepreneurial? *City, 12*(3), 303–320. https://doi.org/10.1080/13604810802479126
Kourtit, K., & Nijkamp, P. (2012). Smart cities in the innovation age. *Innovation: The European Journal of Social Science Research, 25*(2), 93–95. https://doi.org/10.1080/13511610.2012.660331
Kourtit, K., Nijkamp, P., & Arribas, D. (2012). Smart cities in perspective – A comparative European study by means of self-organizing maps. *Innovation: The European Journal of Social Science Research, 25*(2), 229–246. https://doi.org/10.1080/13511610.2012.660330
Lazaroiu, G. C., & Roscia, M. (2012). Definition methodology for the smart cities model. *Energy, 47*(1), 326–332.
Lea, R. (2017). Smart cities: An overview of the technology trends driving smart cities. *IEEE Advancing Technology for Humanity, 3*, 1–16.
Lombardi, P., Giordano, S., Farouh, H., & Yousef, W. (2012). Modelling the smart city performance. *Innovation: The European Journal of Social Science Research, 25*(2), 137–149. https://doi.org/10.1080/13511610.2012.660325
Mahizhnan, A. (1999). Smart cities: The Singapore case. *Cities, 16*(1), 13–18.
Mamlook, R., Al-Jayyousi, O., Alabyooki, S., & Attallah, S. A. A. (2019). Smart city and technology transfer: Towards sustainable infrastructure in Bahrain. In *Research & innovation forum 2019: Technology, innovation, education, and their social impact* (pp. 255–268). Springer International Publishing.

Marsal-Llacuna, M. L., Colomer-Llinas, J., & Melendez-Frigola, J. (2015). Lessons in urban monitoring taken from sustainable and livable cities to better address the Smart Cities initiative. *Technological Forecasting and Social Change, 90*, 611–622.

Mathew, B. P., & Bangwal, D. (2024). People centric governance model for smart cities development: A systematic review, thematic analysis, and findings. *Research in Globalization, 9*, 100237.

Monzon, A. (2015). Smart cities concept and challenges: Bases for the assessment of smart city projects. In *2015 international conference on smart cities and green ICT systems (SMARTGREENS)* (pp. 1–11). IEEE.

Moura, F., & de Abreu Silva, J. (2019). Smart cities: Definitions, evolution of the concept and examples of initiatives. In W. Leal Filho, A. Azul, L. Brandli, P. Ozuyar, & T. Wall (Eds.), *Industry, innovation and infrastructure. Encyclopedia of the UN sustainable development goals*. Springer. https://doi.org/10.1007/978-3-319-71059-46-1

Mutambik, I., Lee, J., Almuqrin, A., & Zhang, J. Z. (2023). Transitioning to smart cities in Gulf Cooperation Council countries: The role of leadership and organisational culture. *Sustainability, 15*(13), 10490. https://doi.org/10.3390/su151310490

Neirotti, P., De Marco, A., Cagliano, A. C., Mangano, G., & Scorrano, F. (2014). Current trends in smart city initiatives: Some stylised facts. *Cities, 38*, 25–36.

Riadh, A. D. (2022). Dubai, the sustainable, smart city. *Renewable Energy and Environmental Sustainability, 7*, 3.

Ringel, M. (2021). Smart city design differences: Insights from decision-makers in Germany and the Middle East/North-Africa region. *Sustainability, 13*(4), 2143. https://doi.org/10.3390/su13042143

Shaaban, K., & Adalbi, M. A. (2021). Smart City transportation system in developing countries: The case of Lusail City, Qatar. In *International conference on applied human factors and ergonomics* (pp. 445–452). Springer International Publishing.

Shokeir, R. G., & Yahia, I. B. (2020). Moving toward smart cities: Insights from the MENA region. *International Journal of Web Based Communities, 16*(1), 92–108.

Talari, S., Shafie-khah, M., Siano, P., Loia, V., Tommasetti, A., & Catalao, J. P. S. (2017). A review of smart cities based on the internet of things concept. *Energies, 10*(4), 421. https://doi.org/10.3390/en10040421

Thite, M. (2011). Smart cities: Implications of urban planning for human resource development. *Human Resource Development International, 14*(5), 623–631. https://doi.org/10.1080/13678868.2011.618349

Thuzar, M. (2011). Urbanization in Southeast Asia: Developing smart cities for the future? In M. Montesano & P. Lee (Eds.), *Regional outlook: Southeast Asia 2011-2012* (pp. 96–100). ISEAS Publishing. https://doi.org/10.1355/9789814311694-022

Winters, J. V. (2011). Why are smart cities growing? Who moves and who stays. *Journal of Regional Science, 51*(2), 253–270.

Zygiaris, S. (2013). Smart city reference model: Assisting planners to conceptualize the building of smart city innovation ecosystems. *Journal of the Knowledge Economy, 4*, 217–231.

Chapter 8
The GCC Green Growth Opportunities and Sustainable Finance Framework: A Healthcare Sector Perspective

Immanuel Azaad Moonesar, Khulood Mohamed Hussain Ibrahim Alsayegh, and Elizabeth Berryman

Abstract The Gulf Co-operation Council (GCC) nations actively seek to transition towards a more sustainable and eco-friendly economic model after the global climate crisis. This chapter explores the potential for green growth opportunities and the implementation of a sustainable finance framework within the healthcare sector across the GCC region. By examining the intricate interplay between environmental sustainability, economic development, and public health, the authors comprehensively analyze the challenges and prospects associated with this transformation.

The book chapter delves into various aspects through a multidisciplinary approach, including regulatory frameworks, financial incentives, technological innovations, and stakeholder collaborations. It highlights the pivotal role of sustainable finance in catalyzing the adoption of green initiatives within the healthcare industry, ultimately leading to improved health outcomes and a reduced environmental footprint.

Drawing upon case studies and best practices from around the world, the book chapter offers practical recommendations and strategies for policymakers, healthcare providers, financial institutions, and other key stakeholders to leverage the potential of green growth and sustainable finance in the GCC healthcare sector. Ultimately,

I. A. Moonesar (✉)
Mohammed Bin Rashid School of Government, Dubai, UAE
e-mail: Immanuel.moonesar@mbrsg.ac.ae

K. M. H. I. Alsayegh
Dubai Health Authority, Dubai, UAE
e-mail: khalsayegh@dha.gov.ae

E. Berryman
Chnnl Ltd, Auckland, New Zealand
e-mail: liz@chnnl.app

I. A. Moonesar
International Vaccine Institute, Seoul, South Korea

H. Miniaoui and M. Bellalah (eds.), *Green Growth Opportunities and Sustainable Finance in the Gulf Cooperation Council Region*, Gulf Studies 24,
https://doi.org/10.1007/978-981-96-8287-4_8

this chapter is a valuable resource for researchers, practitioners, and decision-makers seeking to foster a more sustainable and resilient healthcare system in the region.

Keywords GCC · Green growth · Sustainable finance · Healthcare sector · Environmental sustainability · Economic development

8.1 Introduction

The worldwide climate problem, marked by increasing temperatures, rising sea levels, and severe weather events, presents an unparalleled challenge to humanity (Intergovernmental Panel on Climate Change, 2021, 2024). The Gulf Co-operation Council (GCC) region, which includes Bahrain, Kuwait, Oman, Qatar, Saudi Arabia, and the United Arab Emirates (UAE), is especially susceptible to the effects of climate change. The region is at risk of water scarcity, desertification, and coastal flooding because of its dry environment, limited freshwater supplies, and low-lying coastal areas (Salami & Babatunde, 2024; Tolba & Saab, 2009).

8.1.1 Overview of the Global Climate Crisis and its Impact on the GCC Region

Climate change impacts in the GCC region are extensive and varied. Increasing temperatures and extended periods of high heat can lead to a rise in heat-related illnesses and deaths (Neira et al., 2023; Pal & Eltahir, 2016). Water scarcity, worsened by reduced precipitation and rising evaporation rates, presents a substantial risk to agriculture, food security, and public health (Waha et al., 2017). Coastal cities like Dubai and Abu Dhabi are at high risk of sea-level rise, which may result in flooding, erosion, and saltwater entering freshwater aquifers (Burt et al., 2013), potentially disrupting millions of people's livelihoods.

8.1.2 The Need for a Sustainable and Eco-Friendly Economic Model in the GCC

Due to the seriousness of the climate catastrophe and its possible effects on the area, the GCC nations must promptly shift towards a more sustainable and environmentally friendly economic model. The economies of the GCC have traditionally depended on producing and exporting fossil fuels, which have substantially impacted greenhouse gas emissions (Luomi, 2015). The GCC states must diversify their economies and adopt sustainable development strategies to maintain economic resilience and environmental sustainability as the global community moves towards a low-carbon future.

Multiple GCC nations have acknowledged sustainable development's significance and implemented measures to tackle climate change through national plans, regulations, and initiatives (Reiche, 2010). The UAE has initiated the UAE Vision 2021 and Green Agenda 2030, emphasizing sustainable development and environmental conservation (UAE Government, 2023). Saudi Arabia's Vision 2030 advocates for economic diversification and ecological practices (Vision 2030 Kingdom of Saudi Arabia, 2023). However, more focused actions are needed to make a thorough and organized shift towards a sustainable and environmentally friendly economic model throughout the GCC region.

8.2 Green Growth Opportunities in the GCC Healthcare Sector

8.2.1 Definition and Significance of Green Growth

The notion of green growth aims to harmonize economic expansion with environmental sustainability (OECD, 2011). It focuses on optimizing natural resource utilization, reducing environmental harm, and advocating for low-carbon technology and behaviors. Green growth potential in the healthcare sector can result in substantial economic, environmental, and public health advantages (World Health Organization, 2020).

8.2.2 Potential Areas for Green Growth in the Healthcare Sector

8.2.2.1 Energy Efficiency and Renewable Energy Initiatives

Healthcare facilities are energy-intensive and contribute to greenhouse gas emissions (Alshqaqeeq et al., 2020). Adopting energy-efficient strategies like replacing equipment, improving building envelopes, and encouraging behavior change can lower energy usage and expenses (Liu et al., 2021). Moreover, including renewable energy sources like solar and wind power can help decrease the healthcare sector's carbon footprint (Al-Shetwi, 2022). Implementing energy efficiency measures is essential for lowering energy usage and expenses and reducing the environmental implications. In the following list of terms, "sustainability goals" represents the environmental context, while "sustainable practices" represents the operational context and "sustainable waste management" represents the environmental/operational context.

Healthcare facilities can significantly reduce their energy consumption through several important strategies. The process begins with comprehensive energy audits that identify opportunities for improvement and establish baseline energy usage metrics. A crucial step is upgrading to energy-efficient equipment, including LED lighting systems, modern HVAC units, and medical devices designed for optimal energy performance. Building envelope improvements play a vital role, encompassing better insulation, air leak remediation, and the installation of energy-efficient windows and doors. The implementation of sophisticated energy management systems and building automation technology enables real-time monitoring and control of energy usage throughout the facility. Finally, success depends on engaging both staff and patients in energy conservation efforts, promoting simple but effective habits like turning off unused equipment and lights.

Utilizing renewable energy sources can help decrease healthcare facilities' carbon footprint and be part of implementing energy efficiency measures. Healthcare facilities have several viable options for incorporating renewable energy into their operations. They can generate their own sustainable power through on-site installations such as solar panels or wind turbines, while also having the flexibility to purchase renewable energy directly from utility providers or participate in renewable energy certificate programs. Local conditions and resources may support additional renewable options like geothermal systems or biomass energy solutions. By implementing these renewable energy strategies alongside energy efficiency initiatives, healthcare organizations can achieve multiple benefits, including reduced environmental impact, lower operational costs, and improved efficiency in their facilities (Sherman et al., 2020). Refer to Exhibits 8.1 and 8.2.

8.2.2.2 Sustainable Waste Management Practices

The healthcare industry produces a significant volume of both hazardous and non-hazardous waste, which threatens the environment and public health (Janik-Karpinska et al., 2023). Adopting sustainable waste management strategies such as trash minimization, segregation, recycling, and correct disposal can reduce risks and enhance resource efficiency (Mahjoob et al., 2023). Improper waste management can lead to environmental pollution, the spread of infectious diseases, and potential health risks for healthcare workers and the community. Sustainable waste management practices are crucial to mitigate these risks and promote resource efficiency. There is an increased need to enhance the monitoring of Bio-Medical Waste Incinerators (BMWIs) through exposure and risk assessment studies and calls for action by governments and international regulations (Prathish, 2023).

Effective healthcare waste management relies on several interconnected approaches. At its foundation is waste minimization, which encompasses reusing and recycling products where possible, optimizing procurement, and establishing waste-conscious protocols throughout operations. A crucial component is the implementation of thorough waste segregation systems that effectively separate hazardous materials like infectious, radioactive, and chemical waste from non-hazardous items such as paper, plastics, and food waste directly at their point of origin. Healthcare facilities must also establish comprehensive recycling programs for common materials while exploring material recovery and reuse opportunities. Safely treating and disposing of hazardous waste requires specialized technologies, including autoclaving, incineration, or chemical treatment methods. To maintain and improve these systems, facilities should conduct regular waste audits that identify areas needing improvement, track progress, and ensure compliance with all relevant regulations and standards.

Using sustainable waste management techniques, healthcare providers can reduce their environmental effects, lower health risks, and save costs (Mahjoob et al., 2023). Refer to Exhibits 8.1 and 8.2.

8.2.2.3 Green Building and Infrastructure Development

Building and running healthcare facilities can cause notable environmental effects (Guenther & Vittori, 2013). Implementing green building concepts, like energy-efficient design, water conservation measures, and sustainable materials, can lessen these effects and promote healthier indoor spaces (Karimi et al., 2023). Healthcare facilities' design, construction, and operation can significantly affect the environment through energy consumption, water usage, and material waste (Guenther & Vittori, 2013). Implementing green building concepts and sustainable infrastructure development can reduce these consequences and improve interior conditions for patients and personnel.

Sustainable healthcare facility design incorporates multiple integrated strategies to minimize environmental impact while promoting healing. Energy-efficient design begins with fundamental passive principles, leveraging optimal building orientation, natural airflow, and daylight to reduce the energy demands of heating, cooling, and lighting systems. Water conservation plays a crucial role through efficient fixtures and innovative systems that capture both rainwater and greywater for reuse. The selection of construction materials prioritizes local, recycled, or renewable resources, reducing environmental impact and supporting regional economic development. Green roofs and ecological landscaping serve multiple functions, providing additional insulation, combating urban heat island effects, and creating spaces for biodiversity. On-site renewable energy systems, such as solar panels or wind turbines, generate clean power for facility operations. Throughout the building, careful attention to indoor environmental quality ensures optimal conditions for healing by maintaining superior air quality, comfortable temperatures, and appropriate lighting levels.

Exhibit 8.1 Green Growth and Sustainable Finance in Organ Donation

Organ Donation:

(a) Implementing sustainable transportation strategies like electric or hybrid vehicles, route optimization, and carbon offsetting can lessen the environmental impact of organ procurement and transplantation logistics.
(b) Utilizing energy-efficient technologies and practices for organ preservation and storage, like sustainable cooling systems and optimized storage facilities, can help decrease the energy consumption and carbon footprint of organ transplantation processes.
(c) Using sustainable packing and shipping materials for organ transportation can reduce waste and support sustainability in the organ donation supply chain.
(d) Implementing sustainable procurement practices and green supply chain management strategies can help reduce environmental risks, enhance resource efficiency, and ensure ethical and sustainable sourcing of materials and equipment for organ donation and transplantation.

Healthcare providers can implement green building concepts and sustainable infrastructure development to design energy-efficient, water-efficient, and environmentally friendly buildings and provide healthier indoor environments for patients and staff (Veleva et al., 2015). Refer to Exhibits 8.1 and 8.2.

Promotion of Eco-Friendly Products and Services

The healthcare sector can promote eco-friendly products and services, such as sustainable medical devices, environmentally friendly cleaning supplies, and telemedicine services, reducing resource consumption and greenhouse gas emissions (Weiss et al., 2019; Yao et al., 2023). The industry may support environmental sustainability by advocating for the adoption of environmentally friendly products and services in all its activities.

Healthcare facilities can significantly reduce their environmental footprint through comprehensive sustainable operations strategies. The foundation begins with sustainable procurement practices, prioritizing eco-friendly medical supplies, cleaning products, and certified sustainable items. Modern healthcare delivery has been transformed by telemedicine and digital health services, improving access to care and reducing transportation-related environmental impacts by minimizing unnecessary patient travel. Transportation sustainability extends to staff and visitors by promoting public transit options and carpooling programs and installing electric vehicle charging infrastructure. In food service operations, facilities are implementing sustainable food procurement practices, reducing food waste, and expanding plant-based menu options, to decrease the environmental impact of institutional food services. These efforts are complemented by sustainable packaging and distribution initiatives that move away from single-use plastics in favor of eco-friendly materials and distribution methods, reducing waste and emissions throughout the supply chain.

Healthcare providers can showcase leadership in sustainable practices and encourage other sectors to do the same by endorsing eco-friendly products and services, which can help lessen their environmental impact (Weiss et al., 2019; Yao et al., 2023). Refer to Exhibits 8.1 and 8.2.

A comprehensive and coordinated approach encompassing policymakers, healthcare providers, financial institutions, and other stakeholders is needed to capitalize on green growth potential in the GCC healthcare industry. The healthcare sector may enhance environmental sustainability, operational efficiency, cost reduction, and public health by focusing on energy efficiency, waste management, green infrastructure, and promoting eco-friendly practices.

Exhibit 8.2 Green Growth and Sustainable Finance in Promoting Well-Being Amongst Healthcare Professionals

Enhancing the well-being of healthcare professionals:

(a) Design of healthcare facilities incorporating green and biophilic elements. Implementing biophilic design principles, like natural lighting, indoor plants, and outside access, can enhance the well-being and productivity of healthcare professionals by decreasing stress, enhancing air quality, and establishing a more soothing and rejuvenating atmosphere.
(b) Promoting sustainable commuting options like public transportation, carpooling, cycling, and electric vehicle incentives can help healthcare professionals reduce emissions, increase physical activity, and improve work-life balance.
(c) Implementing sustainable food and nutrition programs in healthcare facilities, like promoting plant-based diets, reducing food waste, and sourcing locally-grown produce, can benefit healthcare professionals' well-being and lessen the environmental impact of food production and consumption.
(d) Financial support for employee wellness programs with a focus on sustainability. Investing in financial resources and establishing sustainable finance mechanisms, like ESG-linked incentives or impact investing, to back employee wellness programs, mental health initiatives, and work-life balance measures can enhance the overall well-being and productivity of healthcare professionals.
(e) Implementing sustainable facilities management practices, like energy-efficient lighting, indoor air quality monitoring, and ergonomic workplace design, can improve the work environment for healthcare professionals by reducing stress and health risks.

8.3 Sustainable Finance Framework for the GCC Healthcare Sector

8.3.1 Introduction to Sustainable Finance and Key Distinctions

Sustainable finance refers to integrating environmental, social, and governance (ESG) considerations into investment decisions and financial products and services (United Nations Environment Programme Finance Initiative (UNEP FI), 2019). It aims to channel financial resources towards sustainable economic activities and promote long-term value creation (Schoenmaker, 2020).

8.3.1.1 Financial Sustainability Mechanisms

- Integration of environmental, social, and governance (ESG) criteria into investment decisions.
- Development of financial products and services that support long-term economic viability.
- Creation of funding structures that enable continuous healthcare system improvements.

8.3.1.2 Environmental Sustainability Outcomes

- Measurable reductions in carbon emissions and waste.
- Improvements in resource efficiency and conservation.
- Enhanced environmental protection through healthcare operations.

8.3.1.3 Operational Sustainability Practices

- Implementation of efficient management systems.
- Development of resilient supply chains.
- Adoption of sustainable clinical and administrative processes.

8.3.2 *Role of Sustainable Finance in Promoting Green Initiatives*

Sustainable finance is essential for the shift towards a sustainable and environmentally friendly healthcare industry (Dasgupta, 2021). Sustainable financing may facilitate the execution of green initiatives, including energy-efficient technologies, renewable energy projects, and sustainable infrastructure development, by offering money and financial incentives (Taghizadeh-Hesary & Yoshino, 2020). This framework supports sustainability through three distinct channels:

8.3.2.1 Financial Mechanisms

- Green bonds and sustainability-linked loans.
- ESG investment criteria.
- Performance-based financing structures.

8.3.2.2 Environmental Impact

- Funding for energy-efficient technologies.
- Investment in renewable energy projects.
- Support for sustainable infrastructure development.

8.3.2.3 Operational Implementation

- Resource allocation for sustainable practices.
- Financial incentives for operational improvements.
- Investment in staff training and development.

8.3.3 *Regulatory Frameworks and Policies*

8.3.3.1 Environmental Regulations and Standards

Stringent environmental rules and standards are crucial for encouraging sustainable practices in the healthcare industry (Malik et al., 2020). Emission limitations, waste management rules, and green building codes can enable healthcare providers to implement environmentally friendly practices and meet sustainability standards.

8.3.3.2 Monetary Rewards and Tax Advantages

Authorities can implement financial rewards and tax advantages to promote the adoption of sustainable finance practices among healthcare providers and investors (Dasgupta, 2021). These may encompass grants, subsidies, tax credits, and favorable loan rates for environmentally friendly enterprises and efforts.

8.3.3.3 Collaborations Between the Public and Business Sectors

Public–private partnerships (PPPs) can help mobilize resources and expertise for sustainable finance efforts in the healthcare sector by fostering collaboration between the public and private sectors (Haran et al., 2020). Public–private partnerships (PPPs) can utilize both sectors' advantages to encourage the sharing of knowledge, risks, and resources for opportunities in green growth.

The summary regulatory framework addresses each sustainability aspect:

Financial Regulations:

- Investment criteria and standards.
- Reporting requirements.
- Risk assessment frameworks.

Environmental Standards:

- Emission limitations.
- Waste management regulations.
- Green building codes.

Operational Guidelines:

- Management system requirements.
- Performance metrics.
- Compliance protocols.

8.4 Framework for Integrated Healthcare Sustainability

8.4.1 Core Sustainability Dimensions and Their Interactions

8.4.1.1 Environmental Sustainability

Environmental sustainability in healthcare encompasses several interconnected components that work together to create a comprehensive approach to ecological responsibility. The foundation rests on energy efficiency paired with renewable energy adoption, alongside robust waste reduction and management systems. Green building infrastructure and resource conservation efforts support these initiatives, while specific measures target the mitigation of environmental impacts. These environmental components create ripple effects across other operational dimensions of healthcare delivery. Financial benefits emerge through reduced operational costs driven by improved efficiency, while operational aspects are enhanced through thoughtful facility design and streamlined clinical processes. The clinical dimension benefits from improved patient care environments that support better health outcomes, and the social dimension is strengthened through contributions to broader community health and wellbeing. Together, these elements form an integrated approach to sustainable healthcare delivery that serves both institutional and community needs.

8.4.1.2 Financial Sustainability

Financial sustainability in healthcare integrates several critical components that work together to ensure long-term viability while supporting environmental goals. The framework begins with sustainable funding mechanisms and cost-effective operations, supported by strategic investments in green infrastructure. These elements are protected through comprehensive risk management strategies and careful resource allocation practices. The financial dimension maintains strong connections with other aspects of healthcare delivery - it enables environmental initiatives through investments in green technologies, enhances operational efficiency through optimal resource utilization, provides the foundation for high-quality clinical care delivery, and helps maintain healthcare accessibility for the broader community. Together, these financial elements create a robust framework that supports both immediate healthcare needs and long-term sustainability goals. Financial sustainability in healthcare integrates several critical components that work together to ensure long-term viability while supporting environmental goals. The framework begins with sustainable funding mechanisms and cost-effective operations, supported by strategic investments in green infrastructure. These elements are protected through comprehensive risk management strategies and careful resource allocation practices.

8.4.1.3 Operational Sustainability

Operational sustainability in healthcare encompasses a set of integrated components designed to ensure smooth and effective service delivery. At its core are efficient management systems working in concert with comprehensive staff training and development programs, while supply chain optimization ensures reliable resource availability. These elements are enhanced through continuous process improvement initiatives and rigorous quality assurance protocols. The operational dimension interacts meaningfully with other aspects of healthcare sustainability - it enables the implementation of environmental practices, supports financial efficiency through optimized resource utilization, facilitates effective healthcare delivery, and maintains consistent service reliability for the community. These operational elements work together to create a foundation for sustainable healthcare delivery that balances efficiency with quality care.

8.4.1.4 Clinical Sustainability

Clinical sustainability in healthcare is built around essential elements that ensure effective patient care and continuous improvement. Quality of care and patient outcomes form the foundation, supported by the implementation of evidence-based practices and ongoing healthcare innovation, while maintaining consistent service continuity. The clinical dimension has significant interactions with other aspects of healthcare sustainability - it drives the creation of healthy environments, provides justification for quality-focused investments, shapes operational process design, and ultimately delivers tangible health benefits to the community. These clinical components work together to create a framework that not only supports immediate patient care needs but also ensures long-term sustainability of healthcare services.

8.4.1.5 Social Sustainability

Social sustainability in healthcare integrates several vital components that address both community and workforce needs. It begins with active community engagement and a fundamental commitment to health equity, while maintaining a strong focus on workforce wellbeing and cultural sensitivity in service delivery. Public health promotion serves as a cornerstone of these efforts, extending the healthcare system's impact beyond individual patient care. The social dimension maintains important connections with other aspects of healthcare sustainability - it strengthens environmental initiatives through public health promotion, directs community investment decisions, shapes how services are designed and delivered, and ensures that clinical care aligns with community needs and cultural values. Together, these social elements create a framework that supports both immediate healthcare delivery and long-term community wellbeing.

8.4.2 System-Wide Integration Points

8.4.2.1 Infrastructure Development

The nexus between environmental and financial sustainability in healthcare creates a powerful dynamic that shapes multiple aspects of service delivery. This integration influences operational efficiency through resource optimization and sustainable practices, while simultaneously affecting how clinical services are delivered and maintained. The combined impact extends to community access, as environmental and financial decisions directly influence the availability and affordability of healthcare services. These interconnected elements work together to create a healthcare delivery system that balances ecological responsibility with financial viability.

8.4.2.2 Resource Management

The connection between financial and operational sustainability in healthcare creates a framework that drives system-wide performance. This linkage has direct implications for environmental impact through resource allocation and efficiency measures, while providing crucial support for clinical service delivery. The integration of these elements helps ensure equitable distribution of healthcare resources and services across communities. These interconnected components work together to create a healthcare system that is both efficient and accessible.

8.4.2.3 Quality Improvement

The intersection of clinical and operational sustainability in healthcare creates a dynamic framework that shapes multiple aspects of service delivery. This connection directly affects financial performance through optimized resource utilization and improved efficiency while influencing environmental practices in healthcare settings. Integrating clinical and operational elements also enhances social outcomes by ensuring consistent, high-quality care delivery that meets community needs. These interconnected components create a system that supports effective patient care and sustainable healthcare operations.

8.4.2.4 Innovation and Technology

The comprehensive integration of all sustainability dimensions in healthcare creates a synergistic system that drives overall performance improvement. This holistic approach enables significant efficiency gains across operations while supporting enhanced patient and organizational outcomes. The interconnected nature of these dimensions promotes greater healthcare accessibility, ensuring that sustainable practices benefit individual patients and the broader community. This integrated framework creates a foundation for healthcare delivery that is both effective and enduring.

This framework demonstrates how different aspects of sustainability are interconnected in healthcare, creating a holistic system where improvements in one area can positively impact others. The success of sustainable healthcare depends on recognizing and leveraging these interactions while maintaining balance among all dimensions.

8.5 Challenges and Prospects

8.5.1 Challenges in Implementing Green Growth and Sustainable Finance

8.5.1.1 Economic Barriers

Implementing eco-friendly development and sustainable finance solutions may need substantial initial investments, which can be a barrier for healthcare providers and investors (Dasgupta, 2021). Restricted funding, anticipated higher costs, and unpredictable investment yields could hinder the adoption of sustainable techniques.

8.5.1.2 Technical and Infrastructural Challenges

The healthcare industry could have technical and infrastructural obstacles when introducing environmentally friendly projects, including a shortage of experienced workers, insufficient technological infrastructure, and a restricted supply of sustainable products and services (Malik et al., 2020).

8.5.1.3 Social and Cultural Factors

Social and cultural problems like resistance to change, lack of information, and conflicting priorities might hinder the implementation of green growth and sustainable financing methods in the healthcare industry (Delmas & Toffel, 2008).

8.5.1.4 Regulatory and Policy Hurdles

Insufficient or inconsistent legislation, policies, and incentives might hinder the implementation of green growth and sustainable finance efforts (Dasgupta, 2021). Challenges may arise due to a lack of harmonization and cooperation among many parties.

8.5.2 Prospects and Opportunities

8.5.2.1 Economic Benefits

Adopting green growth and sustainable finance practices can provide economic benefits to healthcare providers, such as cost savings through improved resource efficiency, increased competitiveness, and access to new markets and financing opportunities (Taghizadeh-Hesary & Yoshino, 2020).

8.5.2.2 Environmental and Public Health Benefits

Green growth initiatives in the healthcare sector can contribute to reducing greenhouse gas emissions, minimizing waste, and promoting a healthier environment, ultimately leading to improved public health outcomes (World Health Organization, 2020).

8.5.2.3 Technological Advancements

Advancements in sustainable technologies, such as renewable energy, energy-efficient equipment, and green building materials, can facilitate the implementation of green growth initiatives in the healthcare sector (Al-Shetwi, 2022).

8.5.2.4 Global Collaborations and Knowledge Sharing

International collaborations and knowledge sharing can accelerate the adoption of best practices, promote capacity building, and foster a global transition towards a sustainable and eco-friendly healthcare sector (United Nations Environment Programme Finance Initiative (UNEP FI), 2019).

8.6 Multidisciplinary Approach

8.6.1 Regulatory Frameworks and Policies

Developing solid regulatory frameworks and regulations is essential for encouraging environmentally friendly growth and sustainable financial practices in the healthcare industry (Malik et al., 2020). These could encompass environmental restrictions, building rules, emission requirements, and financial incentives to promote sustainable behaviors.

8.6.2 Financial Incentives and Investment Strategies

Offering financial incentives like grants, subsidies, tax credits, and preferential loan rates can help promote the implementation of environmentally friendly projects and sustainable financial practices in the healthcare industry (Dasgupta, 2021). Moreover, creating investment plans focusing on sustainable and environmentally friendly enterprises can help gather resources for green growth prospects.

8.6.3 Technological Innovations and Best Practices

Promoting technical advancements and exchanging successful strategies will help facilitate the adoption of energy-efficient solutions, renewable energy technology, sustainable waste management techniques, and green construction principles in the healthcare industry (Al-Shetwi, 2022).

8.6.4 Stakeholder Collaborations and Public Awareness

Facilitating partnerships across various stakeholders such as healthcare providers, policymakers, financial institutions, civil society organizations, and the public is crucial for advancing green growth and sustainable finance (United Nations Environment Programme Finance Initiative (UNEP FI), 2019). Promoting the significance of sustainability and the advantages of eco-friendly practices among the public might increase the desire for sustainable healthcare services.

8.7 Case Studies and Best Practices

8.7.1 Successful Case Studies from Around the World

Numerous case studies worldwide demonstrate the successful implementation of green growth and sustainable finance initiatives in the healthcare sector.

8.7.1.1 Cleveland Clinic, Ohio, USA

Sustainable Practices: The Cleveland Clinic in the United States has implemented sustainable practices like energy conservation, recycling, and green building design, as well as sustainable healthcare products and services (Cleveland Clinic, 2021). The Cleveland Clinic initiated a comprehensive energy efficiency program incorporating LED lighting, energy-efficient HVAC systems, and advanced building management systems. They implemented sustainable procurement techniques and enhanced recycling efforts.

Benefits: Decreased energy usage by 20% and cut millions in yearly energy expenses, proving that sustainable practices can be both environmentally and financially advantageous.

8.7.1.2 John Radcliffe Hospital, Oxford, UK

Sustainable Practices: In 2023, John Radcliffe Hospital erected a combined heat and power (CHP) facility to generate electricity and hot water using waste heat (John Radcliffe Hospital, 2024). The hospital implemented sustainable mobility choices for employees and patients, decreasing its carbon impact.

Benefits: The hospital's efforts have led to substantial decreases in greenhouse gas emissions and energy expenses and enhanced air quality in the surrounding area, which has benefitted public health.

8.7.1.3 Melbourne Health, Australia

Sustainable Practices: Melbourne Health (2023) established an extensive waste management initiative to decrease the use of disposable plastics and encourage recycling (Melbourne Health, Australia, 2023). They implemented a green rooftop garden to enhance patient well-being and utilized energy-efficient lighting options.

Benefits: This initiative demonstrated the wide-ranging benefits of sustainability in healthcare by decreasing waste production, enhancing patient satisfaction, and reducing energy consumption.

8.7.1.4 University Medical Center Utrecht, Netherlands

Sustainable Practices: In 2023, the medical institution adopted a circular procurement policy focusing on reusing and recycling medical equipment and supplies. They also allocated resources to on-site renewable energy sources (University Medical Center Utrecht, Netherlands, 2023).

Benefits: This policy demonstrated the promise of circular economy concepts in healthcare through significant waste reduction, financial savings, and minimized environmental effects.

8.7.1.5 Hospital Lusíadas Lisboa, Portugal

Sustainable Practices: In 2023, Hospital Lusíadas Lisboa established an extensive sustainability program that encompassed energy-efficient building designs, waste reduction programs, and the encouragement of sustainable mobility choices for employees (Hospital Lusíadas Lisboa, Portugal, 2023).

Benefits: The beneficial impacts of sustainability in a healthcare system include decreased greenhouse gas emissions, significant cost savings, and enhanced community and staff well-being.

8.7.2 Best Practices for Implementing Green Growth and Sustainable Finance

Various optimal strategies for incorporating green growth and sustainable finance in the healthcare industry have been identified. These activities involve completing energy audits, installing energy management systems, adopting sustainable procurement methods, establishing waste minimization and recycling programs, and integrating sustainability criteria into decision-making processes (Malik et al., 2020). Engaging stakeholders, boosting capacity building, and fostering partnerships and collaborations will help ensure the successful execution of green growth and sustainable financing projects. For example, refer to Exhibit 8.3 for insights on green growth and sustainable finance in genetics and precision medicine.

Exhibit 8.3 Green Growth and Sustainable Finance in the Field of Genetics and Precision Medicine
The science of genetics and precision medicine has a substantial environmental impact because it heavily relies on data-intensive computation, biobanking,

specialized equipment, and resource usage. Implementing sustainable practices in this field is essential for reducing environmental effects and promoting scientific research and healthcare innovations. Crucial topics for promoting environmentally friendly expansion and stable financial support in genetics/precision medicine encompass:

1. Implementing sustainable data management and computing infrastructure by using green data centers, with energy-efficient storage and processing solutions, and reducing redundant data storage.
2. Sustainable biobanking includes energy-efficient facilities, eco-friendly sample management processes, and alternate preservation technologies with fewer environmental implications.
3. Implementing green procurement and sustainable supply chains by focusing on eco-friendly items, sustainable packaging and transportation, and evaluation of suppliers based on environmental and social responsibility criteria.
4. Sustainable financing and investing structures include green bonds, impact investment funds, and integrating environmental, social, and governance (ESG) criteria into investment decision-making.
5. Ethical considerations, particularly in the realm of Artificial Intelligence (AI), and social responsibility involve creating ethical guidelines, encouraging public involvement and education, and working with various stakeholders to guarantee fair and sustainable execution of precision medicine projects (Manoj Kumar et al. 2022).

The science of genetics and precision medicine can contribute to environmental stewardship, advance scientific knowledge, improve healthcare outcomes, and promote responsible innovation by incorporating sustainable practices and utilizing sustainable finance options.

8.8 Recommendations and Strategies

8.8.1 Recommendations for Policymakers

Policymakers are essential in advancing green growth and sustainable finance in the healthcare industry. Key considerations include developing thorough national policies and action plans for sustainable healthcare, as Malik et al. (2020) suggested: implementing strong regulatory frameworks, environmental standards, and building rules (Dasgupta, 2021); offering financial incentives like grants, subsidies, and tax credits to support green activities (Taghizadeh-Hesary & Yoshino, 2020); encouraging collaborations between the public and commercial sectors to gather resources and knowledge for sustainable initiatives (Haran et al., 2020); and enhancing capacity strengthening, building and facilitating knowledge exchange among healthcare practitioners and stakeholders (United Nations Environment Programme Finance Initiative (UNEP FI), 2019).

8.8.2 Strategies for Healthcare Providers

Healthcare providers should implement the following techniques to support environmentally friendly economic development and stable financial practices: performing energy audits and establishing energy management systems to enhance energy efficiency (Liu et al., 2021); implementing sustainable waste management strategies including trash minimization, segregation, and recycling (Hossain et al., 2024); implementing eco-friendly building principles and sustainable materials in facility design and construction (Karimi et al., 2023); advocating for environmentally friendly products and services, like sustainable medical equipment and telemedicine (Weiss et al., 2019; Yao et al., 2023); and involving stakeholders such as policymakers, financial institutions, and local communities to promote cooperation and exchange of knowledge (United Nations Environment Programme Finance Initiative (UNEP FI), 2019).

8.8.3 Initiatives for Financial Institutions

Financial institutions can have a crucial impact on promoting green growth and sustainable finance in the healthcare sector by implementing the following initiatives: creating sustainable financial instruments, including green bonds, sustainability-linked loans, and ESG-focused investment funds (Dasgupta, 2021); integrating ESG factors into investment decision-making and risk management frameworks (United Nations Environment Programme Finance Initiative (UNEP FI), 2019) and aligning the United Nations 2030 Sustainable Development Goals with company/institutional values (Kango, 2023); for sustainable healthcare projects, offering advantageous financing choices, such as reduced interest rates and extended loan durations (Taghizadeh-Hesary & Yoshino, 2020); and participating in capacity building and knowledge exchange with healthcare practitioners and other stakeholders to encourage sustainable finance practices (United Nations Environment Programme Finance Initiative (UNEP FI), 2019).

8.8.4 Role of Other Stakeholders

Additional stakeholders, including civil society organizations, academic institutions, and research centers, can support green growth and sustainable finance in the healthcare sector by increasing public awareness and advocating for education on sustainability and eco-friendly practices (United Nations Environment Programme Finance Initiative (UNEP FI), 2019); conducting research and producing creative solutions for sustainable healthcare, such as new technology, materials, and practices (Malik et al., 2020); promoting partnerships and exchanging knowledge among stakeholders to support the implementation of optimal practices (United Nations Environment Programme Finance Initiative (UNEP FI), 2019); and advocating for policy changes and regulatory reforms to promote green growth and sustainable finance initiatives (Dasgupta, 2021).

8.9 Conclusion

8.9.1 Summary of Key Findings and Recommendations

This chapter has demonstrated how the GCC healthcare sector intersects with three critical themes: green growth, sustainable finance, and resilience to global shocks. Our analysis reveals that the healthcare sector presents significant opportunities for green growth through energy efficiency measures, renewable energy integration, sustainable waste management, and green construction. These initiatives reduce environmental impact and create new economic opportunities within the healthcare ecosystem.

Our findings regarding sustainable finance highlight the crucial role of innovative funding mechanisms in transforming the healthcare sector. Establishing green bonds, sustainability-linked loans, and environmental, social, and governance (ESG) investment frameworks is essential for financing the transition to sustainable healthcare infrastructure and operations.

The chapter on resilience to global shocks demonstrates how sustainable healthcare practices enhance the sector's ability to withstand environmental, economic, and public health challenges. The COVID-19 pandemic has highlighted the importance of building robust and adaptable healthcare systems that can respond effectively to unexpected crises while maintaining environmental responsibility.

8.9.2 Future Research Directions

The future research agenda for sustainable healthcare encompasses three critical areas of investigation. In green growth, researchers need to develop region-specific sustainability strategies, assess the economic advantages of environmental initiatives, and identify emerging opportunities for green innovation in healthcare delivery. The sustainable finance domain calls for exploring novel funding mechanisms, evaluating various financial instruments, and developing frameworks to measure returns on sustainable healthcare investments. Research into healthcare's response to global shocks focuses on understanding how sustainable practices contribute to system resilience, examining technology's dual role in crisis response and sustainability, and investigating methods to integrate sustainability principles into emergency preparedness planning. Together, these research directions aim to strengthen healthcare's ability to maintain effective service delivery while meeting environmental and social responsibilities.

8.9.3 Measuring the Impact of Sustainable Practices on Health and Well-Being for GCC Residents

A practical measurement framework for sustainable healthcare must incorporate multiple dimensions of performance assessment. It should quantify the economic advantages of green healthcare initiatives while evaluating how sustainable financing mechanisms contribute to improved healthcare outcomes. The framework must also assess the system's capacity to withstand and adapt to various environmental, health, and economic disruptions. These interconnected metrics provide a comprehensive view of healthcare sustainability performance.

These metrics should be integrated into a comprehensive monitoring system that tracks progress across all three themes. This would enable policymakers to assess the success of sustainable healthcare initiatives while ensuring preparedness for future challenges.

8.9.4 Importance of Fostering a Sustainable and Resilient Healthcare System

The transformation of the GCC healthcare sector represents a critical intersection of green growth principles, sustainable finance practices, and resilience building. By embracing sustainable practices, the sector can simultaneously achieve environmental protection, economic prosperity, and enhanced capacity to respond to global shocks. This holistic approach ensures that the healthcare system contributes to sustainability goals and maintains its essential function as a cornerstone of public health and societal well-being in the face of future challenges.

References

Al-Shetwi, A. Q. (2022). Sustainable development of renewable energy integrated power sector: Trends, environmental impacts, and recent challenges. *Science of the Total Environment, 822*, 153645. https://doi.org/10.1016/j.scitotenv.2022.153645

Alshqaqeeq, F., Esmaeili, M. A., Overcash, M., & Twomey, J. (2020). Quantifying hospital services by carbon footprint: A systematic literature review of patient care alternatives. *Resources, Conservation and Recycling, 154*, 104560. https://doi.org/10.1016/j.resconrec.2019.104560

Burt, J. A., Feary, D. A., & Van Lavieren, H. (2013). Persian gulf reefs: An important asset for climate science in urgent need of protection. *Ocean Challenge, 20*, 49–56. Retrieved from https://nyuscholars.nyu.edu/en/publications/persian-gulf-reefs-an-important-asset-for-climate-science-in-urge

Cleveland Clinic. (2021). *Environment: Sustainability & going green*. Retrieved April 12, 2024, from https://my.clevelandclinic.org/about/community/sustainability/sustainability-global-citizenship/environment

Dasgupta, P. (2021). *The economics of biodiversity: The Dasgupta review* (p. 103). HM Treasury. Retrieved April 12, 2024, from https://www.gov.uk/government/publications/final-report-the-economics-of-biodiversity-the-dasgupta-review

Delmas, M. A., & Toffel, M. W. (2008). Organizational responses to environmental demands: Opening the black box. *Strategic Management Journal, 29*(10), 1027–1055. https://doi.org/10.1002/smj.701

Guenther, R., & Vittori, G. (2013). *Sustainable healthcare architecture* (2nd ed., pp. 225–319). John Wiley & Sons.

Haran, M., Neitzert, T., & Shrivastava, A. (2020). *Public-private partnerships for sustainable development in the healthcare sector*. Sustainable Development and World Health Organization. Retrieved from https://apps.who.int/iris/handle/10665/334001

Hospital Lusíadas Lisboa, Portugal. (2023). *Sustainability*. Retrieved April 12, 2024, from https://www.lusiadas.pt/hospitais-clinicas/hospital-lusiadas-lisboa

Hossain, I., Haque, A. M., & Ullah, S. A. (2024). Assessing sustainable waste management practices in Rajshahi City Corporation: An analysis for local government enhancement using IoT, AI, and Android technology. *Environmental Science and Pollution Research*, 1–19.

Intergovernmental Panel on Climate Change. (2021). *Climate change 2021: The physical science basis*. Retrieved April 12, 2024, from https://www.ipcc.ch/report/ar6/wg1/

Intergovernmental Panel on Climate Change. (2024). *Climate change 2024*. Retrieved April 12, 2024, from https://www.ipcc.ch/

Janik-Karpinska, E., Brancaleoni, R., Niemcewicz, M., Wojtas, W., Foco, M., Podogrocki, M., & Bijak, M. (2023). Healthcare waste—A serious problem for global health. *Healthcare, 11*(2), 242. https://doi.org/10.3390/healthcare11020242

John Radcliffe Hospital. (2024). *Sustainability*. Retrieved April 12, 2024, from https://www.ouh.nhs.uk/hospitals/jr/

Kango, R. (2023). *The United Nations SDGs and their impact on ESG investing. SDGs. Sustainability Magazine*. Retrieved from https://sustainabilitymag.com/articles/united-nations-sdgs-and-their-impact-on-esg-investing

Karimi, H., Adibhesami, M. A., Bazazzadeh, H., & Movafagh, S. (2023). Green buildings: Human-centered and energy efficiency optimization strategies. *Energies, 16*(9), 3681. https://doi.org/10.3390/en16093681

Liu, J., Jian, L., Wang, W., Qiu, Z., Zhang, J., & Dastbaz, P. (2021). The role of energy storage systems in resilience enhancement of health care centers with critical loads. *Journal of Energy Storage, 33*, 102086. https://doi.org/10.1016/j.est.2020.102086

Luomi, M. (2015). *The gulf monarchies and climate change: Abu Dhabi and Qatar in an era of natural unsustainability*. Oxford University Press.

Mahjoob, A., Alfadhli, Y., & Omachonu, V. (2023). Healthcare waste and sustainability: Implications for a circular economy. *Sustainability, 15*(10), 7788. https://doi.org/10.3390/su15107788

Malik, N., Shrivastava, A., Neitzert, T., & DiPrete, L. (2020). *Sustainable healthcare in the WHO European region: A focus on environmental sustainability in healthcare.* Sustainable Development and World Health Organization. Retrieved from https://apps.who.int/iris/handle/10665/336119

Manoj Kumar, M. V., Atalla, S., Almuraqab, N., & Moonesar, I. A. (2022). Detection of COVID-19 using deep learning techniques and cost-effectiveness evaluation: A survey. *Frontiers in Artificial Intelligence, 5*, 912022. https://doi.org/10.3389/frai.2022.912022

Melbourne Health, Australia. (2023). *Sustainability*. Retrieved April 12, 2024, from https://vahi.vic.gov.au/hospital-and-health-services/melbourne-health

Neira, M., Erguler, K., Ahmady-Birgani, H., Al-Hmoud, N. D., Fears, R., Gogos, C., & Christophides, G. (2023). Climate change and human health in the eastern Mediterranean and Middle East: Literature review, research priorities, and policy suggestions. *Environmental Research, 216*, 114537. https://doi.org/10.1016/j.envres.2022.114537

OECD. (2011). *Towards green growth: A summary for policymakers*. Organization for Economic co-operation Development. Retrieved April 12, 2024, from https://www.oecd.org/greengrowth/48012345.pdf

Pal, J. S., & Eltahir, E. A. (2016). Future temperature in Southwest Asia projected to exceed a threshold for human adaptability. *Nature Climate Change, 6*(2), 197–200. https://doi.org/10.1038/nclimate2833

Prathish, K. P. (2023). Dioxins emissions from bio-medical waste incineration: A systematic review on emission factors, inventories, trends and health risk studies. *Journal of Hazardous Materials, 465*, 133384. https://doi.org/10.1016/j.jhazmat.2023.133384

Reiche, D. (2010). Renewable energy policies in the Gulf countries: A case study of the carbon-neutral, renewable energy based strategy of Masdar City in Abu Dhabi. *Energy Policy, 38*(1), 378–382. https://doi.org/10.1016/j.enpol.2009.09.028

Salami, A. A., & Babatunde, O. R. (2024). Environmental challenges, the impacts of climate change in North Africa region: A review. In *Natural resources deterioration in MENA region: Land degradation, soil erosion, and desertification* (pp. 281–294). Springer. https://doi.org/10.1007/978-3-031-58315-5_14

Schoenmaker, D. (2020). *The impact economy: Balancing profit and impact*. Working paper, 4. Retrieved from https://www.econstor.eu/bitstream/10419/237617/1/1736696165.pdf

Sherman, J. D., Thiel, C., MacNeill, A., Eckelman, M. J., Dubrow, R., Hopf, H., & Bilec, M. M. (2020). The green print: Advancement of environmental sustainability in healthcare. *Resources, Conservation and Recycling, 161*, 104882. https://doi.org/10.1016/j.resconrec.2020.104882

Taghizadeh-Hesary, F., & Yoshino, N. (2020). Sustainable solutions for green financing and investment in renewable energy projects. *Energies, 13*(4), 788. https://doi.org/10.3390/en13040788

Tolba, M. K., & Saab, N. W. (2009). Arab environment: Climate change. In *Beirut, annual report of the Arab Forum for environment and development*. AFED. Retrieved 13 April 2024, from http://www.afedonline.org/

UAE Government. (2023). *UAE green agenda 2030*. Retrieved 12 April 2024, from https://u.ae/en/about-the-uae/strategies-initiatives-and-awards/strategies-plans-and-visions/environment-and-energy/the-uaes-green-agenda-2030

United Nations Environment Programme Finance Initiative (UNEP FI). (2019). *Industry-based principles*. Retrieved April 12, 2024, from https://www.unepfi.org/about/about-us/

University Medical Center Utrecht, Netherlands. (2023). *Sustainability*. Retrieved April 12, 2024, from https://www.umcutrecht.nl/nl

Veleva, V., Todorova, S., Lowitt, P., Angus, N., & Neely, D. (2015). Understanding and addressing business needs and sustainability challenges: Lessons from Devens eco-industrial park. *Journal of Cleaner Production, 87*, 375–384. https://doi.org/10.1016/j.jclepro.2014.09.014

Vision 2030 Kingdom of Saudi Arabia. (2023). *National transformation program 2023*. Retrieved April 15, 2024, from https://www.vision2030.gov.sa/v2030/vrps/ntp/

Waha, K., Krummenauer, L., Adams, S., Aich, V., Baarsch, F., Coumou, D., Fader, M., Hoff, H., Jobbins, G., Marcus, R., Mengel, M., Otto, I. M., Perrette, M., Rocha, M., Robinson, A., & Schleussner, C. F. (2017). Climate change impacts in the Middle East and northern Africa (MENA) region and their implications for vulnerable population groups. *Regional Environmental Change, 17*(6), 1623–1638. https://doi.org/10.1007/s10113-017-1144-2

Weiss, E., Weiss, G., Weiss, R., Zuzik, J., & Labant, S. (2019). Economic aspects reorganization of accommodation facilities on eco-friendly hotel in Slovakia. *SGEM, 19*(5), 209–216. https://doi.org/10.5593/sgem2019/5.3/S21.027

World Health Organization. (2020). *WHO global strategy on health, environment and climate change: The transformation needed to improve lives and Well-being sustainably through healthy environments*. World Health Organization. Retrieved April 11, 2024, from https://iris.who.int/bitstream/handle/10665/331959/9789240000377-eng.pdf?ua=1

Yao, S., Pan, Y., Wang, L., Sensoy, A., & Cheng, F. (2023). Building eco-friendly corporations: The role of minority shareholders. *Journal of Business Ethics, 182*(4), 933–966. https://doi.org/10.1007/s10551-022-05291-y

Chapter 9
Sustainable Finance in the Gulf Region: Motivations, Risk-Return Analysis, and Government Actions

Inass El Farissi and Mondher Bellalah

Abstract This paper examines the motivations, use, and development of sustainable finance in GCC countries. We study the risk-return of ESG investments by considering cost of debt, investment/financial performance, stock returns, mean-variance optimized equity portfolios, and value. We survey the government actions proposed in the literature. We show that the main motivations behind the employment and the development of sustainable finance in the Gulf region are: the threat of decline in oil prices; the opportunity to use the available receipts to diversify the economies for an energy transition; the shared foundational principles between Shariah-based finance, ESG considerations, and the United Nations' sustainable development goals; and the ambition to improve revenues. Our analysis reveals that superior ESG rating is correlated with: lower cost of debt; inferior short-term financial performance; superior long-term financial performance; better investment performance; lower volatility and lower returns; higher market value; and superior free cash flow and dividend payout ratio. The principal government actions proposed in the literature to develop sustainable finance are to: enhance environmental performance; prepare the environment and employ financial and economic instruments to raise capital, manage risks, and stimulate the private sector; and multiply actions to support the free market economy.

Keywords Sustainable investment · Ethical investment · Green investment · Responsible investment · Socially responsible investment

---

I. El Farissi (✉)
ISCAE, Casablanca, Morocco
e-mail: ielfarissi@groupeiscae.ma

M. Bellalah
ISC, Paris, France

CY Cergy Paris University, Paris, France

M. Bellalah
e-mail: mbellalah@iscparis.com

H. Miniaoui and M. Bellalah (eds.), *Green Growth Opportunities and Sustainable Finance in the Gulf Cooperation Council Region*, Gulf Studies 24,
https://doi.org/10.1007/978-981-96-8287-4_9

9.1 Introduction

Sustainable finance takes environmental, social, and governance (ESG) considerations into account in investment and financing decisions.[1] This area of finance gained prominence at the beginning of the 2000s due to the augmented consciousness of environmental and social issues, the developing importance of corporate social responsibility, and the expansion of new ESG instruments.

ESG investing is an incorporation of ESG criteria into investment decisions. This is at the same time sustainable investing and green financing. Green finance focuses one's attention on providing capital for projects and activities that respect the environment. ESG debt is a bond that considers ESG impact. Most of the ESG investment market is comprised of debt investments. Green bonds are debt instruments of the green finance market that were introduced in 2007 as a financing solution for sustainable investments. Green bond issuers can benefit from the lower cost of capital that is occasioned by the green label. In an *activity-based debt*, funds are destined to finance or refinance specific green projects or activities. In a *behavior-based debt*, funds are not destined to specific uses. The financing is dependent on the company successfully achieving fixed ESG key performance indicators. In a *transition debt*, funds are used in projects or activities that are not classified as green presently, but are involved in the transition to a more sustainable future.

The Gulf Cooperation Council is a political and economic union of Arab states neighboring the Gulf established in 1981. Bahrain, Kuwait, Oman, Saudi Arabia, Qatar, and United Arab Emirates share the same religion, culture, and political systems, and all suffer from exposure to climate change, with persistent problems like water unavailability, food insecurity risk, and social welfare challenges. They have introduced diversification and sustainability strategies to reduce their dependence on oil, which makes their economies vulnerable to falling oil prices. The green bond market in the GCC was born when the first green bond was issued by the National Bank of Abu Dhabi in 2017. UAE and Saudi Arabia are leading the sustainable finance efforts in their region.

The organization of the chapter is as follows. Firstly, we explore the motivations behind the use and the development of sustainable finance. Secondly, we analyze the risk and the return of ESG investments by examining subsequently the cost of debt, investment performance, financial performance, stock returns, mean-variance optimized equity portfolios, and value. Thirdly, we survey the government actions suggested in the literature to develop sustainable finance in the GCC countries.

---

[1] Historically, in the 1990s and 2000s, the appellation "responsible investment" was used. It was replaced, in the 2010s, by "ESG investing."

9.2 Motivations, Use, and Development of Sustainable Finance

The six countries of the GCC acknowledge that they cannot count on hydrocarbons for an unlimited time. They desire to reduce their dependence on oil, which makes their economies vulnerable to falling oil prices. This is why they have introduced diversification and sustainability strategies. The principal motivation for increased renewable energy investment has been the threat of decline in oil prices.

GCC countries believe that the climate program could provide opportunities to diversify their economies for an energy transition. The increase in oil and gas prices provides an unexpected gain that they find crucial to investing in an economic and environmental transition. Prosperous hydrocarbon prices enhanced the receipts, balanced the fiscal state, and reduced the public debt of GCC countries. They also improved their external balances. The primary source of electricity in the GCC has been domestically produced oil or natural gas. For the future, they aim to decrease the rate at which they consume their hydrocarbons. The use of solar and wind energy permits these countries to obtain more revenues by exporting their oil and gas, or to reduce the extraction rate and maintain revenue at the same level.

The World Bank advise the oil-rich countries of the GCC to use the oil windfall to advance diversification and the share of the private sector in their economies. The 2020 report of the World Bank shows that, in 2018, 53% of total exports in Bahrain, 91% in Kuwait, 75% in Oman, 94% in Qatar, 77% in Saudi Arabia, and 74% in the UAE consisted of oil and gas revenues. These countries expect to increase the contribution of the private sector by reducing the participation of the public sector. In 2023, non-oil GDP continued its growth trajectory, reaching 3.9%.

The Islamic religious tradition has fixed as a condition of the validity of a transaction the avoidance of harm to the environment and society. Adil et al. (2023) focus on the shared foundational principles between Shariah-based finance, ESG considerations, and the United Nations' sustainable development goals. The authors point out the role of Islamic capital markets in satisfying the UN SDGs. They illustrate the convergence between Shariah-based finance and ESG through the ESG Sukuk instrument. Mahenthiran et al. (2023) studied 531 listed firms in the six GCC countries from 2010 to 2021. The results show that firm value is dependent on both ESG adoption and Shariah compliance.

The 2021 BP Statistical Review of World Energy report reveals that 99% of the GCC's energy consumption is fossil fuels, the share of renewable energy being 1%. Following Connelly and Xydis (2021), the investments in renewable energy technologies are characterized by long payback periods. Al-Sarihi and Mansouri (2022), among other authors, find that the lack of information and high uncertainty surrounding these technologies discourage investors from initiating investment decisions.

9.3 Risk-Return Analysis of ESG Investments

The GCC region offers an attractive return on investments, and it is a tax haven. Numerous barriers to the entrance of foreign investors have been abolished. As a result, the region attracts foreign capital and international investments. We propose henceforward a defined risk-return analysis by distinguishing the correlation between ESG rating, cost of debt, investment performance, financial performance, stock returns, mean-variance optimized equity portfolios, and value.

Apergis et al. (2022) studied whether, for the period between 2010 and 2019, firms that demonstrate solid ESG scores profit from lower cost of debt and higher bond ratings compared to firms with fragile ESG scores. Firms with low ESG ratings are characterized as riskier, since they encounter additional liabilities associated with ESG considerations that increase their likelihood of failing to honor commitments. The authors demonstrate that the bond spreads for borrowing firms are negatively correlated with their ESG rating. They prove that superior ESG rating is correlated with inferior cost of debt.

Whelan et al. (2021) studied the relationship between ESG and performance in more than a thousand research papers published between 2015 and 2020. They divided the papers into those concentrated on *corporate* financial performance (return on equity, return on assets, or stock price) and those concentrated on *investment* performance (Sharpe ratio), to discern if there was a difference in the results. 58% of the "corporate" studies detect a positive relationship between ESG and financial performance. 8% find a negative relationship and 34% show neutral or mixed results. 59% of the "investment" studies exhibit similar or better investment performance compared to conventional investment approaches. 14% find a negative link. Furthermore, it is proven that the positive effect of the ESG on financial performance is more pronounced over the long term.

Gupta and Chaudhary (2023) measure ESG investment performance in three developed markets and three emerging markets. With daily data, the authors did not find any significant difference in returns between ESG and conventional indices. With annual data, ESG indices reveal positive alpha and offer risk-adjusted returns superior to the overall market indices in most countries. In addition, ESG portfolios offer more protection against downside risk in the majority of countries. Finally, the authors did not find enough evidence that the performance of ESG indices is significantly different between developed and emerging markets.

By studying all GCC countries, Abdmouleh et al. (2015) find that investments in renewable energy projects do not generate satisfactory short-term financial returns. But when considering the increase in energy demand produced by the industrial development in the region, the authors expect the expansion of renewable energy projects.

Alqahtani et al. (2020) evaluate the stock return predictive power of the global and Saudi geopolitical risk indices and crude oil returns in the context of the six GCC countries. With monthly data from February 2007 to December 2019, global and Saudi GPR indices have shown feeble evidence of predictability of excess stock returns. Crude oil prices are shown to be generally a better predictor of GCC stock market returns. The predictability of crude oil returns is particularly efficacious in the Kuwaiti, Omani, Qatari, Saudi, and UAE stock markets.

Cornell (2020) addresses the conceptual issues that arise when attempting to assess the relation between risk, return, and ESG rating. The preference of investors for highly rated ESG companies will lead to higher prices and lower expected returns. But the lower expected returns signify a lower discount rate and higher market values for green companies. Investors' preferences for companies with high ESG ratings can lead to lower costs of equity capital that encourage investment in green technologies. Furthermore, a high ESG rating constitutes a hedge against climate shocks and changes in environmental regulation. Once more, high ESG ratings imply lower expected returns for investors.

El Ghoul and Karoui (2017) study the effect of corporate social responsibility on fund performance. They find that an increase in the level of CSR leads to a reduction in performance. More precisely, the CSR score of the portfolio is negatively related to risk-adjusted performance.

Riedl and Smeets (2017) seek to understand why investors choose socially responsible mutual funds. They find that social preferences as well as social signaling explain socially responsible investment decisions more than financial motivations. The authors explain that investors are disposed to sacrifice financial performance in order to invest in harmony with their social predilections. Socially responsible investors, in this experimental evidence, expect to receive lower returns on socially responsible investment funds than on conventional funds and to pay higher management fees.

Prol and Kim (2022) studied mean-variance optimized equity portfolios between 2018 and 2019. They find that high ESG score portfolios have lower volatility and lower returns than those with lower ESG scores, resulting in lower Sharpe ratios and performances.

Kilic et al. (2022) studied conventional stock market returns and ESG stocks returns with daily data from 2007 to 2021 for 19 developing and 19 developed countries. The authors conclude that there are significant portfolio diversification gains from adding ESG stocks to a portfolio in developed countries, but limited gains in developing countries.

Kibria et al. (2023) measure the relationship between the ESG score, the cost of equity, cost of debt, return on equity, and dividend payout ratio. The sample size consisted of 35 publicly traded banks and 88 general corporates listed on stock exchanges in the GCC countries. The authors detect a positive causality between the environmental score, free cash flow, and profitability. They find a robust relationship between the overall ESG score and the dividend payout ratio.

The recent international regulations required the replacement of the conventional investor wealth maximization by a sustainable value creation model. Fatemi and Fooladi (2013), Schoenmaker and Schramade (2019), and Kurznack, Schoenmaker and Schramade (2021) suggested new long-term value creation models that take into account sustainability strategies.

Starks (2023) highlights the difference between investors' and managers' motivations for ESG decision making. *Values* of an ESG project are non-pecuniary preferences. Numerous motivations can explain the *values approach.* For instance, a preference for avoiding projects with objectionable products or supporting projects in accordance with religious values. Non-pecuniary preferences differ among investors and managers, engendering differences in priorities (respect for natural and human resources). The *value* of an ESG project can be summarized financially as its coupled risk and return. Several motivations can explicate the *value approach.* As an illustration, ESG considerations can clarify financially the most important risks (mainly environmental, climate, and regulatory aspects). This permits the attenuation of risks and improved returns. Social responsibility oriented investors adopt ESG investment first and foremost, following a *values approach.* Financially oriented investors accept ESG investment uniquely following a *value approach.*

Rui et al. (2019) propose an industry equilibrium model, expecting that corporate social responsibility decreases systematic risk and increases firm value. The empirical results support the model's predictions.

Flammer (2021) covered all corporate green bonds issued by public and private companies throughout the world between 2013 and 2017. The author finds that the stock market responds positively to the issuance of green bonds. The cumulative abnormal return around the announcement of green bond issues is 0.67%, meaning that corporate green bonds are value creating. In the years subsequent to the issuance of green bonds, the study finds an increase both in long-term value (Tobin's Q), and in the operating performance (Return on Assets). The empirical evidence suggests, furthermore, that corporate green bonds permit the enhancement of environmental performance (the environmental rating). Finally, the issuance of green bonds allows the firms to captivate an investor clientele interested in the long-term investment horizon and the natural environment.

Financial analysts act as an intermediary channel between the firms and the stakeholders. They reduce the information asymmetry between firms and investors by incorporating firms' corporate social responsibility information into their recommendations. Alazzani et al. (2021) find that, in the six GCC countries, firms with political connections obtain more optimistic recommendations than non-politically connected firms. Authors show that sell-side analysts tend to provide favorable recommendations to firms with political connections. They also find that ESG ratings for firms with connections to royalty are higher than for non-connected firms.

By studying more than 23,000 large firms from 114 countries, Liang and Renneboog (2017) find that a firm's corporate social responsibility (CSR) practice is strongly correlated to its country's legal origin. Firms from common law countries have a lower level of CSR than firms from civil law countries. The authors explain the strong link between legal origins and CSR by differences in ex-post shareholder legal action risk and stakeholder regulations. Legal origins seem to be the solidest forecasters of CSR adoption and performance, solider than social preferences, and a firm's financial and operational performance.

9.4 Government Actions to Develop Sustainable Finance

Ma et al. (2023) investigate the relationship between green finance, renewable energy investment, and environmental performance. Green finance enhances environmental performance more in countries where credit and capital markets are less developed. In developing countries, green financing promotes green innovation. The authors' empirical results recommend that developing countries encourage green innovation and improve environmental performance.

Magazzino (2022) examines fiscal sustainability in the six GCC countries between 1990 and 2017. The results of the tests reveal that Saudi Arabia and Oman "are in a condition of risk," having no choice but to maintain the debt under control. Bahrain and Qatar show signs of the severest difficulties: government revenues, expenditures, and debt. The results confirm the fiscal synchronization: the GCC governments decide on their revenues and expenditures synchronously.

Beyer and Bayoumi (2022) studied how governments can create investment conditions and use financial tools to organize sustainable finance. The authors recommend preparing the environment and employing financial and economic instruments to raise capital, manage risks, and stimulate the private sector.

The preparation of the environment recommended by Beyer and Bayoumi consists of both the definition of a regional sustainability classification and forging an understanding of sustainability's benefits. A harmonized regional sustainability classification is an agreed system that permits recognition of ESG projects and activities. This classification would yield a standard definition of sustainable projects and activities. The regional classification should precede the national classifications, but the GCC countries are starting from the national one. Furthermore, the understanding of sustainability's benefits should motivate change.

The financial and economic instruments suggested by Beyer and Bayoumi are the development of a regional carbon market and the leverage of sovereign wealth funds for transition finance. Market mechanisms are sensitive to economic changes and reduce the cost of emission reduction. The sustainability-linked loans stimulate sustainable investments. They can take the form of direct loans or using financial intermediation. The GCC countries should subsidize consumer energy prices to encourage consumption. By offering guarantees and risk insurance, finance institutions and multilateral banks permit lower financing costs and stimulate private investment. Green bonds and green Sukuk mobilize private capital toward sustainable investments. Debt-for-nature and debt-for-climate swaps are a consensus that diminish a financial debt in favor of an obligation to invest the liberated funds to protect the climate. Sovereign wealth funds are government investment funds with long-term investment strategies in the sustainable transition. State-owned enterprises allow governments to make ESG progress beyond the private sector. By playing the role of regulator, executor, and owner, the government can drive sustainable transition via state-owned enterprises. The government can award these enterprises subsidies, preferred financing, guarantees, and/or exemptions.

Sweidan (2021) finds the institutional actions of the GCC countries insufficient to engender sustainable consumption and production. The state of diversification being still feeble, the main recommendation of the author is that the GCC countries should develop the diversification of the economy, multiplying actions in support of the free market economy, and undertaking more concrete actions toward renewable energy.

9.5 Conclusion

The GCC countries admit that they cannot continue to count on hydrocarbons infinitely. They aim to decrease their dependence on oil and the vulnerability of their economies to fluctuations in oil prices. Sustainable finance provides opportunities to diversify these economies and, at the same time, to develop the part of the private sector. They have the opportunity to use the available receipts for an energy transition and to earn more revenue. Furthermore, the convergence between Shariah-based finance, ESG considerations, and the United Nations' sustainable development goals can enhance the development of sustainable finance.

Approximately 580 published research papers find a positive correlation between ESG and financial performance (ROE, ROA, and Sharpe ratio). Furthermore, it has been proven that lower cost of debt is associated with higher ESG rating. As a consequence, an intuitive and robust relationship exists between the ESG score and the dividend payout ratio. The GCC countries are incentivized to develop ESG investments and enhance environmental performance. They must prepare their environment and develop appropriate financial and economic instruments. However, the absolute majority of the GCC's energy consumption is still fossil fuels today.

References

Abdmouleh, Z., Alammari, R. A. M., & Gastli, A. (2015). Recommendations on renewable energy policies for the GCC countries. *Renewable and Sustainable Energy Reviews, 50*(C), 1181–1191.

Adil, M., Hoogendoorn, H. J., & Jumat, Z. H. (2023). Sukuk innovation: Powering sustainable finance. Palgrave CIBFR studies in Islamic finance. In Z. H. Jumat, S. H. Khateeb, & S. N. Ali (Eds.), *Islamic finance, FinTech, and the road to sustainability* (pp. 145–158). Palgrave Macmillan.

Alazzani, A., Wan-Hussin, W. N., Jones, M., & Al-hadi, A. (2021). ESG reporting and analysts' recommendations in GCC: The moderation role of royal family directors. *Journal of Risk and Financial Management, 14*(2), 72.

Alqahtani, A., Bouri, E., & Vinh Vo, X. (2020). Predictability of GCC stock returns: The role of geopolitical risk and crude oil returns. *Economic Analysis and Policy, 68*, 239–249.

Al-Sarihi, A., & Mansouri, N. (2022). Renewable energy development in the Gulf cooperation council countries: Status, barriers, and policy options. *Energies, 15*(5), 1923.

Apergis, N., Poufinas, T., & Antonopoulos, A. (2022). ESG scores and cost of debt. *Energy Economics, 112*, 106186.

Beyer, J., & Bayoumi, M. (2022). *Financing a green transition in the Middle East*. Mohammed Bin Rashid School of Government.

Connelly, C., & Xydis, G. (2021). Wind energy in the Gulf cooperation council region: Progress, challenges and strategies for development. *Review of Economics and Political Science, 6*, 287–291.

Cornell, B. (2020). ESG preferences, risk and return. *European Financial Management, 27*(1), 12–19.

El Ghoul, S., & Karoui, A. (2017). Does corporate social responsibility affect mutual fund performance and flows? *Journal of Banking and Finance, 77*, 53–63.

Fatemi, A. M., & Fooladi, I. J. (2013). Sustainable finance: A new paradigm. *Global Finance Journal, 24*, 101–113.

Flammer, C. (2021). Corporate green bonds. *Journal of Financial Economics, 142*, 499–516.

Gupta, H., & Chaudhary, R. (2023). An analysis of volatility and risk-adjusted returns of ESG indices in developed and emerging economies. *Risks, 11*(10), 182.

Kibria, M. K., Qazi, H. S., Damilola Shuaib, S. O., Amrullah, A., & Nasriddinov, E. (2023). Assessment of ESG integration in corporate valuation in GCC region. *International Journal of Islamic Economics and Governance, 4*(1), 62–79.

Kilic, Y., Destek, M. A., Cevik, E. I., Bugan, M. F., Korkmaz, O., & Dibooglu, S. (2022). Return and risk spillovers between the ESG global index and stock markets: Evidence from time and frequency analysis. *Borsa Istanbul Review, 22*(2), S141–S156.

Kurznack, L., Schoenmaker, D., & Schramade, W. (2021). A model of long-term value creation. *Journal of Sustainable Finance & Investment, 11*, 1–19.

Liang, H., & Renneboog, L. (2017). On the foundations of corporate social responsibility. *The Journal of Finance, 72*, 853–910.

Ma, M., Zhu, X., Liu, M., & Huang, X. (2023). Combining the role of green finance and environmental sustainability on green economic growth: Evidence from G-20 economies. *Renewable Energy, 207*, 128–136.

Magazzino, C. (2022). Fiscal sustainability in the GCC countries. *International Journal of Economic Policy Studies, 16*, 389–408.

Mahenthiran, S., Gjerde, T., & Ansari, J., (2023). *Importance of ESG ratings for Shariah-compliant companies listed on gulf cooperation council countries stock exchanges*. Working Paper.

Prol, J. L., & Kim, K. (2022). Risk-return performance of optimized ESG equity portfolios in the NYSE. *Finance Research Letters, 50*(December), 103312.

Riedl, A., & Smeets, P. (2017). Why do investors hold socially responsible mutual funds? *Journal of Finance, 72*(6), 2505–2550.

Rui, A., Koskinen, Y., & Zhang, C. (2019). Corporate social responsibility and firm risk: Theory and empirical evidence. *Management Science, 65*, 4451–4469.

Schoenmaker, D., & Schramade, W. (2019). Investing for long-term value creation. *Journal of Sustainable Finance & Investment, 9*, 356–377.

Starks, L. T. (2023). Sustainable finance and ESG issues—value versus values. *The Journal of Finance, 78*(4), 1837–1872.

Sweidan, O. D. (2021). The environmental and energy policies to enable sustainable consumption and production in the Gulf cooperation council countries. *Clean Technologies and Environmental Policy, 23*, 2639–2654.

Whelan, T., Atz, U., Holt, T. V., & Clark, C. (2021). *ESG and financial performance: Uncovering the relationship by aggregating evidence from 1,000 plus studies*. Working Paper, New York University, Center for Sustainable Business.

Chapter 10
Harnessing Sustainable AI: Advancing Green Growth and Innovation in the GCC

Ashhadul Islam, Samir Brahim Belhaouari, and Sami Miniaoui

Abstract By focusing on compressing AI models and making them both energy-efficient and cost-effective, the Gulf Cooperation Council (GCC) can take meaningful steps toward reducing the carbon footprint of its rapidly expanding AI ecosystem. Leveraging sustainable finance, robust regulatory frameworks, and fostering innovation hubs will ensure that AI technologies contribute to green growth, aligning with the region's vision for a sustainable future. This integrated approach not only accelerates the development of low-energy AI solutions but also attracts global investments, positioning the GCC as a leader in sustainable technological innovation. These efforts will solidify the GCC's role in promoting green growth while advancing its economic competitiveness on the global stage.

Keywords Sustainable AI · Network pruning · Sparse matrices

10.1 Energy Efficiency Through AI Model Compression

The rapid development and deployment of Large Language Models (LLMs) and other advanced AI technologies have brought immense computational demands, leading to significant energy consumption (Hadi et al., 2024). This poses a challenge to environmental sustainability, particularly in regions like the GCC, where

A. Islam
KTH Royal Institute of Technology, Stockholm, Sweden
e-mail: aisla@kth.se

S. B. Belhaouari (✉)
Hamad Bin Khalifa University, Doha, Qatar
e-mail: sbelhaouari@hbku.edu.qa

S. Miniaoui
University of Dubai, Dubai, UAE
e-mail: sminiaoui@ud.ac.ae

H. Miniaoui and M. Bellalah (eds.), *Green Growth Opportunities and Sustainable Finance in the Gulf Cooperation Council Region*, Gulf Studies 24,
https://doi.org/10.1007/978-981-96-8287-4_10

energy consumption is directly tied to oil and gas production (Al-Mohannadi & Al-Mohannadi, 2023). As the region strives to balance technological advancement with sustainability, addressing the energy needs of these models is becoming increasingly critical.

Training large AI models, particularly LLMs, requires substantial computational resources, which in turn generates a considerable amount of carbon emissions (Jiang et al., 2024). For instance, the process of training a single large AI model can emit as much CO_2 as five average cars over their lifetime (Delanoë et al., 2023). Given the global push for sustainable development, the GCC, with its vision for a greener future, cannot afford to perpetuate these energy-intensive practices. The region must adopt solutions that align with its sustainability goals while continuing to advance its AI capabilities.

One viable solution lies in the compression of AI models. Techniques such as pruning, quantization, and knowledge distillation can significantly reduce the size and complexity of these models without sacrificing performance (Dantas et al., 2024). These methods work by minimizing the number of operations required during both the training and inference phases, resulting in lower energy consumption. By compressing AI models, the computational power needed for running these models is reduced, thereby lowering operational costs. This also diminishes the energy demands for cooling and maintaining large data centers, further contributing to a reduced carbon footprint (Bose et al., 2021).

The adoption of energy-efficient AI technologies aligns perfectly with the GCC's green growth strategies. As the region continues to invest in sustainable economic development, compressed AI models offer a pathway to achieving greater operational efficiency without exacerbating environmental challenges (Sarwar, 2022). These smaller, more efficient models can be applied across various sectors, including energy, finance, logistics, and transportation, providing governments and businesses with the tools to optimize their operations while adhering to sustainability goals. By prioritizing energy-efficient AI, the GCC can position itself as a leader in both technological innovation and environmental stewardship, ensuring that its AI-driven growth supports its long-term vision for sustainable development.

It is important to note that the energy savings discussed here are primarily a result of reducing the computational intensity of AI models through techniques such as pruning and quantization. These methods optimize model structure, reducing the energy required for both training and inference, particularly in large-scale deployments.

10.2 Experimental Results

As seen in Fig. 10.1, models pruned with our techniques (architecture-modified) (Belhaouari & Islam, 2023; Islam et al., 2024) consistently demonstrate significant advancements in sustainability and efficiency, serving as a pivotal step toward a greener AI ecosystem. By utilizing 70% fewer floating-point operations (FLOPs), emitting 65% less carbon, occupying 68% less storage space, and performing inference 62% faster compared to unpruned models, these techniques address the dual challenges of computational inefficiency and environmental sustainability. These improvements highlight the potential of pruning as a transformative approach to reducing the carbon footprint of AI systems while simultaneously cutting operational costs.

10.3 Sustainable Finance for Green AI Innovation

The development of sustainable AI technologies in the Gulf Cooperation Council (GCC) region demands considerable investment in research, infrastructure, and workforce training. Building and implementing green AI innovations, such as energy-efficient models, involves substantial upfront costs, which can be a barrier for many businesses (Al Bakri & Kisswani, 2024). However, at the same time, sustainable finance is gaining momentum in the GCC as governments and industries look to diversify their economies and reduce reliance on oil-based revenues. This shift toward sustainable finance provides a unique opportunity to fund AI innovations that align with environmental goals and the broader vision for a greener future.

One of the key mechanisms for financing green AI innovation is the issuance of green bonds. These bonds are designed to raise capital for projects that promote environmental sustainability (Chishti et al., 2024). In the context of AI, green bonds could be used to fund research and development in areas such as AI model compression and low-energy AI technologies (Al-Kasasbeh et al., 2024). By channeling funds from green bonds into AI innovation, the GCC can support the development of AI solutions that contribute to reduced energy consumption and lower carbon emissions.

Sustainability-linked loans offer another avenue for supporting green AI efforts. These loans provide companies with financial incentives to meet specific environmental performance targets, such as reducing their carbon footprint (Elias et al., 2024). Companies working on energy-efficient AI models could secure these loans with the promise of lower interest rates if they achieve measurable environmental goals. This financial structure would not only drive innovation in the AI sector but also ensure that sustainability remains at the forefront of technological advancements (Edunjobi, 2024).

Impact investment is a third significant mechanism. AI startups and tech companies focused on developing low-energy AI models are well-positioned to attract impact investors—those who seek both financial returns and positive environmental outcomes. By investing in AI technologies that prioritize sustainability, impact investors can drive advancements in green AI, supporting both economic growth and environmental stewardship in the region (Addy et al., 2024).

The GCC's financial institutions, including its powerful sovereign wealth funds such as the Abu Dhabi Investment Authority, the Kuwait Investment Authority, and Saudi Arabia's Public Investment Fund, are well-placed to play a pivotal role in fostering green AI innovation (Khalil et al., 2023). These funds can invest directly in

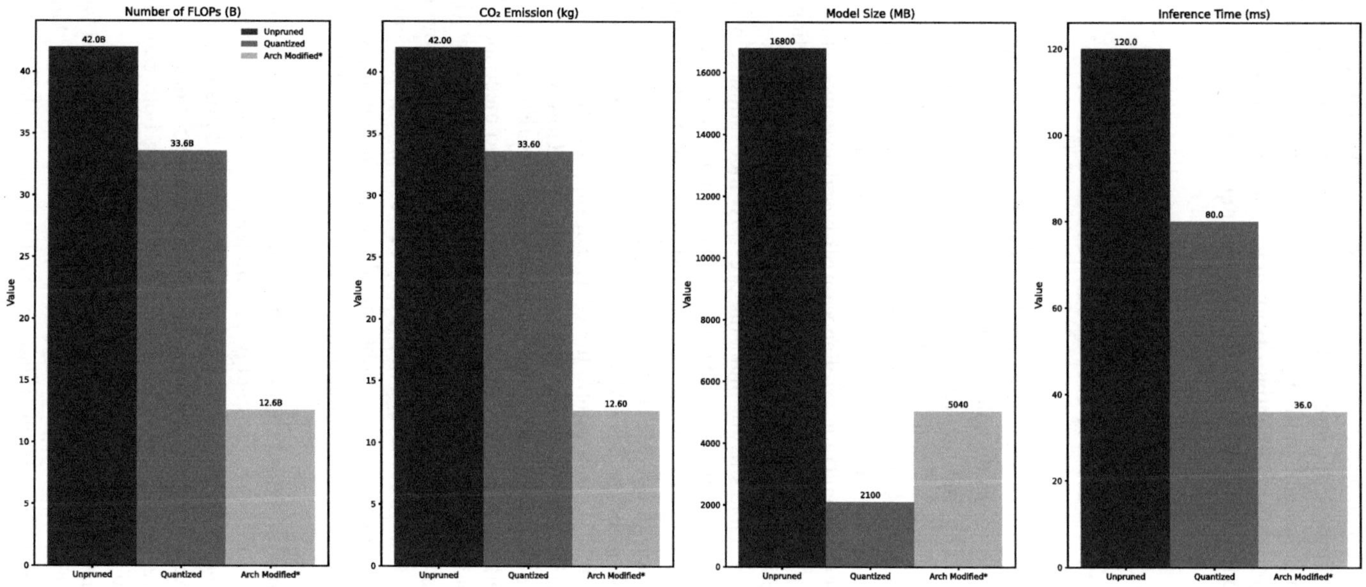

Fig. 10.1 Comparing unpruned models, quantized models, and models modified with our techniques

AI startups and initiatives dedicated to the development of compressed and energy-efficient AI models. By aligning their investment strategies with Environmental, Social, and Governance (ESG) frameworks, these institutions can support the region's digital transformation while contributing to global sustainability goals (Al Bakri & Kisswani, 2024).

Investing in AI innovation through sustainable finance mechanisms not only supports the GCC's environmental objectives but also positions the region as a leader in both green technology and AI development. As global interest in sustainable technology grows, the GCC has the potential to become a hub for cutting-edge AI solutions that drive both economic growth and environmental sustainability (Sadriwala et al., 2024).

10.4 Synergies with GCC Green Growth Strategies

The Gulf Cooperation Council (GCC) is actively pursuing economic diversification, with a strong focus on reducing its dependence on hydrocarbons. Visionary initiatives such as Saudi Vision 2030 (Belaïd et al., 2023) and the UAE's Green Growth Strategy aim to drive development in renewable energy, tourism, and digital innovation. These strategies highlight the region's commitment to fostering sustainable growth while transitioning toward a more diversified and resilient economy. To achieve these goals, emerging technologies like AI will play a critical role in optimizing resources and supporting the region's green growth ambitions (Al-Saidi & Zaidan, 2024).

AI's contribution to green growth in the GCC is multifaceted, particularly in the context of smart city development. AI-powered smart city solutions can significantly improve the management of urban resources, including energy, water, and public transportation. By employing AI technologies to optimize energy grids and reduce water waste, especially important in the region's arid climate, cities can achieve more efficient and sustainable urban growth. Compressed AI models are well-suited for these smart city applications (Ajlan & Al Abed, 2023), as they require fewer computational resources, making them both scalable and environmentally friendly. This enables the GCC to pursue green urban development without imposing additional burdens on energy consumption.

Beyond smart cities, AI can also enhance energy optimization efforts in the GCC. Compressed AI models can improve the performance of renewable energy infrastructures, such as solar farms and wind turbines. Through predictive maintenance and more efficient energy distribution, AI technologies can ensure that renewable energy sources are used to their full potential, supporting the region's shift toward cleaner energy. The energy efficiency of compressed models further contributes to this goal by reducing the overall operational demands of these AI systems (Ghaithan et al., 2024).

Environmental monitoring is another area where AI can contribute to the GCC's green growth strategies. Through integration with Internet of Things (IoT) devices, AI models can track emissions, monitor pollution levels, and optimize the use of natural resources (Alam & Azalie, 2023). The real-time, data-driven insights provided by AI technologies enable faster and more efficient decision-making, which is essential for addressing environmental challenges. Since compressed AI models require less energy to operate, they are especially valuable in such applications, where minimizing energy consumption is key to achieving sustainability objectives (Rashid & Mohsenin, 2024).

By incorporating AI-driven solutions into smart infrastructure, transportation, and resource management, the GCC can ensure that its economic growth is both sustainable and efficient. The use of compressed AI models aligns with the region's energy diversification efforts, as these models allow for widespread AI adoption without significantly increasing energy demands or carbon emissions (Alhammadi et al., 2023). In this way, AI technologies can serve as a vital tool in supporting the GCC's green growth strategies, enabling the region to remain at the forefront of sustainable development while reducing its reliance on traditional energy sources.

10.5 Regulatory Support for Sustainable AI Development

Governments in the GCC are steadily constructing regulatory frameworks to support green finance and broader sustainability initiatives. However, specific regulations that address the environmental impact of AI technologies have yet to be fully developed. While progress is being made in many areas, there remains a gap in ensuring that AI, which is rapidly becoming a cornerstone of innovation in the region, adheres to sustainability standards. The environmental implications of large-scale AI deployments must be carefully managed to align with the region's broader sustainability goals (Al Bakri & Kisswani, 2024; Radhi et al., 2024).

One key step toward sustainable AI regulation could be the introduction of Green AI certifications (Alzoubi & Mishra, 2024). These certifications would establish benchmarks for AI models, ensuring they meet specific sustainability criteria, such as energy efficiency, minimized environmental impact, and optimized resource use (Filho et al., 2024). Compressed AI models, which are designed to operate with reduced computational requirements, are particularly well-suited for such certifications. By setting industry standards for AI sustainability, these certifications could help guide the development of environmentally responsible AI technologies (Is, 2024).

In addition to certifications, governments could offer tax incentives and subsidies to encourage businesses and startups to develop and adopt AI technologies that prioritize sustainability (Ibekwe et al., 2024). Companies that focus on model compression or other energy-saving techniques could benefit from reduced taxes or financial support, accelerating the adoption of sustainable AI practices (Gorina et al., 2024). These incentives would not only promote green AI innovation but also drive economic growth by supporting tech companies that align their goals with the region's sustainability objectives (Zhou et al., 2024).

Public procurement policies can also play a crucial role in advancing sustainable AI technologies (Singh et al., 2024). Governments can lead by example, mandating that AI systems used in public sectors—such as transportation, urban planning, and energy management—comply with established sustainability standards. This would create a demand for energy-efficient AI models in essential services, encouraging the private sector to develop and implement more sustainable AI solutions (Al Bakri & Kisswani, 2024; Timm et al., 2023.

By integrating AI-specific sustainability regulations into existing policies, the GCC can strengthen its commitment to a low-carbon economy. These frameworks will not only help reduce emissions but also foster an innovation ecosystem where AI is developed and deployed with sustainability at its core (Almulhim & Al-Saidi, 2023; Saqib et al., 2023). This approach will allow the GCC to continue advancing its digital transformation while ensuring that AI technologies contribute to its green growth objectives, ultimately positioning the region as a leader in sustainable technological development (Shehri et al., 2023).

10.6 AI-Powered Sustainability Solutions

Industries such as transportation, renewable energy, manufacturing, and water management increasingly rely on AI-driven tools to improve operational efficiency and reduce resource consumption. However, the use of large, power-intensive AI models can sometimes undermine the very sustainability goals they are meant to support (Bassey et al., 2024). The challenge lies in balancing the need for advanced AI technologies with the environmental responsibility of minimizing their energy consumption (Bolón-Canedo et al., 2024).

In the realm of sustainable applications, compressed AI models offer a promising solution. For example, smart energy management systems can benefit greatly from the deployment of compressed AI models (Zhang et al., 2024). These systems are designed to optimize the generation, storage, and consumption of renewable energy sources like solar and wind power. By accurately forecasting energy demands and minimizing wastage, compressed AI models can help ensure that renewable energy infrastructures operate at peak efficiency while consuming fewer resources (Kakolyris et al., 2024).

Another critical application of compressed AI models is in predictive maintenance. In the context of renewable energy infrastructures, such as wind turbines and solar farms, compressed models can be used to predict equipment failures and schedule maintenance before significant issues arise. This not only reduces downtime but also extends the operational lifespan of key assets, contributing to more sustainable energy production and reducing the need for frequent repairs or replacements (Assareh et al., 2025).

AI can also play a pivotal role in carbon tracking. By optimizing logistics, supply chains, and various industrial processes, AI technologies can help industries track and reduce their carbon footprints. These optimizations lead to more efficient use of resources and lower emissions, contributing to a reduction in overall environmental impact (Gaur et al., 2023; Lei, 2024). Compressed AI models, with their reduced energy requirements, are particularly suited for such tasks, as they enable these optimizations without adding to the energy burden.

The application of compressed AI models across multiple industries in the GCC is essential for improving energy efficiency and minimizing resource waste. For example, AI-powered systems can optimize the operations of desalination plants, which are crucial for the region's water management efforts (Sherif et al., 2023). These technologies not only contribute to more sustainable water use but also support the broader climate goals of the GCC. As the region continues to pursue green growth strategies, compressed AI models will play an integral role in enhancing industrial competitiveness while ensuring that sustainability remains at the forefront of development initiatives (Alloghani, 2024; Dsilva et al., 2024).

The energy savings achieved by AI applications in sectors such as energy planning or transport optimization differ from those resulting from model compression. While compressed models focus on reducing the operational energy demands of AI systems, AI applications in these sectors optimize real-world systems, such as energy distribution grids or traffic flow, to achieve broader sustainability goals.

10.7 Innovation Hubs for Sustainable AI

The GCC is positioning itself as a leading hub for innovation, with a particular focus on sectors such as fintech, artificial intelligence, and sustainable energy (Khalil et al., 2023). While significant progress has been made, many of the AI innovations emerging from the region could benefit from a stronger emphasis on sustainability. As AI becomes an increasingly important tool for development across industries, the need to integrate sustainable practices into AI technology has become more pressing (Al-Kasasbeh et al., 2024; Fardan et al., 2024).

One approach to fostering this integration is the establishment of AI and sustainability incubators. These incubators would focus on supporting the development of energy-efficient AI solutions, such as compressed models (Sargiotis, 2024). By providing startups with access to funding, infrastructure, and technical expertise, these incubators can accelerate the creation of AI technologies that align with sustainability goals. This type of specialized support system would not only encourage innovation but also ensure that sustainability remains a priority in the development of new AI solutions (Saba & Hadidi, 2025).

Collaboration between academia, industry, and government (Ahmed et al., 2023) is another key element in promoting sustainable AI innovation. By forming partnerships between these sectors, the GCC can leverage diverse expertise and resources to drive the development of sustainable AI technologies (Hojeij, 2024). These collaborations (Sulthan et al., 2023) can lead to the creation of AI models that are not only energy-efficient but also scalable and applicable across a variety of industries, such as energy, transportation, and manufacturing (Shwedeh et al., 2024).

The GCC's ongoing innovation initiatives, particularly in countries like the UAE and Saudi Arabia, provide an ideal platform for advancing sustainable AI technologies (Hafner et al., 2023). By focusing on the development of low-energy AI models, the region can lead the global movement toward green AI solutions. These innovation hubs will not only help the GCC meet its climate and sustainability goals but also drive economic growth by positioning the region at the forefront of cutting-edge AI technologies (Sanfilippo et al., n.d.). Through these efforts, the GCC can emerge as a global leader in both technological innovation and sustainability (Zaidan, 2023).

10.8 Conclusion

In conclusion, the Gulf countries stand at the crossroads of transformative change, moving toward sustainable energy practices while maintaining economic resilience. The integration of compressed AI models and sustainable technologies provides a critical pathway to achieving the region's dual goals of reducing carbon emissions and fostering economic diversification. By addressing the immense computational demands of AI through model optimization and leveraging sustainable finance mechanisms, such as green bonds and impact investments, the GCC can align its technological advancements with its sustainability objectives.

Furthermore, these efforts prepare the region to respond effectively to global shocks, such as climate change and fluctuating energy markets, by reducing dependence on hydrocarbons and strengthening its green growth strategies. Compressed AI models not only enhance the efficiency of renewable energy and resource management but also contribute to broader applications in transportation and urban development, reinforcing the GCC's position as a global leader in sustainable innovation.

By fostering collaboration between academia, industry, and government, and establishing innovation hubs dedicated to green AI, the GCC can accelerate the adoption of sustainable practices across industries. This integrated approach ensures that the region's digital transformation aligns with its economic and environmental priorities, enabling it to thrive in an increasingly competitive global landscape. Ultimately, the GCC's commitment to sustainable AI exemplifies how technological innovation can serve as a powerful tool for green growth, positioning the region as a key contributor to the global sustainability agenda.

References

Addy, W. A., Ofodile, O. C., Adeoye, O. B., Oyewole, A. T., Okoye, C. C., Odeyemi, O., & Ololade, Y. J. (2024). Data-driven sustainability: How fintech innovations are supporting green finance. *Engineering Science & Technology Journal, 5*(3), 760–773.

Ahmed, E., Shaukat, A., & AlDhaen, E. (2023). A review of the recent developments in the higher education sector globally and in the GCC region. In *Artificial intelligence and transforming digital marketing* (pp. 635–651). Springer.
Ajlan, A. M., & Al Abed, A. M. (2023). Transformation model towards sustainable smart cities: Riyadh, Saudi Arabia as a case study. *Current Urban Studies, 11*(1), 142–178.
Al Bakri, A., & Kisswani, M. (2024). The role of artificial intelligence in supporting green economy in GCC: Its economic, managerial and necessary legal framework impact. *Artificial Intelligence: Legal and Economic Prospects and Challenges, 66*(3), 759–778.
Alam, M., & Azalie, I. A. N. (2023). Greening the desert: Sustainability challenges and environmental initiatives in the GCC states. In *Social change in the Gulf region: Multidisciplinary perspectives* (pp. 493–510). Springer.
Alhammadi, A., Alsyouf, I., Semeraro, C., & Obaideen, K. (2023). The role of industry 4.0 in advancing sustainability development: A focus review in The United Arab Emirates. *Cleaner Engineering and Technology, 18*, 100708.
Al-Kasasbeh, O., Al-Khazaleh, S. M., & Alsheikh, G. (2024). The dynamic impact of environmental sustainability, green finance, and FinTech on energy efficiency in middle eastern economies. *International Journal of Energy Economics and Policy, 14*(4), 574–579.
Alloghani, M. A. (2024). *Artificial intelligence and sustainability* (pp. 1–210). Springer.
Al-Mohannadi, S. A., & Al-Mohannadi, D. M. (2023). Qatar in the energy transition: Low carbon economy challenges and opportunities. In *Sustainable Qatar* (p. 109). Springer.
Almulhim, A. I., & Al-Saidi, M. (2023). Circular economy and the resource nexus: Realignment and progress towards sustainable development in Saudi Arabia. *Environmental Development, 46*, 100851.
Al-Saidi, M., & Zaidan, E. (2024). Smart cities and communities in the GCC region: From top-down city development to more local approaches. *Frontiers in Built Environment, 10*, 1341694.
Alzoubi, Y. I., & Mishra, A. (2024). Green artificial intelligence initiatives: Potentials and challenges. *Journal of Cleaner Production, 468*, 143090.
Assareh, E., Keykhah, A., Bedakhanian, A., Agarwal, N., & Lee, M. (2025). Optimizing solar photovoltaic farm-based cogeneration systems with artificial intelligence (AI) and cascade compressed air energy storage for stable power generation and peak shaving: A Japan-focused case study. *Applied Energy, 377*, 124468.
Bassey, K. E., Juliet, A. R., & Stephen, A. O. (2024). AI-enhanced lifecycle assessment of renewable energy systems. *Engineering Science & Technology Journal, 5*(7), 2082–2099.
Belaïd, F., Amine, R., & Massie, C. (2023). Smart cities initiatives and perspectives in the MENA region and Saudi Arabia. In *Smart cities: Social and environmental challenges and opportunities for local authorities* (pp. 295–313). Springer International Publishing.
Belhaouari, S. B., & Islam, A. (2023). Neural network optimization with weight evolution. In *ICML 2023 workshop neural compression: From information theory to applications*. ICML.
Bolón-Canedo, V., Morán-Fernández, L., Cancela, B., & Alonso-Betanzos, A. (2024). A review of green artificial intelligence: Towards a more sustainable future. *Neurocomputing, 599*, 128096.
Bose, R., Roy, S., Mondal, H., Chowdhury, D. R., & Chakraborty, S. (2021). Energy-efficient approach to lower the carbon emissions of data centers. *Computing, 103*(8), 1703–1721.
Chishti, M. Z., Dogan, E., & Binsaeed, R. H. (2024). Can artificial intelligence and green finance affect economic cycles? *Technological Forecasting and Social Change, 209*, 123740.
Dantas, P. V., da Silva, S., Cordeiro, L. C., & Carvalho, C. B. (2024). A comprehensive review of model compression techniques in machine learning. *Applied Intelligence, 54*, 11804–11844.
Delanoë, P., Tchuente, D., & Colin, G. (2023). Method and evaluations of the effective gain of artificial intelligence models for reducing CO_2 emissions. *Journal of Environmental Management, 331*, 117261.
Dsilva, J., Shaikh, M. I., & Usmani, F. Z. A. (2024). Net zero policies and climate resilience: A focus on emerging technologies and implementation strategies in the Middle East. In *Transition towards a sustainable future* (pp. 119–138). Springer.

Edunjobi, T. E. (2024). Sustainable supply chain financing models: Integrating banking for enhanced sustainability. *International Journal for Multidisciplinary Research Updates, 7*(2), 1–11.

Elias, O., Awotunde, O. J., Oladepo, O. I., Azuikpe, P. F., Samson, O. A., Oladele, O. R., & Ogunruku, O. O. (2024). The evolution of green fintech: Leveraging AI and IoT for sustainable financial services and smart contract implementation. *World Journal of Advanced Research and Reviews, 23*(1), 2710–2723.

Fardan, S., Hamdan, A., & Kanan, M. (2024). Investing the role of artificial intelligence in advanced sustainable development goals: Evidence from banks in Bahrain. In *Business development via AI and digitalization: Volume 1* (pp. 925–934). Springer.

Filho, W. L., Kautish, S., Wall, T., Rewhorn, S., & Paul, S. K. (2024). Optimizing green power and green energy through digital technologies. In *Digital technologies to implement the UN sustainable development goals* (pp. 275–304). Springer.

Gaur, L., Afaq, A., Arora, G. K., & Khan, N. (2023). Artificial intelligence for carbon emissions using system of systems theory. *Ecological Informatics, 76*, 102165.

Ghaithan, A. M., Mohammed, A., Al Hanbali, A., & Alshibani, A. (2024). A mixed integer linear programming model for optimal sizing of hydrogen refueling station powered using a PV-grid system. *Arabian Journal for Science and Engineering, 50*, 11129–11144.

Gorina, L., Korneeva, E., Kovaleva, O., & Strielkowski, W. (2024). *Energy-saving technologies and energy efficiency in the post-COVID era.* Sustainable Development.

Hadi, M. U., Al Tashi, Q., Shah, A., Qureshi, R., Muneer, A., Irfan, M., & Shah, M. (2024). Large language models: A comprehensive survey of its applications, challenges, limitations, and future prospects. Authorea Preprints.

Hafner, M., Raimondi, P. P., & Bonometti, B. (2023). Low-carbon energy strategies in mena countries. In *The energy sector and energy geopolitics in the MENA region at a crossroad: Towards a great transformation?* (pp. 175–261). Springer.

Hojeij, Z. (2024). An overview of university-industry collaboration in the Arab world. *Journal of Innovation and Entrepreneurship, 13*(1), 40.

Ibekwe, K. I., Umoh, A. A., Nwokediegwu, Z. Q. S., Etukudoh, E. A., Ilojianya, V. I., & Adefemi, A. (2024). Energy efficiency in industrial sectors: A review of technologies and policy measures. *Engineering Science & Technology Journal, 5*(1), 169–184.

Is, H. (2024). Strategic approaches to eco-efficient computing in institutional environments. In *Proceedings of the cognitive models and artificial intelligence conference* (pp. 186–190). IEEE.

Islam, A., Belhaouari, S., & Bermak, A. (2024). Sustainable AI: Efficient pruning of large language models in resource-limited environments. In *Workshop on machine learning and compression.* NeurIPS.

Jiang, P., Sonne, C., Li, W., You, F., & You, S. (2024). Preventing the immense increase in the life-cycle energy and carbon footprints of LLM-powered intelligent Chatbots. *Engineering, 40*, 202–210.

Kakolyris, A. K., Masouros, D., Vavaroutsos, P., Xydis, S., & Soudris, D. (2024). SLO-aware GPU frequency scaling for energy efficient LLM inference serving. arXiv preprint arXiv:2408.05235.

Khalil, R. G., Damrah, S., Bajaher, M., & Shawtari, F. A. (2023). Unveiling the relationship of ESG, fintech, green finance, innovation and sustainability: Case of gulf countries. *Environmental Science and Pollution Research, 30*(54), 116299–116312.

Lei, J. (2024). Efficient strategies on supply chain network optimization for industrial carbon emission reduction. arXiv preprint arXiv:2404.16863.

Radhi, S. Q., Hamdan, A., & Amir, D. (2024). The impact of ESG (environmental, social, and governance) disclosures on banks in GCC (gulf cooperation council). In *Business development via AI and digitalization: Volume 1* (pp. 513–523). Springer.

Rashid, H. A., & Mohsenin, T. (2024). TinyM^2 net-V3: Memory-aware compressed multimodal deep neural networks for sustainable edge deployment. arXiv preprint arXiv:2405.12353.

Saba, D., & Hadidi, A. (2025). Opportunities, challenges, and future directions of digital innovations for renewable energy and conservation. In *Digital innovations for renewable energy and conservation* (pp. 393–424). Springer.

Sadriwala, K. F., Shannaq, B., & Sadriwala, M. F. (2024). GCC cross-national comparative study on environmental, social, and governance (ESG) metrics performance and its direct implications for economic development outcomes. In *The AI revolution: Driving business innovation and research: Volume 2* (pp. 429–441). Springer.

Sanfilippo, A., Benito Bermudez, V., & Vermeersch, M. (n.d.). Energy transition strategies in the Arabian gulf.

Saqib, N., Mahmood, H., Murshed, M., Duran, I. A., & Douissa, I. B. (2023). Harnessing digital solutions for sustainable development: A quantile-based framework for designing an SDG framework for green transition. *Environmental Science and Pollution Research, 30*(51), 110851–110868.

Sargiotis, D. (2024). Smart infrastructure for sustainable energy consumption: Leveraging AI and big data.

Sarwar, S. (2022). Impact of energy intensity, green economy and blue economy to achieve sustainable economic growth in GCC countries: Does Saudi vision 2030 matters to GCC countries. *Renewable Energy, 191*, 30–46.

Shehri, T. A., Braun, J. F., Howarth, N., Lanza, A., & Luomi, M. (2023). Saudi Arabia's climate change policy and the circular carbon economy approach. *Climate Policy, 23*(2), 151–167.

Sherif, M., Liaqat, M. U., Baig, F., & Al-Rashed, M. (2023). Water resources availability, sustainability and challenges in the GCC countries: An overview. *Heliyon, 9*, e20543.

Shwedeh, F., Salloum, S. A., Aburayya, A., Fatin, B., Elbadawi, M. A., Al Ghurabli, Z., & Al Dabbagh, T. (2024). AI adoption and educational sustainability in higher education in the UAE. In *Artificial intelligence in education: The power and dangers of ChatGPT in the classroom* (pp. 201–229). Springer.

Singh, S., Singh, G., Singh, S., & Misra, S. C. (2024). Understanding green procurement dynamics: An assessment framework for public sector organizations. *Journal of Environmental Management, 351*, 119756.

Sulthan, N., AlGahtani, F., Al-Surf, M., & Navas, S. (2023). Physician entrepreneurs and AI technology: An in-depth study of knowledge, competence, adoption, and sustainability in the GCC region. *Emirati Journal of Business, Economics, & Social Studies, 2*(2), 28–38.

Timm, J. F. G., Maciel, V. G., & Passuello, A. (2023). Green public procurement model for environmental assessment of constructive systems. *International Journal of Construction Management, 23*(4), 718–728.

Zaidan, E. (2023). From planning cities to sustaining communities: Smart and sustainable urban strategies for the post-carbon transition in the Gulf states. In *Sustainable Qatar* (p. 283). Springer.

Zhang, Y., Li, P., Hong, J., Li, J., Zhang, Y., Zheng, W., & Chen, T. (2024). Revisiting zeroth-order optimization for memory-efficient LLM fine-tuning: A benchmark. arXiv preprint arXiv:2402.11592.

Zhou, W., Zhuang, Y., & Chen, Y. (2024). How does artificial intelligence affect pollutant emissions by improving energy efficiency and developing green technology. *Energy Economics, 131*, 107355.

Chapter 11
FinTech Landscape and Sustainable Economy in the GCC Region

Faten Ben Bouheni

Abstract This chapter examines the promising transition toward sustainability facilitated by financial technology (FinTech), which serves to advance the development of a sustainable economy. A sustainable economy emphasizes the importance of societal initiatives and the utilization of technological advancements to promote environmentally responsible economic growth. Green finance plays a pivotal role by deploying innovative FinTech solutions that prioritize environmental, social, and governance (ESG) responsibilities.

The chapter highlights the proactive measures taken by the member countries of the Gulf Cooperation Council (GCC) to reform policies and implement regulations for building a more sustainable economy. These nations are leading the way in integrating technological solutions and promoting the growth of non-oil sectors. They recognize the potential benefits of these innovations in enhancing operational efficiency and fostering green growth. This transition presents a valuable opportunity for oil-dependent economies to innovate and evolve, paving the way toward a more sustainable and prosperous future.

Keywords Fintech · Financial inclusion · Green finance · Sustainable economy · ESG · Regulation · Policy

F. Ben Bouheni (✉)
Menlo College, Atherton, CA, USA
e-mail: faten.benbouheni@menlo.edu

H. Miniaoui and M. Bellalah (eds.), *Green Growth Opportunities and Sustainable Finance in the Gulf Cooperation Council Region*, Gulf Studies 24,
https://doi.org/10.1007/978-981-96-8287-4_11

11.1 Introduction

Green finance has become a crucial focus in recent years due to growing concerns about sustainability, climate change, and the urgent need to achieve the United Nations' Sustainable Development Goals. In this light, the concept of green finance has emerged as a significant topic of discussion (Sachs et al., 2019).

In 2024, global energy investment is expected to surpass USD 3 trillion for the first time, with USD 2 trillion allocated to clean energy technologies and infrastructure. Investment in clean energy has been accelerating since 2020, and spending on renewable power, grids, and storage now exceeds total spending on oil, gas, and coal (International Energy Agency, 2024).

Nonetheless, technological advancements are becoming pivotal in supporting the Sustainable Development Goals and facilitating the growth of green finance. These innovations not only optimize resource efficiency but also promote the integration of environmentally sustainable practices across various sectors (e.g. the growth of the electric vehicles market; the rise of FinTech solutions, etc.). As a result, economists perceive opportunities, including the development of renewable energy sources, the adoption of sustainable business practices, and the creation of eco-friendly infrastructure projects.

However, these opportunities come with considerable challenges. For instance, economies contend with the initial costs associated with the implementation of new technologies, the demand for skilled labor to manage these innovations, and the potential economic disparities from varying adaptation rates among regions. The effective balance of immediate economic needs with long-term sustainability goals requires meticulous planning and collaborative efforts among governments, businesses, and communities.

Recent academic studies are focused on enhancing the understanding of the interconnections and synergies between financial technology (FinTech) and sustainable economic development.

In fact, the digital economy and FinTech are transforming the financial sector by introducing new cultures, innovative business models, and practices that disrupt traditional financial services. This phenomenon is highly relevant not only for companies in the financial sector but also for external stakeholders such as investors or researchers (Firmansyah et al., 2023; Siddiqui & Rivera, 2022; Suryono et al., 2020; Trotta et al., 2024).

In investigating the sustainable economy and the transition to greener solutions by leveraging technological advancements, this chapter aims to address the following questions:

1. How are technological solutions fostering financial inclusion?
2. What major initiatives have been undertaken by GCC member countries to ensure a green transition while promoting sustainable finance growth?
3. How does FinTech promote environmental, social, and governance (ESG) principles?

4. What are the global and local policies and reforms necessary for creating an inclusive business environment in the FinTech sector?
5. What are the major areas for development in the GCC region?

The chapter discusses the member countries of the Gulf Cooperation Council (GCC). These nations have experienced significant growth in FinTech and sustainable investment strategies, which present both opportunities and challenges.

FinTech and sustainability have become increasingly relevant in today's fast-paced and interconnected world. As the financial industry continues to embrace technological advancements, there is a growing emphasis on ensuring that these innovations are aligned with sustainable practices. This includes considerations of environmental impact, social responsibility, and long-term economic stability. The intersection of FinTech and sustainability presents both challenges and opportunities for businesses, regulators, and consumers as they navigate toward a more sustainable and inclusive financial ecosystem. In a recent study, Ben Bouheni et al. (2023) employ advanced econometric modeling to examine the relationship between the adoption of FinTech and the operating performance of Islamic banks in the Middle East and find a significant impact of technology on Islamic banks' profitability and growth. The study attributes the financial growth of Islamic banks to technology implementation and internet coverage.

The transition away from oil and gas toward sustainable energy is a major focus for countries in the GCC. Consequently, the Environmental, Social and Governance (ESG) agenda is becoming the core topic in the Middle East and a competitive advantage. In a Bloomberg report (2021)[1] of expert insights, the conclusion is that "*All panel members agreed that ESG has quickly moved from the periphery of company and shareholder concern to center stage and that the time to disclose data is now. The speakers also agreed that strong action around ESG is likely to lower risk and bring returns, particularly in the medium- to long-term.*"

The growth of awareness of sustainable practices and responsible business strategies lead the future of economies in the region. In this chapter, we provide an overview on the finance landscape in the GCC and investigate driving forces and challenges behind the growth of green finance.

This chapter includes five sections. Section 11.2 discusses the relationship between FinTech and financial inclusion, as well as challenges and opportunities. Section 11.3 presents the linkages between FinTech and each component of the Environmental, Social and Governance (ESG) agenda. Section 11.4 tracks the progress in regulating the FinTech and technology markets. Section 11.5 discusses the green finance initiatives to support sustainable economies in the GCC region. And Sect. 11.6 discuses areas for development in supporting the green transition in the GCC region.

---

[1] The Bloomberg report is available at https://www.bloomberg.com/professional/blog/expert-insights-into-data-and-esg-in-the-middle-east/.

11.2 Financial Technology and Financial Inclusion

In today's interconnected world, characterized by widespread internet access and a tech-savvy population, the influence of financial technology, or FinTech, continues to reshape traditional banking systems. By providing innovative and cost-effective solutions, FinTech firms are revolutionizing the way consumers manage their finances. These firms are introducing new methods of conducting transactions and digitally managing money. The global presence of FinTechs is on the rise, with significant growth observed across all regions. For instance, in Europe, the Middle East, and Africa (EMEA), the number of FinTech firms surged from 3581 in 2018 to 9681 in 2023. Similarly, in the Americas, encompassing North America, South America, Central America, and the Caribbean, the number of FinTech firms experienced substantial growth, increasing from 5686 in 2018 to 11,651 in 2023, as depicted in Fig. 11.1. This growth underscores the appeal of the FinTech industry and the potential it holds for promoting financial inclusion.

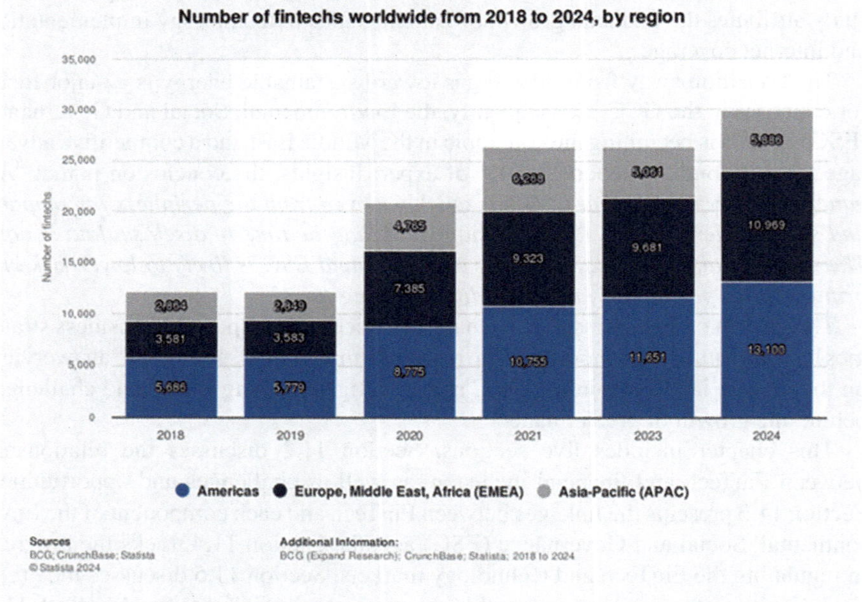

Fig. 11.1 Number of FinTech firms worldwide from 2018 to 2024 by region. Source: Statista (2024a). Retrieved from https://www.statista.com/statistics/893954/number-fintech-startups-by-region/

11.2.1 Financial Inclusion Challenges

Despite significant global initiatives to promote financial technology literacy, there are still notable disparities in financial inclusion. According to a comprehensive summary of the status of financial inclusion in Africa by Demirgüç-Kunt and Klapper (2012), based on Global Findex and survey data, many individuals and firms continue to be excluded from accessing financial services. The main hurdles identified include the cost, distance, and documentation requirements. Similarly, an analysis of financial inclusion in China by Fungáčová and Weill (2015) revealed that higher income, better education, being male, and older age are correlated with increased use of formal accounts and formal credit. These researchers attribute the influence of income and education to financial inclusion in China.

A study of populations in the Middle East and North Africa (MENA) by Lyons and Kass-Hanna (2021) suggests that economically vulnerable populations are less likely to benefit from financial inclusion. Their research indicates that households in MENA countries with higher levels of financial literacy are more likely to engage in positive savings behaviors and less likely to resort to borrowing, especially from informal sources.

In recent research on the Gulf Cooperation Council (GCC) economies, there has been a focus on promoting more inclusive finance. Niankara (2023) conducted a study on the impact of financial inclusion on digital payment solutions in GCC countries using the Global Findex surveys. The study suggests that implementing complementary financial inclusion policies can help improve the use of digital payment solutions, leading to more effective financial inclusion and equality in the region.

In another study, Uddin et al. (2023) analyzed the determinants of financial inclusion in GCC countries and found that between 2011 and 2017, there was a 16% increase in the number of individuals reporting having a formal financial institution account. However, the study also revealed that formal savings and borrowings were lower in GCC countries compared to the average of high-income countries. Additionally, the research found that financial inclusion was higher for males, older individuals, more educated individuals, and those with higher incomes, while women and younger people were less likely to benefit from financial inclusion in GCC countries.

The level of financial inclusion is significantly influenced by gender, as demonstrated by the statistics in Saudi Arabia as of January 2024. While 74% of the total population has a bank account, there is a disparity between male and female account holders, with 82% being male and 63% female. Additionally, the use of mobile payments for sending money or paying bills is predominantly male-driven. When it comes to sending money via mobile, 64% of users are male and 47% are female, and for paying bills, 71% of users are male compared to 49% female (see Fig. 11.2).

In the GCC countries, the widespread adoption of mobile technology has the potential to advance financial inclusion and foster equality. In 2016, the average mobile usage across the GCC countries was 42%, with Kuwait leading at 47% and Qatar trailing at 37%. Furthermore, the distribution of mobile technology by type indicates a shift toward faster and more reliable mobile technology, particularly

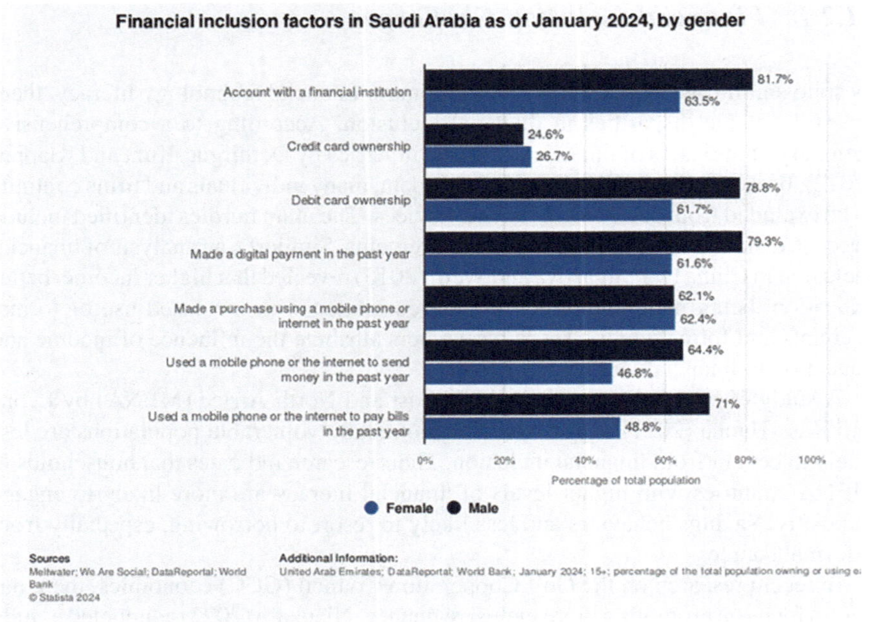

Fig. 11.2 Financial inclusion factors in Saudi Arabia as of January 2024 by gender. Source: Statista (2024b). Retrieved from https://www.statista.com/statistics/1394148/saudi-arabia-distribution-of-financial-inclusion-factors/

the 5G connection. In 2019, 5G accounted for only 1% of mobile connections, but this is projected to increase to 21% by 2025 in the GCC countries. Meanwhile, 4G remains the most prevalent connection for mobile users, standing at 51% in 2019 and anticipated to reach 70% in 2025 (see Fig. 11.3).

11.2.2 FinTech Ecosystem

In the context of financial technology and financial inclusion, the GCC countries are making ongoing efforts to integrate financial technology and promote financial inclusion. In 2015, PwC reported a positive outlook for the FinTech ecosystem in the GCC, emphasizing the involvement of governments, financial institutions, and entrepreneurs, and expressing optimism about the future of FinTech in the region.

In the United States, the FinTech ecosystem is particularly conducive for young entrepreneurs, with New York leading as the fastest-growing FinTech landscape in the nation. The government provides support through tax credits for business research and development (R&D), as well as patents, and has established incubators and accelerators. In terms of investment, the United States accounted for 83% of global FinTech investments in 2013, with over US $485 billion spent on innovating financial

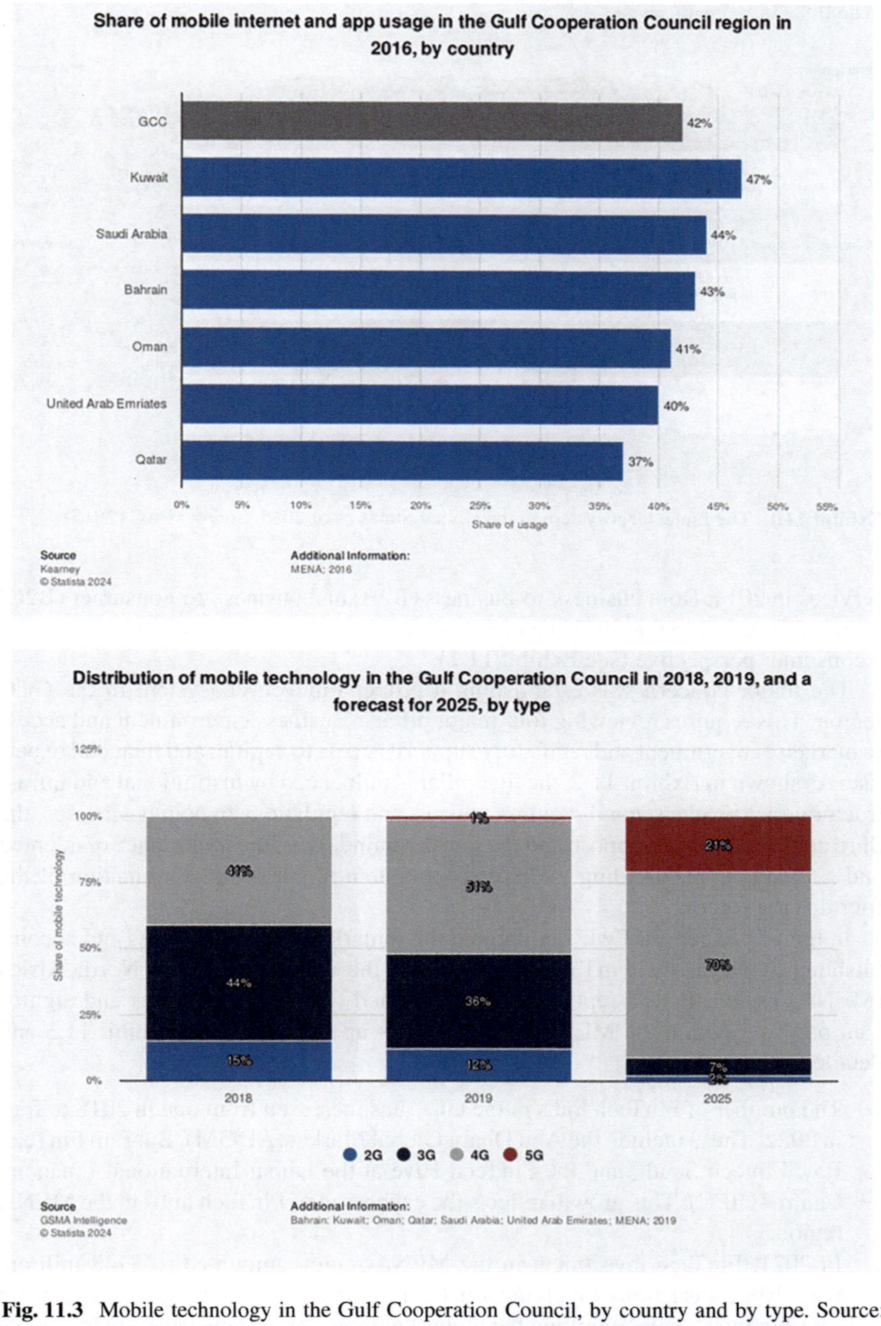

Fig. 11.3 Mobile technology in the Gulf Cooperation Council, by country and by type. Source: Statista (2024c). Retrieved at https://www.statista.com/statistics/669529/gcc-mobil-internet-and-app-usage-by-country/ by country and at https://www.statista.com/statistics/1017359/gcc-mobile-internet-technology/ by type

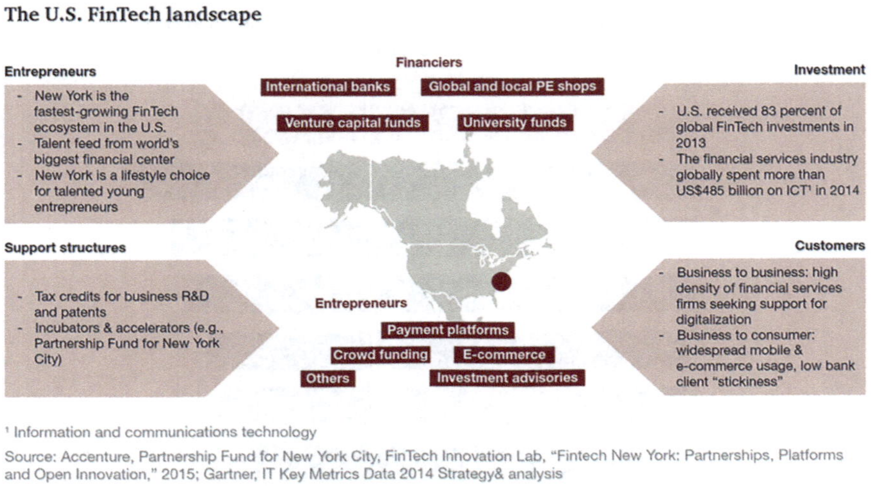

Exhibit 11.1 The Fintech ecosystem in the United States as of 2015. Source: PwC (2015)

services in 2014. Both business-to-business (B2B) and business-to-consumer (B2C) segments are embracing digitalization and seeking more innovative solutions from a consumer perspective (see Exhibit 11.1).

The major concern was establishing a proper FinTech ecosystem in the GCC region. This requires reviewing four major pillars: Business environment and access to markets; government and regulatory support; access to capital; and financial expertise. As shown in Exhibit 11.2, the first pillar is influenced by institutional and infrastructural factors, the second requires policies and regulations to help businesses, the third calls for funding support, and the fourth reminds us of the importance of talented and skilled people. Building a FinTech ecosystem requires the combination of the four driving forces.

In their 2022 report, PwC highlighted the remarkable advancements and accomplishments within the FinTech ecosystem in the Middle East and North Africa (MENA) region. PwC's report for 2022 outlined four key milestones and significant progress within the MENA FinTech landscape, as shown in Exhibit 11.3 and detailed below.

1. The number of FinTech hubs in the GCC has increased from one in 2018 to four in 2022. These include the Abu Dhabi Global Market (ADGM), Bahrain FinTech Bay, Fintech Saudi, and the FinTech Hive at the Dubai International Financial Centre (DIFC). This growth reflects the expansion of FinTech hubs in the MENA region.
2. In 2021, FinTech investment in the MENA region amounted to $448 million, with 108 transactions involving FinTech companies. Additionally, there were four FinTech "exits," marking the highest number for a single year. FinTech exits

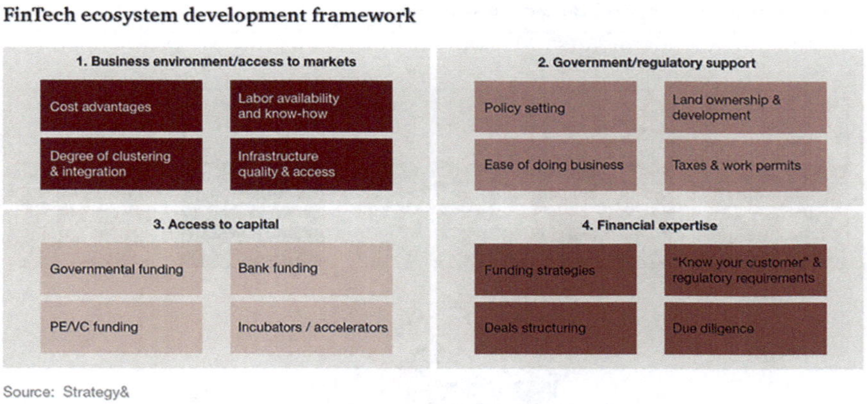

Exhibit 11.2 FinTech ecosystem development framework. Source: PwC (2015)

encompass IPOs, special purpose acquisition companies (SPACs), mergers, and acquisitions.

3. Furthermore, the region saw its first FinTech unicorn in August 2020, when Egypt's *Fawry*, a digital transformation and e-payment platform, achieved a valuation exceeding \$1 billion. Subsequently, other notable companies such as *stc pay* have also attained "unicorn" status, with many more in the region rapidly approaching this milestone.
4. Notable developments have emerged in Saudi Arabia, driven by the recent implementation of the National Fintech Strategy 2030. This strategy aims to establish the country as one of the top 10 global FinTech hubs, with 500 active FinTech companies by 2030.

The rapid growth of the FinTech industry in the Middle East is driven by various factors, with supportive regulation playing a key role. Countries such as Bahrain, Qatar, Saudi Arabia, and the UAE have each formulated comprehensive national FinTech strategies and established government-backed accelerators and incubators to foster the growth of the financial technology industry.

11.3 Initiatives on Green Finance to Support a Sustainable Economy in the GCC Region

The financial sector contributes significantly to the GDP of GCC countries, where asset quality is strong, banking sectors remain well-capitalized and liquid, and financial markets perform well amid geopolitical turbulence (IMF, 2024). Meanwhile, the GCC countries embrace technological changes and innovation, and prioritize a sustainable economy, in globally competitive and interconnected markets.

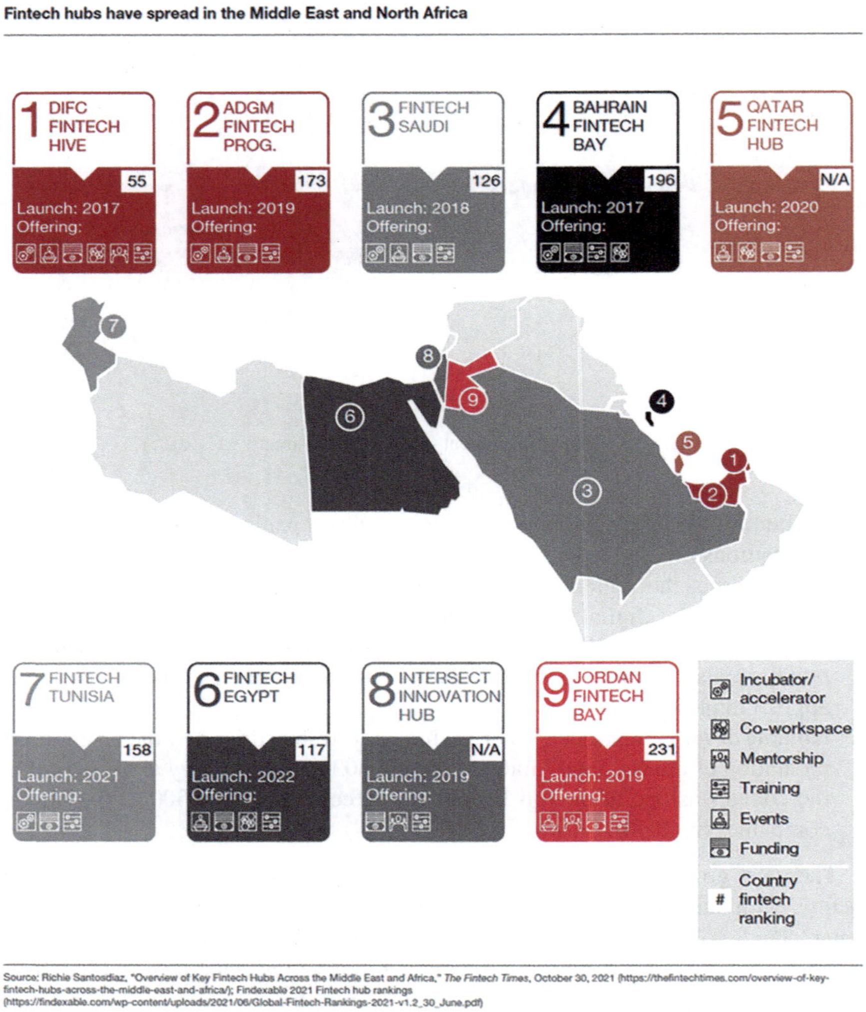

Exhibit 11.3 FinTech hubs in the Middle East and North Africa [MENA] as of 2022. Source: PwC (2022)

In fact, green finance has become a prominent issue in recent years, offering new perspectives to support sustainable investments and economic development. For instance, in 2010, the Green Climate Fund (GCF) was established by the United Nations Framework Convention on Climate Change (UNFCCC), which involves

194 countries. In response to climate change, the GCF aims to support societal transformations and finance green projects.[2]

The GCC region sees opportunities to boost sustainable finance and continues its reform efforts to support long-term visions. The Gulf Cooperation Council (GCC) countries have embarked on several key initiatives aimed at fostering economic growth and sustainability. These initiatives include:

1. Building a Culture of Sustainable Business: The GCC nations are pursuing a vision of cultivating a business environment that prioritizes sustainability. This involves encouraging companies to adopt environmentally friendly practices, invest in renewable energy, and support sustainable supply chains. The representation of the Gulf Cooperation Council (GCC) countries at the United Nations Climate Change Conferences is noteworthy. In December 2023, the United Arab Emirates hosted the 28th Conference of the Parties (COP28) on Climate Change in Dubai. At this conference, representatives from 198 governments reached a significant agreement to expedite the transition from fossil fuels to renewable energy sources. Furthermore, the Arab Coordination Group (ACG) announced a commitment of $10 billion to support this green initiative through 2030.

   In the context of the building sector within the Kingdom of Saudi Arabia and the United Arab Emirates, a study conducted by Asif (2016) documents the increasing trend of sustainable building practices in the GCC region. This research underscores the policies and strategies aimed at promoting sustainable real estate markets.
2. Supporting Islamic Green Finance: The GCC nations are also working to develop Islamic finance models that align with green principles. This includes offering financial products that fund environmentally sustainable projects while adhering to Shariah law, thereby attracting investments that support both economic growth and sustainable development. Islamic green finance promotes global financial stability by integrating Islamic finance principles with sustainable and ecologically friendly projects and encouraging responsible investments that support sustainability and inclusion. The core principles of Islamic finance highlight transparency, good corporate governance, risk and profit sharing, and capital allocation toward socially and environmentally responsible investments (Sidaoui et al., 2022).

> *There has been a growing consensus that we need to urgently redefine the profit motive. Profits may no longer be the sole objective and more attention should be given to sustainability and long-term social impact. In this context, it seems apt that Islamic finance contributes significantly toward this redefinition. The Shariah, which is the foundation of Islamic finance, shares common principles and values on the preservation of the environment and society. With these similarities, they provide tremendous opportunity*

[2] According to the United Nations (2010), "*It allocates its resources to low-emission and climate-resilient projects and programmes in developing countries.*"

for Islamic finance to capitalize on the strong growth of green finance and investment segments across the globe." (World Bank Group, 2019)[3]

3. Leading a NetZero Agenda: The GCC countries are committed to achieving zero netemissions. This initiative involves implementing policies aimed at reducing carbon footprints, investing in clean energy technologies, and promoting energy efficiency across various sectors. For example, the United Arab Emirates (UAE) and Oman are committed to NetZero by 2050, the Kingdom of Saudi Arabia (KSA) is committed to NetZero by 2060, and the KSA is committed to leading sustainable efforts through the Middle East Green Initiative (MGI). The Nationally Determined Contributions (NDCs) of the United Arab Emirates estimate that approximately $36 billion will be required for investments from 2023 to 2030 to achieve climate goals. These goals include reducing greenhouse gas emissions and increasing the use of renewable energy sources.[4]

   On a global scale, the International Energy Agency reported that investments in renewable energy reached record highs of $1.7 trillion in 2022 and $1.8 trillion in 2023, a significant increase from $1.1 trillion in 2015.[5] This steady rise underscores the crucial role of the private sector in facilitating the transition to sustainable energy and illustrates a global commitment to tackling climate change.
4. Attracting Foreign Investments: The Gulf Cooperation Council (GCC) countries rely heavily on the oil and gas industries, making them particularly vulnerable to fluctuations in commodity prices. In response to this challenge, these nations have recognized the necessity for economic diversification and are increasingly focused on attracting foreign direct investment (FDI) to help achieve this goal. FDIs enable the GCC countries to maintain a robust growth of the non-oil sector. According to the World Bank Group (2024b), the economic growth in the GCC region *continues to be driven by the non-oil sector which has shown robust growth of 3.7%, mainly driven by the ongoing diversification efforts and ambitious reforms throughout the region.*[6]

   In a recent report examining investment trends within the GCC region, Moody (2024) provided a comprehensive analysis highlighting several key conclusions. The report delves into various aspects of the GCC investment landscape, including economic growth drivers, sector-specific opportunities, and the impact of regional policies on investment flows. Moody's presents the following major conclusions: (1) GCC countries offer a favorable investment environment with

[3] World Bank Group (2019). https://documents.worldbank.org/en/publication/documents-reports/documentdetail/591721554824346344/islamic-green-finance-development-ecosystem-and-prospects.

[4] World Bank Group (2024a). https://blogs.worldbank.org/en/arabvoices/embracing-climate-opportunities-greener-gcc.

[5] International Energy Agency (2024). https://www.iea.org/reports/world-energy-investment-2024/overview-and-key-findings.

[6] World Bank Group (2024b). https://www.worldbank.org/en/news/press-release/2024/12/01/non-oil-sectors-drive-robust-growth-in-gcc-countries.

strong business conditions and a positive macroeconomic outlook, prioritizing foreign direct investment (FDI) through incentives and diversification efforts; (2) In 2023, Saudi Arabia led in capital investment project (CIP) inflows, particularly in green technology, while cross-border investments diversified, especially in the manufacturing sector. Significant investments were also made in IT infrastructure due to policies requiring regional headquarters for government contracts; (3) North America was a key investor in the GCC, and the region invested in Asia, though investments from Europe and Asia were low due to economic challenges; (4) Despite international risks, GCC countries are positioned to attract and invest in global markets, while the region should continue the reform efforts.

5. Reforming Policies: The GCC countries are reviewing and revising existing regulations to ensure they align with modern economic realities and global best practices, thereby enhancing their competitiveness and adaptability. These initiatives reflect the GCC's commitment to a sustainable and inclusive future while positioning the region for long-term economic success.

With the increasing global awareness of the need for more sustainable financial systems, PwC (2025a) has recently proposed a common sustainable finance framework for the GCC region. By unifying their economic and monetary systems, the region has the potential to enhance economic resilience and ensure a sustainable future. The PwC reports that *in the GCC region, there have been various efforts to establish sustainable finance frameworks at individual country level—for example, the UAE Sustainable Finance Framework 2021–2031, that aims to boost the demand and supply of sustainable finance and strengthen the climate and green investments' ecosystem through robust collaboration among stakeholders. The Public Investment Fund's Green Finance Framework in the Kingdom of Saudi Arabia (KSA) also aims to promote sustainable and environmentally friendly projects... the implementation of a common sustainable finance framework within the GCC can bolster economic growth, empower businesses, establish robust sustainability credentials and facilitate tailored financing solutions, aligned with the region's transition needs.*

In 2023, the World Bank director for the GCC region highlighted a series of legal reforms implemented by the Gulf Cooperation Council (GCC) countries to enhance economic resilience. For instance, between 2019 and 2020, Saudi Arabia introduced a new set of legal reforms, including the civil status law, labor law, enforcement law, and laws regarding access to finance. As a result of these reforms, women's labor force participation increased from 22% in 2019 to nearly 37% in 2023, surpassing the 30% target set by Saudi Vision 2030.

Similarly, the United Arab Emirates (UAE) has eliminated legal barriers to women's employment, moving toward a gender-neutral ecosystem and workplace. The UAE has created numerous social, educational, and economic opportunities for women and boasts one of the highest women's labor force participation rates in the MENA region at 55% compared to the regional average of 19%.

Bahrain has also made significant strides to facilitate women's involvement in economic activities. Additionally, Kuwait passed laws in 2020 and 2021 aimed at combating domestic violence and prohibiting gender-based discrimination and sexual harassment in the workplace.

From a fiscal standpoint, research indicates the critical importance of fiscal stability in facilitating a successful transition to greener economies within the GCC region (Alnafrah & Mouselli, 2024). In addition, fiscal stability enhances investor confidence, thereby attracting both domestic and international investments in green technologies (Shmelev & Speck, 2018).

The green transition from oil-based economies to non-oil alternatives within the GCC region is gaining significant momentum, presenting numerous opportunities for sustainable growth and development. This transformation is not merely a response to global environmental concerns, but a strategic initiative driven by the GCC countries' commitment to diversifying their economies.

In recent years, these nations have taken proactive steps to create a conducive environment for both local and foreign investors. This includes implementing favorable policies, enhancing infrastructure, and fostering innovation to attract investments in renewable energy technologies, green manufacturing, and sustainable agriculture.

Looking ahead to 2025, the GCC is poised to navigate the challenges posed by declining oil revenues, which are a pressing concern for oil-dependent economies worldwide. While diminished oil income could impact fiscal stability, the anticipated surge in revenues from non-oil sectors—such as tourism, finance, real estate, and technology—offers a vital counterbalance.[7]

These emerging sectors are not only projected to help offset potential deficits but also contribute to job creation, economic resilience, and a sustainable future for the region. By embracing this diversification strategy, the GCC aims to build a more robust and sustainable economic landscape that is less vulnerable to fluctuations in global oil prices.

11.4 FinTech and ESG

Amid the COVID-19 pandemic and escalating geopolitical tensions in key oil-producing regions, countries around the world are placing an increasing emphasis on fostering sustainable, environmentally friendly economic growth. This shift is being facilitated through the strategic application of technology aimed at creating new opportunities for sustainable development. Economist Paul Romer's influential work has underscored the pivotal role of technology as an intrinsic driver of long-term, sustainable economic growth.

[7] According to PwC (2025b), in the GCC region, *although oil revenues are anticipated to decline, non-oil revenues are set to benefit from continued growth in the non-oil sector, projected to expand by 4.2% in 2025, up from 1.8% in 2024, driven by sectors such as tourism, trade, real estate and manufacturing.*

Notably, the Gulf region, which has historically been a significant player in the production and importation of non-renewable oil energy, is now actively transitioning toward embracing renewable energy sources and solutions. This shift reflects the region's dedicated commitment to advancing sustainability and promoting environmentally conscious practices.

11.4.1 How Does FinTech Support the Environment?

Recent advancements in technology have paved the way for the development of innovative digital solutions that not only meet our needs but also have a positive impact on the environment. One notable example is the increasing use of digital documents and bills across various sectors such as business, education, and daily life. This shift toward digitalization not only offers convenience but also significantly reduces the environmental impact associated with the production and disposal of tangible paper products.

Changing consumer behaviors, particularly driven by the e-commerce revolution, have created a growing demand for sustainable products. This shift presents increasing opportunities for businesses to engage in the production and trading of environmentally friendly and sustainable products, meeting the evolving consumer preferences.

Empirical research has shed light on the positive environmental implications of financial technology. In a recent study by Udeagha and Muchapondwa (2023) on the BRICS nations, it was concluded that FinTech plays a crucial role in enhancing environmental quality in both the short and long run. This highlights the potential of FinTech to contribute to environmentally sustainable practices and initiatives.

Recent studies have highlighted the potential of integrating financial technologies to address environmental challenges and promote sustainability. For example, the International Institute for Sustainable Development (IISD) released an article in 2018 which showcased the successful implementation of FinTech solutions in water science and management projects. These solutions were sourced from a diverse range of local and international organizations, signaling a promising trend toward leveraging financial technologies for environmental sustainability.

11.4.2 How Does FinTech Promote Social Equity?

The emergence and widespread adoption of financial technology (FinTech) solutions have played a pivotal role in advancing the concept of financial inclusion. Financial inclusion refers to the extent to which individuals and communities have access to and can affordably use essential financial services in daily life. These services encompass a broad spectrum, including but not limited to opening and maintaining bank accounts, engaging in deposit and loan, transferring funds, and conducting various financial transactions, whether through digital channels or traditional in-person methods.

It is worth noting that the global support for more financial inclusion has proven instrumental in addressing the significant disparities that exist between individuals who benefit from comprehensive access to financial services and those who are excluded from such opportunities. This progress has been achieved despite the persistent presence of institutional, regulatory, and cultural barriers that have historically hindered extensively the financial inclusion process.

Bulger and Rouen (2022) assert that the rise of FinTech companies is partly fueled by the potential for FinTech to have a significant social impact. FinTech leaders pledge to enhance financial inclusion for those without access to traditional banking services, as well as to improve financial well-being and bolster digital security. Major FinTech companies such as PayPal, Mastercard, Visa, and Shopify are championing this pledge by framing their products and services as instruments for promoting financial inclusion and fostering fair economic expansion. Investors are also recognizing FinTech's potential for social impact, with the industry attracting around a quarter of all investment focused on creating positive societal change. This amounts to almost $250 billion in assets under management, making FinTech the top recipient of impact-oriented investment compared to any other industry (Bulger & Rouen, 2022).

11.4.3 How Is FinTech Linked to Governance?

The intersection of FinTech, the environment, and society has been the subject of extensive exploration in scholarly research and media coverage. Despite this, the connection between FinTech and corporate governance has received limited attention and is not well substantiated.

In a recent study, Almubarak and Aljughaiman (2024) delved into the relationship between governance and innovation in the banking sector of Saudi Arabia. Their findings suggest that governance may have an impact on both the quantity and quality of innovation. The authors specifically explored the link between the size of the board of directors, the proportion of independent directors on the board, and the provision of FinTech services. Their research revealed a significant negative correlation between board size and a bank's FinTech score. Additionally, they found that the presence of independent board members positively influences a bank's performance by facilitating the introduction of innovative FinTech services.

In another recent empirical study, Dicuonzo et al. (2024) investigated the relationship between the adoption of FinTech and ESG (Environmental, Social, and Governance) improvements in a sample of 180 banks based in Europe and the United States. Their research revealed that the presence of FinTech in the banking sector contributes to enhanced ESG performance. Additionally, they found that factors such as board independence and gender diversity play a supportive role in promoting ESG initiatives. Overall, their findings indicate that a higher level of FinTech adoption is linked to better ESG scores.

Wang et al. (2022) conducted a study using an emerging economy to examine how the development of financial technology affects the environmental, social, and governance (ESG) performance of Chinese listed companies. Their findings show that the easing of corporate financing constraints is one of the ways FinTech development influences corporate ESG performance. In summary, the paper demonstrates the positive impact of FinTech on corporate ESG performance.

11.5 Policy and Framework

Technology has no boundaries. Various new technologies such as higher-speed internet, smartphones, internet of things (IoT), big data, artificial intelligence (AI), generative AI, blockchain systems, and cybersecurity technologies are shaping the modern world.

In the financial industry, the term FinTech encompasses a wide range of operations, including payment transactions, robot advisors, digital currencies, deposit and loan products, insurance, regulatory compliance, and more. This has led to the development of new institutions and organizations, products, business models, policies, and regulations.

The ever-accelerating pace of technological progress necessitates the continuous review and adaptation of policies and regulations, not only on a local scale but also on a global level.

11.5.1 Global Regulatory Environment

In a summary of global regulations in May 2024, Bloomberg provided an overview of recent regulatory agreements. The United States and the United Kingdom signed a Memorandum of Understanding (MOU) on advanced AI (Artificial Intelligence) models. The MOU is intended to support the two countries in aligning their scientific approaches and to "accelerate and rapidly iterate robust suites of evaluations for AI models, systems, and agents."

Furthermore, the countries have plans to develop a common approach to AI safety testing and to share their existing capabilities to address safety concerns. The U.S. and U.K. AI Safety Institutes also plan to conduct at least one joint testing exercise on a publicly available model, while also exploring the exchange of personnel. The MOU agreement goes into effect immediately and reflects the urgency around AI safety expressed by governments worldwide.

In the near future, the focus will be on the following questions:

- What specific AI safety testing methods will the United States and U.K. develop as part of their common approach?
- Which publicly available model will be used for the joint testing exercise by the U.S. and U.K. AI Safety Institutes?

11.5.2 How Will the Exchange of Personnel Between the U.S. and U.K. AI Safety Institutes Be Facilitated and Regulated?

In May 2024, Bloomberg reported also that the Hong Kong Monetary Authority (HKMA) released a report on how financial institutions are using machine learning and other tools of regulation technology (RegTech) to improve their anti-money-laundering systems and enhance risk management. The report outlined the processes that firms use to design, implement, and optimize transaction monitoring systems. These processes include management oversight and governance, risk assessment of the monitoring system, selection of detection scenarios, identification of critical data elements, data quality testing, customer segmentation, threshold setting, functional testing, periodic review, and optimization using RegTech, including AI.

11.5.3 The GCC Regulatory Environment

The global impact of the COVID-19 pandemic has accelerated the widespread adoption of financial technology (FinTech) solutions. This rapid expansion has prompted regulatory authorities in various regions to take proactive measures to address and regulate the FinTech industry. The primary goal of these regulations is to ensure the stability, security, and integrity of the FinTech sector amidst its rapid growth.

CCAF (2021) conducted a comparison of surveys on how regulators view FinTech and their objectives in the Middle East and North Africa (MENA) and around the world. The study revealed that 85% of regulators in the MENA region consider FinTech to be supportive of their goals, compared to the global average of 61%. Additionally, in terms of promoting competition, 62% of regulators in MENA hold this view, in contrast to 47% globally.

The regulatory landscape in the region reflects an encouraging level of enthusiasm and support from regulators. One notable example is the Dubai International Financial Center (IFC) and the United Arab Emirates more generally, where there are five concrete regulatory initiatives in place focusing on payments, remittances, equity crowdfunding, e-money, and peer-to-peer (P2P) lending. Additionally, Abu Dhabi Global Market, Bahrain, Israel, and Saudi Arabia have each implemented four regulatory initiatives, demonstrating a strong commitment to the FinTech regulatory frameworks within their respective financial sectors. These initiatives illustrate a

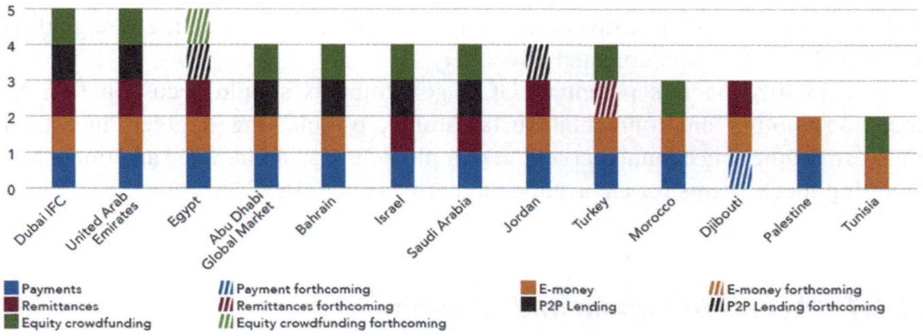

Fig. 11.4 Regulatory frameworks and initiatives in the MENA region as of 2021. Source: CCAF (2021)

proactive approach toward fostering innovation and growth within the financial and FinTech industries in the region (see Fig. 11.4).

Fintech Saudi, a prominent example of financial technology innovation in the Gulf Cooperation Council (GCC) countries, was established in April 2018 through a collaboration between the Saudi Central Bank and the Capital Market Authority. The organization serves as a pivotal entity driving the growth and advancement of the FinTech industry in Saudi Arabia. Leveraging its role as a catalyst, Fintech Saudi plays a vital part in shaping the policy environment and fostering the development of financial technology within the country.[8]

The regulatory framework has been established, along with payment laws, licensing policies, and a new commercial register that features FinTech activity. The number of licensed FinTech firms increased from 10 in 2018 to 82 in 2021. Additionally, several measures were undertaken, including the launch of the first Saudi FinTech accelerator program, the introduction of an instant payments system, and the development of a strategy for open banking. There was a 37% increase in the number of FinTech firms between 2020 and 2021 (PwC, 2022).

11.6 Areas for Development: GCC Region

The Gulf Cooperation Council (GCC) has seen significant investments in environmental sustainability from both governments and businesses, as highlighted in PwC (2022) analysis. However, the financial sector in the region has not fully embraced green finance, which integrates environmental considerations with financial returns. It is crucial to develop a robust framework for green finance in the GCC, as it has the potential to unlock substantial economic growth, creating over a million jobs and generating $2 trillion by 2030. Additionally, green finance can contribute to the

[8] Fintech Saudi, "About Fintech Saudi." Retrieved on June 4, 2024, from https://fintechsaudi.com/about/.

region's goals of economic diversification, job creation, and attracting foreign direct investment (FDI) if implemented effectively.

To capitalize on this potential, GCC governments should focus on four key areas: Promoting environmental sustainability, establishing a green investment entity, strengthening capital markets, and implementing standardized and transparent reporting mechanisms for environmental performance (PwC, 2022).

11.6.1 Access to Capital and Liquidity

Access to capital and liquidity is crucial for the success of the start-up ecosystem. Having readily available funds and access to liquidity enables start-ups to cover their initial expenses, invest in research and development, and navigate through early stages of growth. This financial support plays a pivotal role in fostering innovation, driving job creation, and fueling economic growth within the start-up community. The Middle East and North Africa (MENA) region shows great potential for growth compared to other emerging markets.

According to McKensey (2023), in Southeast Asia and Latin America, venture capital (VC) investments as a share of GDP are at 0.8% and 0.4%, respectively. In comparison, MENA and Pakistan are at 0.1%, indicating a potential for significant growth in the coming years. The region has also demonstrated relative resilience amidst the recent global slowdown in venture activity. In the fourth quarter of 2022, MENAP funding remained at nine times the levels of the first quarter of 2017 (peaking in the first quarter of 2022 at 13 times), while global VC funding slowed to roughly twice the levels of the first quarter of 2017, down from a peak of 5.7 times in the fourth quarter of 2021.

Despite this recent momentum, many VC investors are still cautious in the region. Investors tend to focus on the same top ten start-ups, which account for 35–55% of funding each year, as opposed to around 10% in the United States. Additionally, the same eight start-ups have made multiple appearances in the top ten over the past 5 years (McKensey, 2023).

In recent years, the countries in the GCC region have been increasingly emphasizing the importance of fostering a culture of innovation and entrepreneurship. This shift reflects a growing recognition of the role that innovative thinking and entrepreneurial endeavors can play in driving economic growth and diversification in the region. The growing need for venture capital funds and investors can be attributed to the escalating desire of entrepreneurs to expand and grow their innovative ideas into successful ventures. Consequently, strengthening capital markets and continuing the efforts to attract venture capitalists constitute a fundamental area for development in the region.

11.6.2 Entrepreneurship and Diversity

In comparing entrepreneurial behaviors and attitudes based on five indicators in the United Arab Emirates (UAE), and globally, the 2022 Global Entrepreneurship Monitor (GEM) scorings show higher indexes for the UAE than the global average in terms of:

1. Perceived opportunities: 72.16 for the UAE versus 55.30 global average.
2. Perceived capabilities: 71.66 for the UAE versus 59.09 global average.
3. Entrepreneurial intention: 44.46 for the UAE versus 22.33 global average.
4. Total early-stage entrepreneurship activity: 25.52 for the UAE versus 12.94 global average.
5. Entrepreneurship as a good career choice: 82.80 for the UAE versus 67.36 global average.

However, in 2022, when comparing the established business ownership rate, the global average rate was 7.00, while the rate in the United Arab Emirates was 4.53. This indicates that the business ownership rate in the UAE was lower than the global average in the specified year.

Moreover, the worldwide average rate of female and male entrepreneurship combined was 0.74, with the United Arab Emirates (UAE) recording a rate of 0.71. When focusing on the impact and business services sector, the global average stood at 21.08, while the UAE's rate was slightly lower at 19.50.

11.6.3 Shortage in Specialized Skills

In the GCC FinTech ecosystem, a significant challenge lies in ensuring the availability of talented individuals with the requisite skill sets. This encompasses the need for professionals proficient in areas such as technology, finance, and regulatory compliance, which are vital for the growth and sustainability of the sector.

According to a recent market study conducted by PwC in 2022, a significant percentage of employers in the Gulf Cooperation Council (GCC) countries expressed concerns about the shortage of individuals with specialized skills. Specifically, 75% of the respondents from Kuwait, 60% from Qatar, 58% from Saudi Arabia, and 46% from the UAE believed that their respective countries were facing a scarcity of individuals with specialized skills.

Furthermore, the study highlighted that the GCC respondents showed more confidence than the global survey average in their employers' commitment to prioritizing skills improvement. This suggests that organizations in the GCC region are placing a higher emphasis on enhancing the skill sets of their workforce compared to the global average.

11.7 Conclusion

In recent years, Gulf Cooperation Council (GCC) countries have increasingly recognized the importance of economic diversification in addressing various challenges. They have begun investing in renewable energy projects and other sustainable initiatives, such as Saudi Arabia's Vision 2030 plan, which includes the Saudi Green Initiative, and the UAE Energy Strategy 2050 for renewable energy. These efforts demonstrate a growing awareness among GCC leaders of the necessity of preparing their economies for a sustainable future beyond oil dependency.

The non-oil sectors in the region are experiencing steady growth each year, providing a promising opportunity to enhance overall economic resilience. As these sectors continue to develop, they are expected to effectively offset any deficits in oil revenue, contributing to a more balanced and sustainable economy.

The FinTech ecosystem in the GCC countries stands out as one of the most dynamic sectors, closely aligned with regulatory objectives and the vision of regional leaders. This vibrant ecosystem offers a wealth of opportunities while also presenting certain challenges, laying a solid groundwork for both growth and enhancement. The swift evolution of FinTech has encouraged regulatory authorities across different regions to take proactive measures to effectively oversee and guide the industry. The primary aim of these regulations is to foster stability, security, and integrity within the rapidly expanding FinTech sector, ensuring its sustainable development moving forward.

By prioritizing policy reforms, promoting fiscal responsibility, implementing tax adjustments, fostering advancements in artificial intelligence (AI), enhancing FinTech, and facilitating urban development, the GCC is strategically positioned to embrace and effectively respond to significant economic transformations by 2025.

References

Almubarak, A. I., & Aljughaiman, A. A. (2024). Corporate governance and FinTech innovation: Evidence from Saudi banks. *Journal of Risk and Financial Management, 17*(2), 48. https://doi.org/10.3390/jrfm17020048

Alnafrah, I., & Mouselli, S. (2024). From petrodollars to green power: Fiscal stability and green transition in GCC economies. *Journal of Environmental Management, 370*, 122784. https://doi.org/10.1016/j.jenvman.2024.122784

Asif, M. (2016). Growth and sustainability trends in the buildings sector in the GCC region with particular reference to the KSA and UAE. *Renewable and Sustainable Energy Reviews, 55*, 1267–1273. https://doi.org/10.1016/j.rser.2015.05.042

Ben Bouheni, F., Tewari, M., Sidaoui, M., & Hasnaoui, A. (2023). An econometric understanding of fintech and operating performance. *Review of Accounting and Finance, 22*(3), 329–352. https://doi.org/10.1108/raf-10-2022-0290

Bloomberg. (2021). *Expert insights into data and ESG in the Middle East*. Retrieved on May 28, 2024, from https://spotlight.bloomberg.com/story/implementing-successful-data-strategies/page/6

Bulger, Z., & Rouen, E. (2022). How fintech can deliver on its social impact promises. *Harvard Business Review*. Retrieved from https://hbr.org/2022/07/how-fintech-can-deliver-on-its-social-impact-promises

CCAF. (2021). *FinTech regulation in the Middle East and North Africa, Cambridge Centre for Alternative Finance at the University of Cambridge, Judge Business School*. Retrieved June 4, 2024, from https://www.jbs.cam.ac.uk/faculty-research/centres/alternative-finance/publications/fintech-regulation-in-the-middle-east-and-north-africa/

Demirgüç-Kunt, A., & Klapper, L. F. (2012). *Financial inclusion in Africa: An overview*. World Bank Policy Research Working Paper No. 6088. Retrieved from https://ssrn.com/abstract=2084599

Dicuonzo, G., Palmaccio, M., & Shini, M. (2024). ESG, governance variables and Fintech: An empirical analysis. *Research in International Business and Finance, 69*, 102205. https://doi.org/10.1016/j.ribaf.2023.102205

Firmansyah, E. A., Masri, M., Anshari, M., & Besar, M. H. A. (2023). Factors affecting fintech adoption: A systematic literature review. *FinTech, 2*(1), 21–33.

Fungáčová, Z., & Weill, L. (2015). Understanding financial inclusion in China. *China Economic Review, 34*, 196–206. https://doi.org/10.1016/j.chieco.2014.12.004

IMF. (2024). *Gulf Cooperation Council (GCC)—Pursuing visions amid geopolitical turbulence-Economic prospects and policy priorities for the GCC countries*. Retrieved January 10, 2025 from https://www.imf.org/en/Publications/Policy-Papers/Issues/2024/12/19/Gulf-Cooperation-Council-Pursuing-Visions-Amid-Geopolitical-Turbulence-Economic-Prospects-559734

International Energy Agency. (2024). *World energy investment 2024*. Retrieved January 20, 2025, from https://www.iea.org/data-and-statistics/charts/global-investment-in-clean-energy-and-fossil-fuels-2015-2024

Lyons, A. C., & Kass-Hanna, J. (2021). Financial inclusion, financial literacy and economically vulnerable populations in the Middle East and North Africa. *Emerging Markets Finance and Trade, 57*(9), 2699–2738. https://doi.org/10.1080/1540496X.2019.1598370

McKensey. (2023). *Is MENAP's budding start-up ecosystem ready to blossom?* Retrieved May 30, 2024, from https://www.mckinsey.com/industries/technology-media-and-telecommunications/our-insights/is-menaps-budding-startup-ecosystem-ready-to-blossom#/.

Moody. (2024). *GCC investment: Capitalising on diversification*. Retrieved January 24, 2025, from https://www.moodys.com/web/en/us/site-assets/gcc-cip.pdf

Niankara, I. (2023). The impact of financial inclusion on digital payment solution uptake within the Gulf cooperation council economies. *International Journal of Innovation Studies, 7*(1), 1–17. https://doi.org/10.1016/j.ijis.2022.09.004

PwC. (2015). *Developing a fintech ecosystem in the GCC: Let's get ready for take off*. Retrieved May 29, 2024, from https://www.strategyand.pwc.com/m1/en/reports/2015/developing-fintech-ecosystem-gcc.html

PwC. (2022). *Fintech in the Middle East: Building on the momentum*. Retrieved May 29, 2024, from https://www.strategyand.pwc.com/m1/en/strategic-foresight/sector-strategies/financial-sector-consulting/fintechbuildingmomentum.html

PwC. (2025a). The case for a GCC common sustainable finance framework. Retrieved January 28, 2025 from https://www.pwc.com/m1/en/publications/case-for-gcc-common-sustainable-finance-framework.html

PwC. (2025b). *Five GCC economic themes to watch in 2025*. Retrieved January 29, 2025, from https://www.pwc.com/m1/en/blog/five-economic-themes-to-watch-2025-gcc.html

Sachs, J. D., Woo, W. T., Yoshino, N., & Taghizadeh-Hesary, F. (2019). Importance of green finance for achieving sustainable development goals and energy security. In J. Sachs, W. Woo, N. Yoshino, & F. Taghizadeh-Hesary (Eds.), *Handbook of green finance. Sustainable development*. Springer. https://doi.org/10.1007/978-981-13-0227-5_13

Shmelev, S. E., & Speck, S. U. (2018). Green fiscal reform in Sweden: Econometric assessment of the carbon and energy taxation scheme. *Renewable and Sustainable Energy Reviews, 90*, 969–981. https://doi.org/10.1016/j.rser.2018.03.032

Sidaoui, M., Ben Bouheni, F., Arslankhuyag, Z., & Mian, S. (2022). Fintech and Islamic banking growth: New evidence. *The Journal of Risk Finance, 23*(5), 535–557. https://doi.org/10.1108/JRF-03-2022-0049

Siddiqui, Z., & Rivera, C. A. (2022). Fintech and fintech ecosystem: A review of literature. *Risk Governance & Control: Financial Markets & Institutions, 12*, 1.

Statista. (2024a). *Number of fintechs worldwide from 2018 to 2024, by region.* Retrieved May 28, 2024, from https://www.statista.com/statistics/893954/number-fintech-startups-by-region/

Statista. (2024b). *Financial inclusion in Saudi Arabia as of January 2024, by gender*. Retrieved May 28, 2024 from https://www.statista.com/statistics/1394148/saudi-arabia-distribution-of-financial-inclusion-factors/

Statista. (2024c). *Mobile technology in the Gulf Cooperation Council, by country and by type.* Retrieved May 28, 2024, from https://www.statista.com/statistics/669529/gcc-mobil-internet-and-app-usage-by-country/

Suryono, R. R., Budi, I., & Purwandari, B. (2020). Challenges and trends of financial technology (Fintech): A systematic literature review. *Information, 11*(12), 590.

Trotta, A., Rania, F., & Strano, E. (2024). Exploring the linkages between FinTech and ESG: A bibliometric perspective. *Research in International Business and Finance, 69*, 102200. https://doi.org/10.1016/j.ribaf.2023.102200

Uddin, M.-A., Ahmad, F., Dar, A.-B., & Khan, M.-A.-I. (2023). Determinants of financial inclusion in gulf cooperation council countries. *International Journal of Applied Economics, Finance and Accounting, 16*(2), 141–157. https://doi.org/10.33094/ijaefa.v16i2.940

Udeagha, M. C., & Muchapondwa, E. (2023). Green finance, fintech, and environmental sustainability: Fresh policy insights from the BRICS nations. *International Journal of Sustainable Development & World Ecology, 30*(6), 633–649. https://doi.org/10.1080/13504509.2023.2183526

United Nations. (2010). *Green climate fund.* Retrieved January 10, 2025 from https://www.unep.org/about-un-environment/funding-and-partnerships/green-climate-fund#:~:text=Created%20by%20the%20United%20Nations,and%20programmes%20in%20developing%20countries

Wang, D., Peng, K., Tang, K., & Yewei, W. (2022). Does Fintech development enhance corporate ESG performance? Evidence from an emerging market. *Sustainability, 14*(24), 16597. https://doi.org/10.3390/su142416597

World Bank Group. (2019). *Islamic green finance: Development, ecosystem and prospects.* Retrieved January 20, 2025, from https://documents.worldbank.org/en/publication/documents-reports/documentdetail/591721554824346344/islamic-green-finance-development-ecosystem-and-prospects

World Bank Group. (2024a). *Embracing climate opportunities for a greener GCC*. Retrieved January 23, 2025, from https://blogs.worldbank.org/en/arabvoices/embracing-climate-opportunities-greener-gcc

World Bank Group. (2024b). *Non-oil sectors drive robust growth in GCC countries.* Retrieved January 23, 2025, from https://www.worldbank.org/en/news/press-release/2024/12/01/non-oil-sectors-drive-robust-growth-in-gcc-countries

Chapter 12
Exploring AI Adoption for Sustainable Development Goals: Theoretical Frameworks and Influencing Factors in Higher Education Within the Gulf Region

Adel Bessadok and Amen Alrobai

Abstract This study investigates the relationship between sustainable development and the adoption of artificial intelligence (AI) in higher education. The primary objective is to identify the key factors representing sustainable development and to integrate these factors into a structural equation model to understand their impact on AI adoption. The research framework incorporates elements from the Sustainable Development Goals (SDGs), focusing on quality education, innovation, reduced inequalities, and responsible consumption and production. These elements are aligned with theoretical models such as the DeLone and McLean Information System Success Model, the Technology Acceptance Model (TAM), and the Unified Theory of Acceptance and Use of Technology (UTAUT).

Data was collected through a questionnaire administered to students from various higher education institutions in the Gulf region. A sample of 549 students participated in the survey, providing insights into their perceptions, motivations, and expectations regarding AI in education. The structural equation modeling was employed to explore the collected data and test the proposed hypotheses.

Preliminary results indicate that innovation and quality education are the most influential factors driving AI adoption in higher education. Students who perceive AI as a means to enhance educational quality and foster innovative learning environments show a higher propensity to adopt AI technologies. Additionally, factors related to reduced inequalities and responsible consumption and production also

A. Bessadok (✉) · A. Alrobai
Faculty of Arts and Humanities, Department of Information Science, King Abdulaziz University, Jeddah, Saudi Arabia
e-mail: abessadok@kau.edu.sa

A. Alrobai
e-mail: aaaalrobai@kau.edu.sa

H. Miniaoui and M. Bellalah (eds.), *Green Growth Opportunities and Sustainable Finance in the Gulf Cooperation Council Region*, Gulf Studies 24,
https://doi.org/10.1007/978-981-96-8287-4_12

play significant roles, but to a lesser extent. The study reveals that while all sustainable development factors contribute to AI adoption, the emphasis on innovation and quality education significantly outweighs other factors.

This research contributes to the ongoing discourse on sustainable education by elucidating the interplay between sustainable development principles and AI adoption. The findings offer valuable insights for researchers, policymakers, and educational institutions aiming to promote AI integration in higher education. By understanding the key drivers of AI adoption, stakeholders can develop targeted strategies to foster a more inclusive, innovative, and sustainable educational landscape.

Keywords Sustainable development goals · Artificial intelligence · Technology acceptance model · Structural equation modeling

12.1 Introduction

Sustainable development has emerged as a crucial goal for nations worldwide, particularly in the Gulf region, which faces unique economic, environmental, and social challenges (Zaidan et al., 2019). The Gulf countries, characterized by their rapid economic growth driven by oil and gas revenues, are increasingly recognizing the need to diversify their economies and promote sustainable practices to ensure long-term prosperity and stability (Alkhaldi & Abou Elseoud, 2024). Achieving sustainable development in this region is vital not only for maintaining economic resilience but also for addressing environmental concerns and social inequalities (Al-Zubairi et al., 2024). Among various sectors, higher education plays a pivotal role in this transformation by equipping individuals with the knowledge and skills necessary to drive sustainable development (Leal Filho, Lange Salvia, et al., 2024).

Higher education institutions in the Gulf region are at the forefront of fostering sustainable development (Fathelrahman & AlBreiki, 2024). They serve as hubs for innovation, research, and the dissemination of knowledge, thereby influencing societal values and practices. The integration of sustainable development principles into higher education can enhance the quality of education, promote social equity, and encourage responsible resource management. By embedding sustainability in their curricula and operations, universities can produce graduates who are not only proficient in their fields but also conscious of the environmental and social implications of their actions.

The adoption of artificial intelligence (AI) in higher education presents a significant opportunity to advance sustainable development goals. AI technologies can transform educational practices by enhancing learning experiences, personalizing education, improving administrative efficiencies, and fostering innovative research. These advancements can contribute to sustainable development by making education more accessible and inclusive, promoting lifelong learning, and reducing the environmental footprint of educational institutions through optimized resource use (Shwedeh et al., 2024).

The Sustainable Development Goals (SDGs) established by the United Nations offer a thorough blueprint for tackling global issues (Carlsen & Bruggemann, 2022). Four key SDGs—quality education, innovation, reduced inequalities, and responsible consumption and production—are particularly relevant to the adoption of AI in higher education.

1. Quality education (SDG 4): AI can enhance educational quality by providing personalized learning experiences, enabling adaptive learning technologies, and facilitating efficient educational management systems.
2. Innovation (SDG 9): The integration of AI fosters a culture of innovation within educational institutions, encouraging the development of cutting-edge research and new pedagogical methods.
3. Reduced inequalities (SDG 10): AI can help bridge educational disparities by providing tailored support to diverse learners, ensuring that students from various backgrounds have equal access to high-quality education.
4. Responsible consumption and production (SDG 12): AI can optimize resource utilization in educational institutions, promoting sustainable practices and reducing waste.

These factors are integral to understanding how AI adoption can be aligned with sustainable development goals in higher education.

The integration of AI into higher education holds transformative potential, offering a path to revolutionizing learning systems while advancing the Sustainable Development Goals. By enhancing the quality of education and stimulating innovation, AI is emerging as a cornerstone for promoting green growth through resource-efficient learning environments, reducing environmental impact, and promoting responsible consumption. Moreover, AI-powered educational tools contribute to sustainable finance by reducing costs and improving resource utilization, thereby democratizing access to education and promoting equity. In times of global shocks, such as pandemics or economic disruptions, AI-powered adaptive learning systems serve as resilient frameworks, ensuring uninterrupted, flexible, and inclusive delivery of education. These interconnected benefits underscore the pivotal role that AI adoption in higher education plays as a strategic enabler of sustainable progress, societal resilience, and innovation in an ever-evolving global landscape.

This chapter aims to explore the relationship between sustainable development and AI adoption in higher education by addressing the following research questions:

1. What are the key factors representing sustainable development that influence the adoption of AI in higher education?
2. How do these factors integrate into a structural equation model to assess their impact on AI adoption?
3. Which sustainable development factors have the most significant influence on students' motivation and expectations regarding AI adoption in higher education?
4. What are the implications of AI adoption for promoting sustainable development within higher educational institutions in the Gulf region?

By answering these questions, the study seeks to provide valuable insights into the interplay between sustainable development principles and AI adoption, guiding researchers, policymakers, and educational institutions in fostering a more inclusive, innovative, and sustainable educational landscape.

12.2 Literature Review

12.2.1 Sustainable Development Goals in Higher Education

Sustainable development in higher education involves the incorporation of practices and principles that promote enduring environmental, social, and economic well-being. This comprehensive strategy guarantees the fulfillment of present educational requirements while preserving the capacity of future generations to satisfy their own needs (Bessadok & Bardesi, 2023). Adopted in 2015, the United Nations' Sustainable Development Goals (SDGs) offer a detailed roadmap for achieving global sustainability by 2030. Higher education institutions are pivotal in promoting these goals through their educational programs, research initiatives, community involvement, and sustainable practices (Chankseliani & McCowan, 2021).

Higher education institutions are uniquely positioned to drive sustainable development by fostering critical thinking, innovation, and knowledge dissemination. These institutions educate future leaders and professionals, conduct cutting-edge research, and often lead by example in sustainable practices. Incorporating sustainable development into higher education involves reorienting curricula to include sustainability concepts, promoting interdisciplinary research on sustainability issues, and implementing campus-wide initiatives that reduce environmental impacts and promote social equity (Serafini et al., 2022).

The United Nations' SDGs encompass 17 interconnected goals designed to address global challenges such as poverty, inequality, climate change, environmental degradation, peace, and justice. Higher education institutions can directly influence several of these goals through their core activities.

1. Quality education (SDG 4): This goal strives to ensure inclusive, equitable quality education and to promote lifelong learning opportunities for all. Higher education institutions play a vital role by delivering high-quality education accessible to diverse populations. They foster lifelong learning through continuous education programs and integrate sustainability into their curricula, preparing students to tackle intricate global challenges (Ashida, 2022).
2. Industry, innovation, and infrastructure (SDG 9): Focused on building resilient infrastructure, promoting inclusive and sustainable industrialization, and fostering innovation, this goal sees universities as key contributors. They drive technological and industrial advancements through research, develop sustainable practices via industry partnerships, and provide the necessary infrastructure to support advanced research and innovation (Griebeler et al., 2022).

3. Reduced inequalities (SDG 10): Aiming to decrease inequality within and among countries, higher education institutions promote social equity by ensuring equal access to education for underrepresented groups. They offer scholarships and financial aid and create inclusive campus environments that support the success of all students, regardless of background (Omotosho et al., 2023).
4. Responsible consumption and production (SDG 12): This goal underscores the importance of sustainable management and efficient use of natural resources. Universities lead by example through sustainable operational practices such as waste reduction, energy efficiency, and sustainable sourcing. They also educate students on sustainable consumption and production and conduct research on innovative resource management solutions (Leal Filho et al., 2023).

Higher education institutions are essential in advancing the SDGs through their educational, research, and operational activities. By integrating sustainability into their curricula, universities can equip students with the knowledge and skills needed to address sustainability challenges. Research conducted at universities can contribute to the development of new technologies and approaches that support sustainable development. Furthermore, by implementing sustainable practices on campus, higher education institutions can serve as models for the broader community, demonstrating the feasibility and benefits of sustainable development (Heleta & Bagus, 2021).

The alignment of higher education with the SDGs is crucial for achieving sustainable development on a global scale. By focusing on quality education, innovation, reduced inequalities, and responsible consumption and production, higher education institutions can make significant contributions to the SDGs. These efforts not only enhance the sustainability of the institutions themselves but also prepare students to become leaders and innovators in a sustainable future.

12.2.2 Artificial Intelligence and Its Adoption by Students in Higher Educational Institutions

Artificial intelligence (AI) has emerged as a transformative technology with profound effects on numerous sectors, including higher education. In higher education the adoption of AI encompasses the use of advanced algorithms, machine learning, natural language processing, and other AI-driven technologies to enhance educational processes. This section reviews the literature on the adoption of AI by students in higher educational institutions, discussing its benefits and examining the relationship between students' perceptions of sustainable development goals (SDGs) and their willingness to adopt AI in their studies.

AI in higher education offers numerous advantages, including personalized learning, efficient administrative processes, and improved research capabilities. Personalized learning is perhaps the most notable benefit, as AI can tailor educational content to individual student needs, learning styles, and paces. Adaptive learning platforms powered by AI can identify areas where students struggle and provide targeted resources to address these gaps, thereby enhancing learning outcomes (Qahman et al., 2023).

AI also streamlines administrative tasks such as enrollment, grading, and student support, allowing educators to focus more on teaching and mentoring. Automated grading systems, for instance, can provide immediate feedback to students, which is crucial for the learning process (Holmes et al., 2019). Additionally, AI-driven analytics can help institutions make data-informed decisions to improve educational quality and efficiency.

In research, AI facilitates the analysis of large datasets, enabling researchers to uncover patterns and insights that would be impossible to detect manually. This capability proves exceptionally valuable in disciplines like genomics, climate science, and social sciences, where the sheer magnitude of data is overwhelming (Sharma & Garg, 2021).

For students, adopting AI in their studies can lead to enhanced educational experiences and better preparation for future careers. AI tools can provide personalized tutoring, support collaborative learning environments, and offer simulations and virtual labs that enrich practical learning (Luckin et al., 2016). These technologies not only make learning more interactive but also equip students with essential critical thinking and problem-solving skills.

Moreover, familiarity with AI technologies prepares students for a labor market increasingly dominated by AI and automation. By integrating AI into the curriculum, educational institutions can ensure that students are proficient in using these technologies and understand their ethical implications, thereby enhancing their employability and societal contribution.

Incorporating AI into education represents an innovative approach to delivering personalized content and supporting adaptive learning methods (Stolpe & Hallström, 2024). Grasping the factors that affect student receptiveness and AI adoption is pivotal for the effective deployment of these technologies. To fully leverage AI's advantages in the educational realm, it is imperative to identify and understand the precursors to its adoption within educational systems.

Exploring users' intentions to embrace artificial intelligence (AI) technology can be achieved through a variety of frameworks, from straightforward to intricate and comprehensive models. Parasuraman introduced the Technology Readiness Index (TRI) (Parasuraman, 2000), a comprehensive scale designed to assess users' preparedness for adopting new technologies. In a similar vein, Davis introduced the Technology Acceptance Model (TAM), which is based on the theories of reasoned action (TRA) (Ajzen & Fishbein, 1980) and planned behavior (TPB) (Ajzen, 1985), to clarify and forecast how users embrace new technologies. TAM remains the predominant theoretical model for evaluating the factors influencing technology adoption (Fig. 12.1).

Building on TAM, Venkatesh and Davis introduced TAM2 (Venkatesh et al., 2003), which explains perceived usefulness and usage intentions in terms of social influence and cognitive instrumental processes. Venkatesh and Bala then developed

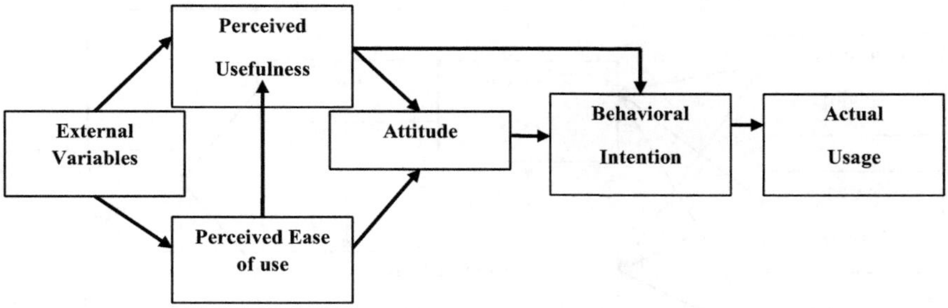

Fig. 12.1 Technology acceptance model (Davis, 1989)

TAM2 further into TAM3 (Venkatesh & Bala, 2008), delving into the common determinants of perceived usefulness and ease of use. Venkatesh et al. developed the Unified Theory of Acceptance and Use of Technology (UTAUT), a highly validated model aimed at explaining user behavior regarding technology adoption in organizational settings. The enhanced version, UTAUT2 (Fig. 12.2), expands this framework by examining technology usage from the users' individual perspectives.

To pinpoint the most crucial aspects of an information system affecting user acceptance and benefits, DeLone and McLean created the Information System Success (ISS) Model (Fig. 12.3).

Researchers have addressed AI adoption by applying these models in various adapted forms, either independently or in combination. For instance, Ma and Lei (2024) assessed students' willingness to use AI technologies through the TAM model. Du and Lv (2024) utilized the UTAUT and the Task-Technology Fit (TTF) Theory

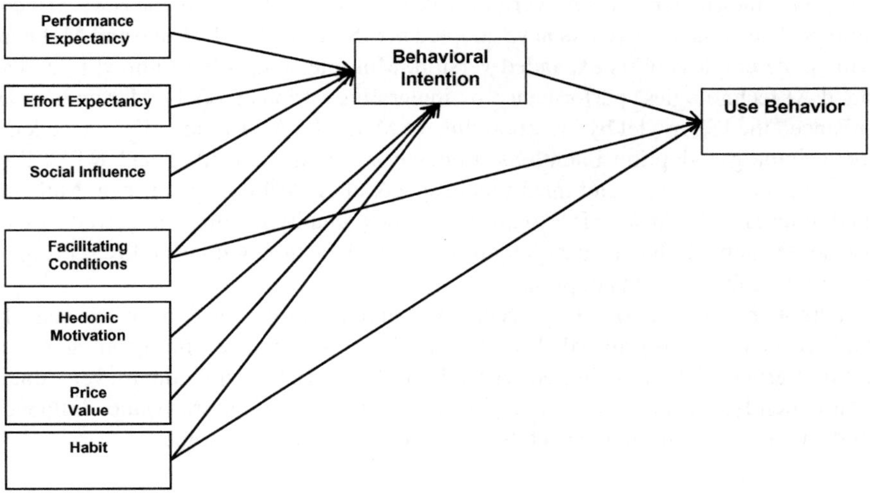

Fig. 12.2 Unified theory of acceptance and use of technology model (Venkatesh et al., 2012)

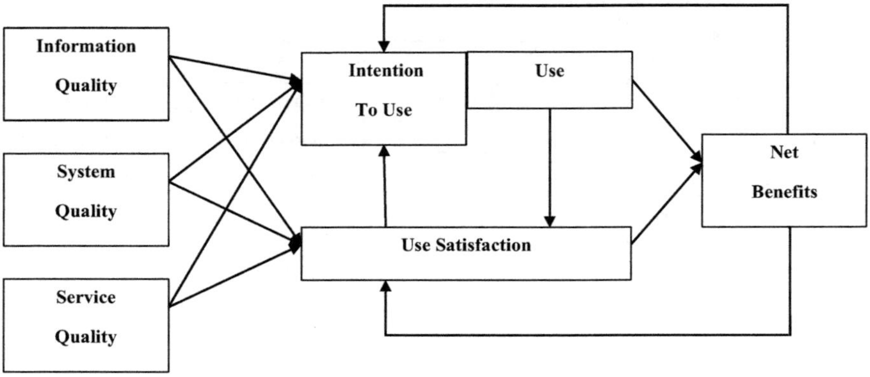

Fig. 12.3 Information system success model (DeLone & McLean, 2003)

to identify the key factors influencing student adoption of AI. Grassini et al. (2024) employed UTAUT2 to uncover the factors affecting AI acceptance, facilitating its use among students. Multiple studies have aimed to understand AI technologies' performance by integrating various models. Gado et al. (2022) enhanced the UTAUT model by integrating TAM and TRA models to identify antecedents for technology adoption intentions. Naidoo (2023) integrated TAM with ISS to investigate how quality features, perceived ease of use, and benefits influence user intentions and satisfaction, with usability serving as a mediating factor in AI adoption. Qahman et al. (2023) emphasized the importance of specific factors for AI-enabled acceptance and effectiveness in learning management.

The scholarly focus on evaluating AI technologies' effectiveness has surged recently, which underscores AI's critical role in higher education. To measure AI technologies' success, researchers have adopted the ISS model by DeLone and McLean, with some employing its expanded version. Multiple studies have aimed to understand AI technologies' performance by integrating various models. Mardiana et al. enhanced the ISS model by incorporating TAM and UTAUT to identify antecedents for technology adoption intentions. Lopes et al. combined TAM and UTAUT to examine the behavioral and intentional aspects of AI technologies usage. Mohammadi integrated TAM with ISS to investigate how quality features, perceived ease of use, and benefits influence user intentions and satisfaction, with usability serving as a mediating factor in AI adoption.

Although AI technologies' acceptance and adoption have been extensively studied, the focus has primarily been on technology acceptance within homogeneous cultural groups. There is limited research on the impact of individual-level values. Consequently, recent studies have aimed to explore how cultural values influence students' motivations and expectations for adopting AI technologies.

The perception of sustainable development goals can significantly influence students' attitudes toward adopting AI in their education. Understanding this relationship can provide insights into how to effectively promote AI adoption while advancing sustainable development objectives.

1. Quality education (SDG 4): Students who value quality education are likely to embrace AI technologies that enhance learning experiences. AI-driven personalized learning aligns with the goal of providing high-quality education by addressing individual learning needs and improving academic outcomes (UNESCO, 2020).
2. Innovation (SDG 9): Students who recognize the importance of innovation for sustainable development may be more inclined to adopt AI, as it represents a cutting-edge technological advancement. AI fosters a culture of innovation within educational institutions, encouraging students to explore new ideas and solutions (Bakhshi et al., 2017).
3. Reduced inequalities (SDG 10): AI can help reduce educational inequalities by providing personalized support to students from diverse backgrounds. Students who are aware of this potential are more likely to support the integration of AI in education, seeing it as a tool to promote social equity (Holmes et al., 2019).
4. Responsible consumption and production (SDG 12): The adoption of AI can lead to more efficient use of resources in educational institutions. Students who understand the importance of responsible consumption and production may appreciate AI's role in optimizing resource allocation and reducing waste (Van den Bergh, 2013).

The adoption of AI by students in higher education holds substantial promise for enhancing educational quality, fostering innovation, reducing inequalities, and promoting responsible resource use (Leal Filho, Ribeiro, et al., 2024). The relationship between students' perceptions of the SDGs and their willingness to adopt AI in their studies underscores the potential of AI to contribute to sustainable development. By aligning AI adoption with the goals of quality education, innovation, reduced inequalities, and responsible consumption and production, higher education institutions can advance both educational and sustainability objectives. Further research is needed to explore these dynamics and develop strategies to effectively integrate AI in ways that support sustainable development.

12.2.3 AI Adoption in Higher Education: A Catalyst for Green Growth, Sustainable Finance, and Global Shock Responses

The integration of artificial intelligence (AI) in higher education presents a unique opportunity to align educational advancements with broader sustainable development goals, particularly in the context of green growth, sustainable finance, and global shock responses. While the adoption of AI is often seen through the lens of technological innovation and educational enhancement, its role in fostering sustainability and resilience is equally pivotal. This subsection explores the multiple dimensions in which AI adoption in the Gulf region's higher education sector can contribute to these global imperatives.

12.2.3.1 AI as an Enabler of Green Growth in Education

One of the most pressing challenges in the global education sector is the need to reduce its environmental impact while simultaneously increasing access to quality learning. AI, as a transformative technology, holds the potential to address both concerns, thereby contributing to green growth. The Gulf region, which is investing heavily in sustainable development and environmental innovations, stands to benefit from AI's ability to optimize resource usage in educational institutions.

For instance, AI can be used to enhance smart campus initiatives, which reduce energy consumption, water usage, and waste generation (Tyagi et al., 2024). By leveraging machine learning algorithms, universities can optimize heating, cooling, and lighting systems, minimizing their carbon footprint (Chagnon-Lessard et al., 2021). Moreover, AI-powered digital learning platforms can reduce the reliance on paper-based resources and physical infrastructures, thus decreasing educational institutions' overall environmental footprint (Shenkoya & Kim, 2023).

12.2.3.2 AI and Sustainable Finance in Higher Education

The financial sustainability of educational institutions is another crucial factor in ensuring that AI adoption is effectively integrated into the broader educational landscape. Sustainable finance, which refers to investments that consider both financial returns and environmental, social, and governance (ESG) factors, has emerged as a key strategy in fostering long-term development (Alshubiri, 2021). By adopting AI, universities can improve their financial efficiency, thus contributing to their own sustainability and enhancing their appeal to green investors and ESG funds.

12.2.3.3 AI as a Response to Global Shocks

The capacity of AI to mitigate the effects of global shocks—such as economic recessions, health crises, and climate-related disasters—has become especially pertinent in recent years. The COVID-19 pandemic, for example, underscored the vulnerability of traditional education systems and the urgent need for more resilient and flexible frameworks. AI can help create adaptive educational systems capable of withstanding such shocks by offering remote learning solutions and ensuring continuity in education, regardless of external disruptions (Azevedo et al., 2021).

In the Gulf region, AI has already demonstrated its potential to support crisis management in education. For instance, AI-based platforms can facilitate personalized learning pathways, ensuring that students continue their educational journey even when face-to-face learning is disrupted. AI tools can also analyze student performance data to identify learning gaps, enabling universities to deploy targeted interventions that address individual needs—an essential aspect of resilience during crises (Bessadok, 2022).

12.2.3.4 Linking AI Adoption to the Sustainable Development Goals (SDGs)

The intersection of AI adoption in higher education and sustainable development is most clearly seen through the lens of the Sustainable Development Goals (SDGs). Specifically, AI adoption contributes to several key SDGs, including quality education (SDG 4), reduced inequalities (SDG 10), responsible consumption and production (SDG 12), and innovation (SDG 9).

1. Quality education (SDG 4): AI can enhance educational quality by personalizing learning experiences, increasing access to knowledge, and improving the overall efficiency of educational systems (UNESCO, 2020). This aligns with the SDG target of ensuring inclusive and equitable quality education and promoting lifelong learning opportunities for all.
2. Reduced inequalities (SDG 10): AI has the potential to democratize education, offering marginalized communities access to cutting-edge learning tools. By delivering high-quality educational content via digital platforms, AI helps bridge educational gaps caused by socio-economic disparities, especially in regions where access to quality education is limited (Tang et al., 2024).
3. Responsible consumption and production (SDG 12): AI's role in optimizing resource use within educational institutions promotes responsible consumption and production. AI can optimize energy usage, reduce waste, and contribute to sustainable campus management, all of which align with the SDG target to reduce the per capita environmental impact of consumption and production (UNDP, 2022).[1]
4. Innovation (SDG 9): AI inherently fosters innovation by enabling new teaching methods, research opportunities, and technological breakthroughs. It also helps universities build a culture of innovation, driving advancements that can be applied in various sectors, including green technologies and sustainable business practices.

[1] https://www.undp.org/publications/undp-annual-report-2022

12.3 Research Model

Our focus in this investigation is directed toward understanding how sustainable development goals (SDGs) influence students' behavioral intentions to engage with artificial intelligence (AI). The research framework, depicted in Fig. 12.4, integrates external variables associated with selected SDGs, key adoption factors, and the behavioral patterns of students regarding AI utilization. This comprehensive model synthesizes elements from the Technology Acceptance Model (TAM), the Unified Theory of Acceptance and Use of Technology (UTAUT), and the DeLone and McLean Information Systems Success (ISS) Model, providing a robust foundation for exploring the interplay between sustainability objectives and AI adoption among students.

12.3.1 Hypotheses

Given the current research objectives, we have devised a series of hypotheses rooted in the essential constructs of the Technology Acceptance Model (TAM). These hypotheses aim to provide a detailed explanation of how the integration of TAM, the Unified Theory of Acceptance and Use of Technology 2 (UTAUT2), and the DeLone and McLean Information Systems Success Model was achieved. The hypotheses are structured as follows:

H1a: Quality education is positively correlated with perceived ease of use.

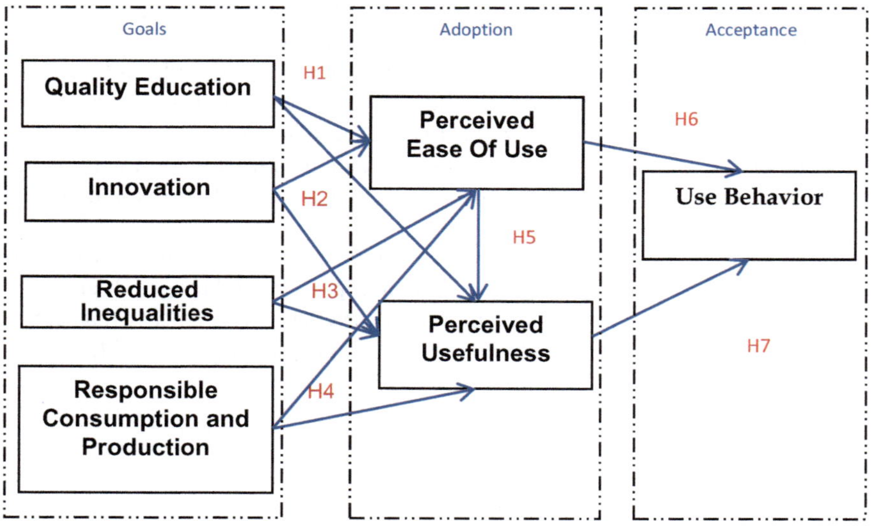

Fig. 12.4 Research model

– This hypothesis suggests that improvements in the quality of education will enhance students' perceptions of how easily AI technologies can be used.

H1b: Innovation is positively correlated with perceived usefulness.

– This hypothesis proposes that advancements and innovations in AI will increase students' perceptions of the technology's utility and benefits.

H2a: Quality education is positively correlated with perceived ease of use.

– Reinforcing H1a, this hypothesis further underscores the connection between quality education and the perceived ease of using AI technologies.

H2b: Innovation is positively correlated with perceived usefulness.

– Echoing H1b, this hypothesis emphasizes the critical role of innovation in shaping perceptions of AI's usefulness.

H3a: Reduced inequality is positively correlated with perceived ease of use.

– This hypothesis explores the idea that efforts to reduce inequality will positively influence students' perceptions of how easy it is to use AI.

H3b: Reduced inequality is positively correlated with perceived usefulness.

– This hypothesis examines the relationship between reduced inequality and perceived usefulness, suggesting that equitable access enhances the perceived benefits of AI.

H4a: Responsible consumption and production are positively correlated with perceived ease of use.

– This hypothesis posits that responsible consumption and production practices will improve students' perceptions of AI's ease of use.

H4b: Responsible consumption and production are positively correlated with perceived usefulness.

– This hypothesis suggests that sustainable practices in consumption and production will increase the perceived usefulness of AI technologies.

H5: The perceived ease of use exerts a positive influence on the perceived usefulness of AI technologies.

– This hypothesis asserts that if students find AI technologies easy to use, they are more likely to perceive these technologies as useful.

H6: The perceived usefulness exerts a positive influence on attitudes toward the use behavior of AI technologies. Perceived usefulness has a positive impact on attitudes toward the use behavior of AI.

– This hypothesis proposes that students who perceive AI as useful will have more positive attitudes toward using AI technologies.

H7: The perceived ease of use exerts a positive influence on attitudes toward the use behavior of AI technologies.

– This hypothesis suggests that the easier students find AI technologies to use, the more favorable their attitudes will be toward engaging with these technologies.

These hypotheses collectively aim to elucidate the interplay between sustainable development goals and AI adoption among students. By integrating the theoretical constructs from TAM, UTAUT2, and the DeLone and McLean Information Systems Success Model, the proposed research model offers a comprehensive framework for analyzing students' behavioral intentions and usage patterns regarding AI technologies. This model, illustrated in the accompanying figures, captures the multifaceted influences of education quality, innovation, inequality reduction, and sustainable practices on AI acceptance and utilization.

12.4 Methods

The conceptual framework, depicted in Fig. 12.4, seeks to explore the determinants influencing students' acceptance of AI technologies. Specifically, it examines how fundamental human values impact students' aspirations through their motivations and expectations. The model employs quantitative methods, including exploratory factor analysis (EFA) and confirmatory factor analysis (CFA), within a structural equation modeling (SEM) context. This methodological approach facilitates a thorough investigation of the causal relationships among the model's constructs, providing a detailed understanding of how these factors interrelate and influence AI adoption.

12.4.1 Sample and Procedures

The study engaged students from three prominent universities within the Gulf region: King Abdulaziz University in Jeddah, Saudi Arabia; Lusail University in Qatar; and the College of Engineering and Information Technology at the University of Dubai. Data collection was facilitated through an online survey platform, Google Forms, ensuring wide accessibility and ease of participation.

The survey was meticulously structured into three primary sections: socio-demographic information, Sustainable Development Goals (SDGs), and constructs related to adoption and acceptance of AI. This comprehensive structure was designed to capture a broad spectrum of relevant data for in-depth analysis.

The questionnaire, translated into Arabic to ensure clarity and comprehension among participants, was distributed to students registered at the three aforementioned institutions. To gauge participants' responses, a 5-point Likert scale was employed, with options ranging from "strongly agree" to "strongly disagree." The introduction to the questionnaire emphasized the voluntary nature of participation, along with

Table 12.1 Participants' demographic information

Demographic	Group	Frequency	Percentage (%)
Gender			
	Woman	228	41.6
	Man	321	58.4
Age			
	<18	14	2.5
	18–23	357	65.0
	24–27	99	18.1
	>27	79	14.4
Education			
	Bachelor's	369	62.2
	Master's	101	17.5
	Ph.D.	79	11.2

assurances of confidentiality and anonymity, aiming to foster honest and uninhibited responses.

For further clarity on the metrics used, additional details are provided in Appendix A. The analysis of the collected data was conducted using IBM SPSS Version 26 and IBM AMOS Version 24, tools known for their robust statistical analysis capabilities. The survey period spanned from September 2023 to March 2024, during which a total of 549 completed questionnaires were collected. The demographic details of the participants, providing insights into the sample's composition, are summarized in Table 12.1.

12.4.2 Structural Equation Modeling

As depicted in Fig. 12.4, the proposed research model integrates a complex web of relationships among various variables. Given the intricate nature of these relationships, structural equation modeling (SEM) was chosen as the ideal method for analysis due to its remarkable flexibility and robustness. SEM is a sophisticated multivariate statistical technique that amalgamates Confirmatory Factor Analysis (CFA) and multiple regression analysis to explore and validate the theoretical constructs and their interconnections. The framework of Structural Equation Modeling (SEM) is delineated through a series of fundamental steps, as illustrated in Fig. 12.5.

Two primary types of SEM are widely utilized: Partial Least Squares SEM (PLS-SEM) (Lohmöller, 1989) and Covariance-Based SEM (CB-SEM) (Jöreskog & van Thillo, 1972). Despite their different approaches, both methodologies are grounded in the same theoretical principles (Wold et al., 1984), aiming to provide comprehensive insights into the data. The implementation of SEM is conducted in two distinct

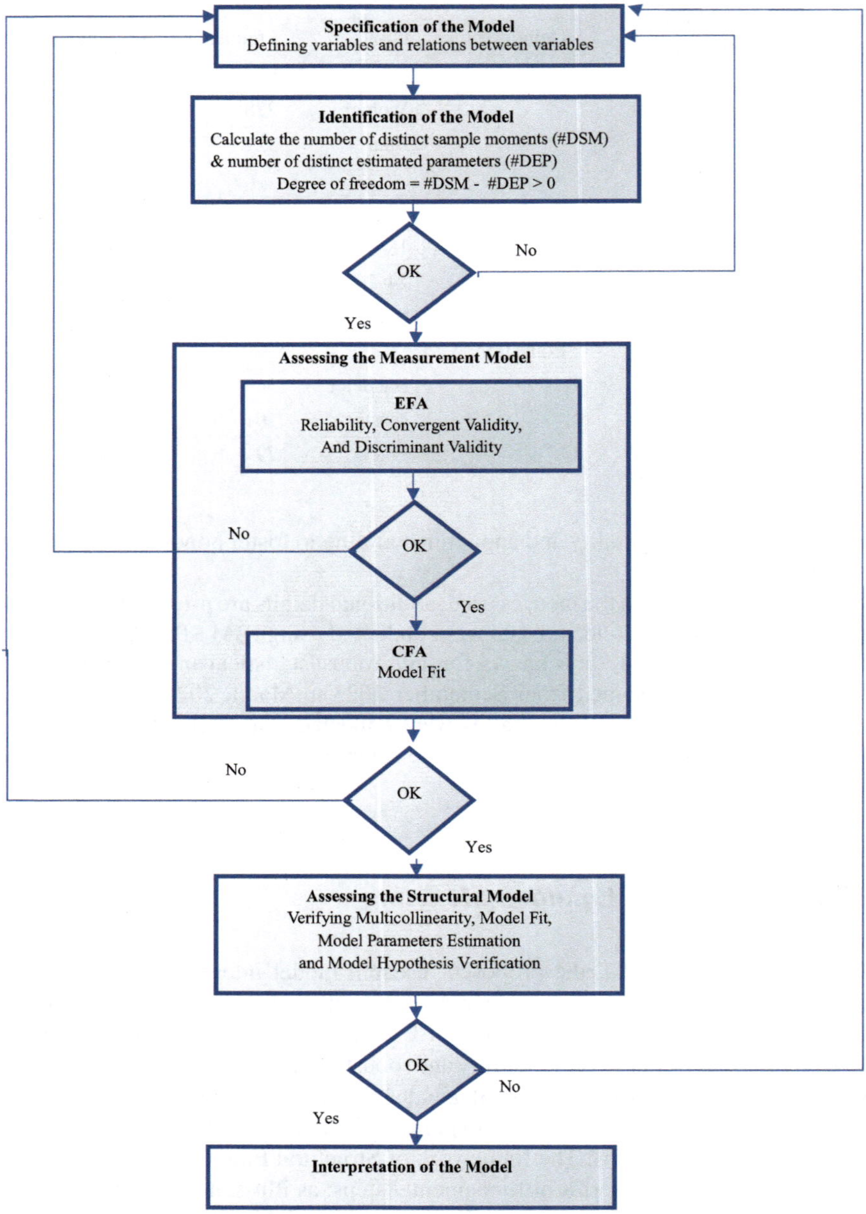

Fig. 12.5 SEM procedure steps (Bessadok, 2022)

phases: the evaluation of the measurement model and the evaluation of the structural model.

1. Evaluation of the measurement model: This initial phase focuses on how well the indicator variables represent the latent constructs. It involves assessing the reliability and validity of the constructs, ensuring that the observed variables accurately measure the underlying theoretical concepts. Techniques such as reliability tests, convergent validity, and discriminant validity are employed to validate the measurement model.
2. Evaluation of the structural model: The second phase examines the hypothesized relationships among the constructs. This stage involves testing the proposed dependencies and interconnections, essentially mapping out how the constructs influence one another. Multiple dependency relationships are analyzed to understand the direct and indirect effects within the model, providing a detailed view of the causal pathways.

Through these stages, SEM facilitates a thorough and nuanced analysis of the research model, enabling researchers to verify the theoretical assumptions and uncover the underlying dynamics of the constructs in question (Hair et al., 2006). This methodological approach ensures a comprehensive understanding of the complex relationships within the proposed model, offering valuable insights into the factors influencing students' acceptance of AI technologies.

12.5 Results

12.5.1 Measurement Model Phase

The measurement model phase in SEM critically assesses the validity and reliability of the indicators for each construct. This process begins with Exploratory Factor Analysis (EFA) to identify the dimensionality of the constructs and guide the development of the scale. EFA is particularly useful in the early stages when the factor structure is not yet established. As the scale and factor structure become more refined and grounded in prior research, Confirmatory Factor Analysis (CFA) is employed for a more precise evaluation.

To ensure the accuracy of the indicators in measuring the constructs, EFA was conducted multiple times to verify the consistency of each factor. Items that had factor loadings lower than 0.5, as suggested by Bagozzi and Yi (1988), were methodically eliminated from the analysis. This iterative process continued until a clear and robust relationship between the factors emerged, ensuring that only strong indicators remained. The final set of indicators for each construct is detailed in Table 12.2.

The measurement model parameter estimation is illustrated in Fig. 12.6.

To further validate the measurement model, the Kaiser-Meyer-Olkin (KMO) test and Bartlett's test of sphericity were applied. The KMO measure, which ranges from

Table 12.2 Loading factors of the measurement model[a]

Item	1	2	3	4	5	6	7
SDG4_1				0.816			
SDG4_2				0.851			
SDG4_3				0.815			
SDG4_4				0.779			
USEB_1							0.606
USEB_2							0.761
USEB_3							0.828
USEB_4							0.730
PEUS_1					0.788		
PEUS_2					0.638		
PEUS_3					0.809		
PEUS_4					0.881		
PEOU_1						0.854	
PEOU_2						0.922	
PEOU_3						0.791	
PEOU_4						0.755	
SDG9_1	0.844						
SDG9_2	0.898						
SDG9_3	0.919						
SDG9_4	0.899						
SDG12_1			0.795				
SDG12_2			0.925				
SDG12_3			0.887				
SDG12_4			0.836				
SDG10_1		0.877					
SDG10_2		0.925					
SDG10_3		0.872					
SDG10_4		0.869					

Note: Extraction method: maximum likelihood. Rotation method: promax with Kaiser normalization. Item: explained in the appendix
[a] Rotation converged in 7 iterations

0 to 1, indicates the adequacy of the sample size for factor analysis. A KMO value near 1.0 is ideal, while values below 0.5 are considered inadequate (Hutcheson & Sofroniou, 1999). For our analysis, a KMO value of at least 0.80 was achieved, which is generally sufficient for proceeding with factor analysis. These values are presented in Table 12.3.

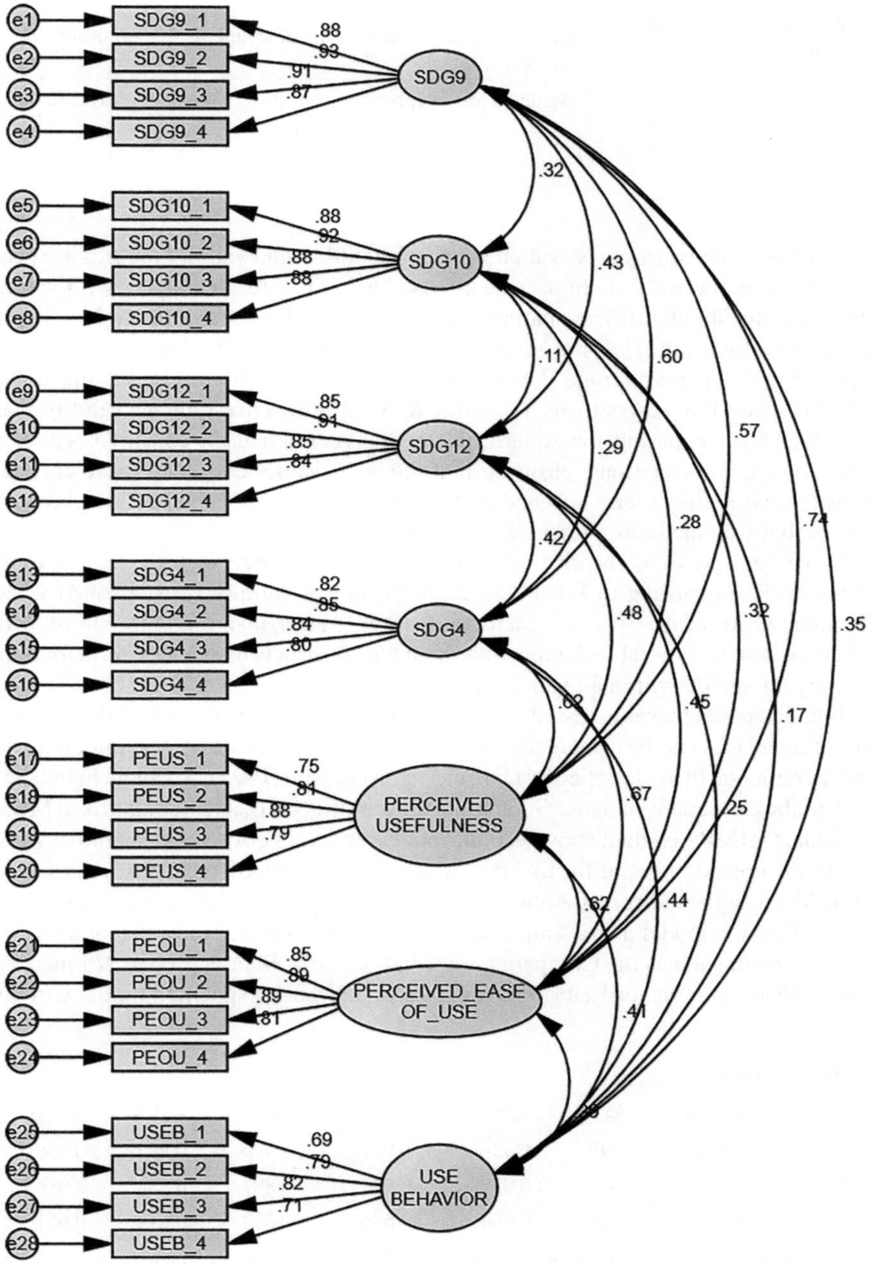

Fig. 12.6 Measurement model result

Table 12.3 Tests of KMO and Bartlett

Kaiser–Meyer–Olkin measure of sampling adequacy		0.877
Bartlett's test of sphericity	Approx. Chi-square	16,272.699
	df	780
	Sig.	0.000

Construct robustness was evaluated using reliability and validity metrics as specified by Fornell et al. Reliability was assessed using Cronbach's alpha, with values above the threshold of 0.7 indicating satisfactory reliability according to Jöreskog's guidelines (Jöreskog, 1993). Convergent validity was evaluated through Composite Reliability (CR) and Average Variance Extracted (AVE), with minimum thresholds set at 0.7 and 0.5, respectively (Bagozzi & Yi, 1988). Discriminant validity was examined by comparing the square root of the AVE for each construct with the inter-construct correlations, ensuring that the AVE values exceeded these correlations following the criteria outlined by Fornell et al. (1982). Table 12.4 displays the results that validate these measures of validity.

Table 12.5 provides the cross-correlation estimates along with the square roots of the AVE. According to Fornell et al., discriminant validity is established when the square root of the AVE for each construct is greater than the correlations with other constructs. This table demonstrates that the constructs meet the criteria for both convergent and discriminant validity.

Finally, model fit was assessed using various indices, as explained by Kline (2016). These indicators can be categorized into three groups: absolute fit, parsimonious fit, and incremental fit as classified by Gerbing and Anderson (1992). Absolute fit indices, such as the chi-square statistic, Goodness-of-Fit Index (GFI), and Standardized Mean Residual (SRMR), evaluate how well the model replicates observed data. Indices that assess the compactness of fit, like the Adjusted Goodness-of-Fit Index (AGFI) and Root Mean Square Error of Approximation (RMSEA), take into consideration the intricacy of the model as parsimonious fit. On the other hand, indices that measure improvement, such as the Comparative Fit Index (CFI), Bentler-Bonett Normed Fit Index (NFI), and Tucker-Lewis Index (TLI), juxtapose the specified model with an

Table 12.4 Validity measures[a]

Measure	SDG 4	SDG 9	SDG 10	SDG 12	PEOU	PEUS	USEB
Alpha	0.894	0.943	0.936	0.919	0.920	0.883	0.838
CR	0.895	0.944	0.937	0.920	0.889	0.884	0.910
AVE	0.687	0.879	0.664	0.688	0.672	0.642	0.651

Alpha Cronbach's alpha, *CR* composite reliability, *AVE* average variance extracted

[a] Constructs abbreviated as follows—Quality Education (SDG 4), Innovation (SDG 9), Reduced Inequalities (SDG 10), Responsible Consumption and Production (SDG 12), perceived usefulness (PEUS), perceived ease of use (PEOU), and use behavior (USEB)

Table 12.5 Discriminant validity of the measurement model[a]

Measure	SDG 4	SDG 9	SDG 10	SDG 12	PEOU	PEUS	USEB
SDG 4	0.826[b]						
SDG 9	0.108	0.911[b]					
SDG 10	0.560	0.543	0.902[b]				
SDG 12	0.681	0.467	0.481	0.832[b]			
PEOU	0.371	0.573	0.365	0.529	0.891[b]		
PEUS	0.345	0.463	0.436	0.463	0.536	0.853[b]	
USEB	0.632	0.419	0.329	0.356	0.418	0.397	0.721[b]

[a] Constructs abbreviated as follows—quality education (SDG 4), innovation (SDG 9), reduced inequalities (SDG 10), responsible consumption and production (SDG 12), perceived usefulness (PEUS), perceived ease of use (PEOU), and use behavior (USEB)
[b] The value of the square roots of the AVE

Table 12.6 Fit indices of the measurement model

Fit indices	Value	Recommended value
Chi-square/degree of freedom	2.920	<3
GFI	0.949	>0.9
CFI	0.886	>0.8
NFI	0.924	>0.9
TLI	0.941	>0.9
RMSEA	0.059	<0.06
SRMR	0.053	<0.06

alternate benchmark model. These indices are summarized in Table 12.6, confirming that the model fit indices support the validity of the measurement model.

By thoroughly evaluating the measurement model through these steps, we ensure the accuracy and robustness of our constructs, laying a solid foundation for the subsequent analysis of the structural model.

12.5.2 Structural Model Phase

Following the validation of the measurement model, the focus shifts to the structural model to estimate the parameters of the research model. This stage of SEM analysis involves assessing the causal relationships among the constructs as specified in the research model. The structural model evaluation was applied to all hypothesized causal links to determine the strength and significance of these relationships.

To ensure the structural model's appropriateness, a variety of fit indices were utilized, as recommended by Hair et al. (2014). These indices, summarized in

Table 12.7 Fit indices of the structural model

Fit indices	Value	Recommended value
Chi-square/degree of freedom	2.067	<3
GFI	0.917	>0.9
CFI	0.917	>0.8
NFI	0.917	>0.9
TLI	0.907	>0.9
RMSEA	0.053	<0.06
SRMR	0.056	<0.06

Table 12.7, include measures such as the Chi-square statistic, Goodness-of-Fit Index (GFI), Comparative Fit Index (CFI), and Root Mean Square Error of Approximation (RMSEA). These indicators help confirm that the model adequately fits the data and that the hypothesized relationships are supported.

The structural model assessment reveals that most of the relationships between the constructs align with the hypothesized directions, although not all were statistically significant. Specifically, Fig. 12.7 shows that the relationships involving self-transcendence, motivation, expectation, and perceived usefulness did not reach statistical significance.

Despite this, the direction of these relationships was consistent with the initial hypotheses, suggesting that the overall conceptual framework of the research model remains intact. Detailed results of these assessments, including path coefficients and significance levels, are provided in Table 12.8.

This comprehensive evaluation of the structural model ensures that the proposed research model is robust and that the causal pathways between constructs are well-defined. By rigorously testing these relationships, we gain deeper insights into the factors influencing students' acceptance and adoption of AI within the context of sustainable development goals. This process ultimately enhances our understanding of the underlying mechanisms driving technology acceptance and adoption behaviors in educational settings.

12.6 Discussion

This study explored the intricate relationship between sustainable development and the adoption of artificial intelligence (AI) in higher education, focusing on key factors represented by the United Nations' Sustainable Development Goals (SDGs). The primary objective was to identify these factors and integrate them into a structural equation model to assess their impact on AI adoption.

The findings indicate that innovation (SDG 9) and quality education (SDG 4) are the most influential factors driving AI adoption in higher education. Students perceive

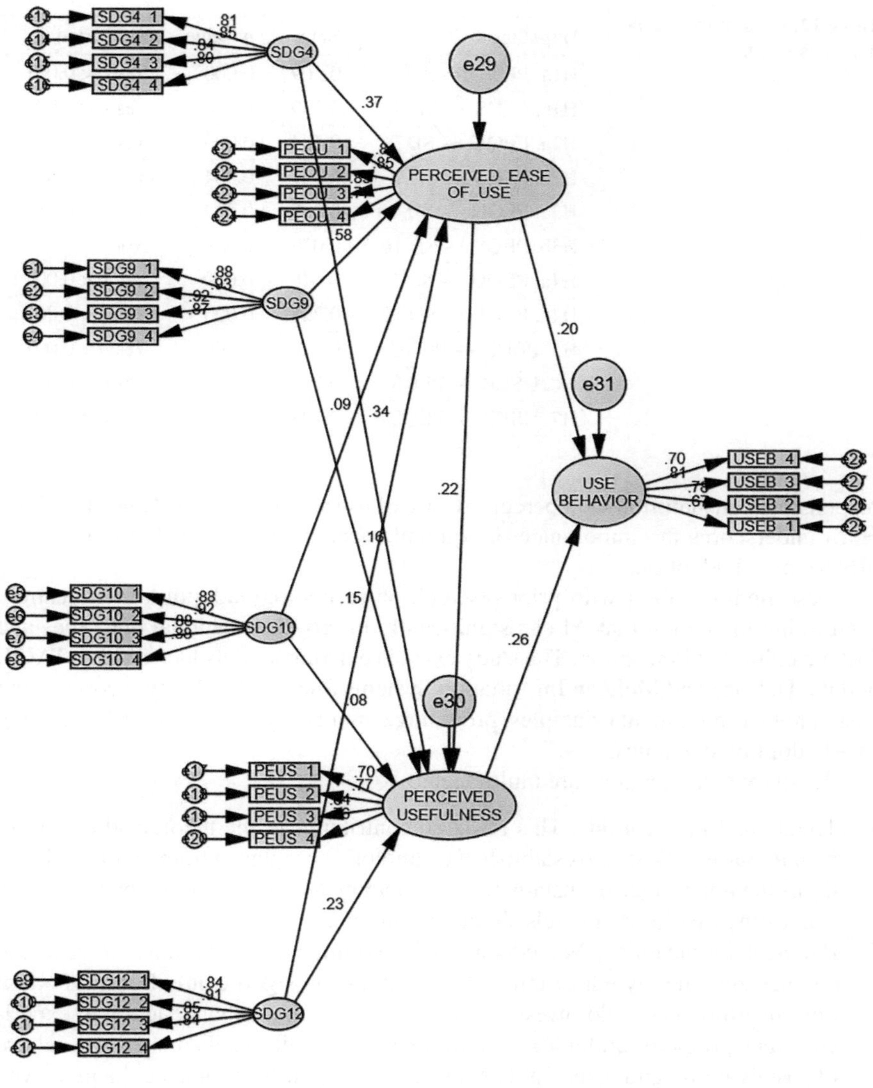

Fig. 12.7 Research model result

AI as a tool to enhance educational quality and foster innovative learning environments. Additionally, factors related to reduced inequalities (SDG 10) and responsible consumption and production (SDG 12) also play significant roles, although to a lesser extent.

The results confirm the hypotheses that quality education and innovation positively influence perceived ease of use and perceived usefulness of AI technologies (H1a, H1b, H2a, H2b). The positive correlation between reduced inequalities and

Table 12.8 Summary of the hypothesis tests

Hypothesis	Beta	p-Value	Supported
H1a. PEOU ← SD 4	0.369	0.000	Yes $p < 0.001$
H1b. PEUS ← SD 4	0.339	0.000	Yes $p < 0.001$
H2a. PEOU ← SD 9	0.584	0.000	Yes $p < 0.001$
H2b. PEUS ← SD 9	0.161	0.004	Yes $p < 0.05$
H3a. PEOU ← SD 10	0.088	0.013	Yes $p < 0.05$
H3b. PEUS ← SD 10	0.079	0.053	Not
H4a. PEOU ← SD 12	0.146	0.000	Yes $p < 0.001$
H4b. PEUS ← SD 12	0.228	0.000	Yes $p < 0.001$
H5. PEUS ← PEOU	0.218	0.000	Yes $p < 0.001$
H6. USEB ← PEUS	0.256	0.000	Yes $p < 0.001$
H7. USEB ← PEOU	0.198	0.000	Yes $p < 0.001$

responsible consumption with perceived ease of use and usefulness (H3a, H3b, H4a, H4b) underscores the importance of equitable access and sustainable practices in promoting AI adoption.

These findings align with prior research on technology adoption in education, reinforcing the notion that AI can significantly improve educational outcomes and foster a culture of innovation. The study extends existing models like TAM, UTAUT, and the DeLone and McLean Information Systems Success Model by incorporating sustainable development principles, providing a more comprehensive understanding of AI adoption dynamics.

The study's implications are multifaceted:

1. Theoretical implications: This research contributes to the theoretical discourse by integrating SDGs into established technology acceptance frameworks. It highlights the importance of sustainability in shaping technology adoption behaviors, suggesting that future models should consider such factors.
2. Practical implications: For educational institutions, the findings suggest that emphasizing quality education and innovation can significantly boost AI adoption. Institutions should invest in AI technologies that enhance learning experiences and promote equitable access to education. Additionally, fostering a culture of sustainability and responsible resource use can further enhance the perceived value and usefulness of AI technologies.
3. Policy implications: Policymakers should support initiatives that integrate AI in education to advance sustainable development. Policies promoting innovation, educational quality, and equitable access can drive the adoption of AI technologies, contributing to broader sustainable development goals.

12.7 Conclusions and Future Research Directions

This study aimed to investigate the relationship between sustainable development and the adoption of artificial intelligence (AI) in higher education within the Gulf region. By integrating key sustainable development factors into a structural equation model, this research sought to understand how these elements influence students' adoption of AI technologies. The Sustainable Development Goals (SDGs), particularly those focused on quality education, innovation, reduced inequalities, and responsible consumption and production, provided the foundational framework for this analysis.

The findings from our questionnaire-based study, which included responses from 600 students across various higher education institutions in the Gulf countries, indicate that innovation and quality education are the most influential factors driving AI adoption. Students who perceive AI as a tool to enhance educational quality and foster innovative learning environments show a higher propensity to adopt AI technologies. This aligns with SDG 4 (Quality Education) and SDG 9 (Industry, Innovation, and Infrastructure), highlighting the critical role of AI in advancing educational practices and supporting innovative research.

Moreover, the study revealed that while factors related to reduced inequalities (SDG 10) and responsible consumption and production (SDG 12) also contribute to AI adoption, their influence is less pronounced compared to the factors of quality education and innovation. This suggests that while students recognize the broader benefits of AI in promoting social equity and sustainable resource use, their primary motivation for adopting AI is its potential to directly enhance their learning experiences and academic outcomes.

The implications of these findings are significant for policymakers, educational institutions, and researchers in the Gulf region. To foster a more inclusive, innovative, and sustainable educational landscape, it is essential to develop targeted strategies that emphasize the enhancement of educational quality and innovation through AI. Institutions should focus on integrating AI technologies that provide personalized learning experiences, support collaborative environments, and offer practical applications such as simulations and virtual labs. Additionally, efforts to raise awareness about the broader social and environmental benefits of AI can further encourage its adoption among students.

In conclusion, this research underscores the importance of aligning AI adoption in higher education with sustainable development principles. By understanding the key drivers of AI adoption and their relation to the SDGs, stakeholders in the Gulf region can create more effective and sustainable educational practices. Future research should continue to explore the interplay between AI technologies and sustainable development, particularly in different regional contexts and across various educational settings. This ongoing investigation will be crucial in ensuring that AI integration in higher education not only enhances learning outcomes but also contributes to the broader goals of sustainability and equity.

Despite its contributions, this study has several limitations. The sample was restricted to students from higher education institutions in the Gulf region, which may limit the generalizability of the findings to other contexts. Additionally, the study relied on self-reported data, which may be subject to biases. Future research should consider longitudinal studies to assess the long-term impact of sustainable development factors on AI adoption.

Future research should explore the impact of sustainable development factors on AI adoption in diverse educational contexts. Investigating the role of other SDGs and their interactions with technology adoption could provide a more holistic understanding. Additionally, studies focusing on faculty and administrative perspectives could offer complementary insights into AI integration in higher education.

Ultimately, this study underscores the significant role of sustainable development goals, particularly quality education and innovation, in driving the adoption of AI in higher education. By aligning AI initiatives with SDGs, educational institutions can not only enhance their technological capabilities but also contribute to a more inclusive, innovative, and sustainable educational landscape. Understanding the key drivers of AI adoption provides valuable insights for stakeholders aiming to foster a transformative and sustainable educational environment.

In conclusion, AI adoption in higher education offers much more than technological advancement; it serves as a strategic tool for advancing green growth, enhancing sustainable finance, and strengthening educational systems' resilience to global shocks. Through its ability to drive innovation, improve efficiency, and support sustainability, AI aligns closely with the Sustainable Development Goals, making it a key component in achieving a more sustainable and resilient future for the Gulf region and beyond. By harnessing the potential of AI, higher education institutions can play a central role in shaping a global education system that is both inclusive and capable of responding to the challenges of the twenty-first century.

References

Ajzen, I. (1985). From intentions to actions: A theory of planned behavior. In J. Kuhl & J. Beckmann (Eds.), *Action control* (pp. 11–39). Springer. https://doi.org/10.1007/978-3-642-69746-3_2

Ajzen, I., & Fishbein, M. (1980). *Understanding attitudes and predicting social behavior*. Pearson.

Alkhaldi, F. K., & Abou Elseoud, M. S. (2024). Assessing the sustainability of GCC economic growth: A proposed theoretical framework. In *Technological innovations for business, education and sustainability* (pp. 203–221). Emerald Publishing Limited. https://doi.org/10.1108/978-1-83753-106-620241014

Alshubiri, F. N. (2021). Analysis of financial sustainability indicators of higher education institutions on foreign direct investment: Empirical evidence in OECD countries. *International Journal of Sustainability in Higher Education, 22*(1), 77–99. https://doi.org/10.1108/IJSHE-10-2019-0306

Al-Zubairi, A., Al Akheli, A., & ELfarra, B. (2024). The impact of financial development, renewable energy and political stability on carbon emissions: Sustainable development prospective for Arab economies. *Environment, Development and Sustainability*. https://doi.org/10.1007/s10668-024-04703-5

Ashida, A. (2022). The role of higher education in achieving the sustainable development goals. In *Sustainable development disciplines for humanity: Breaking down the 5Ps—People, planet, prosperity, peace, and partnerships* (pp. 71–84). Springer. https://doi.org/10.1007/978-981-19-4859-6_5

Azevedo, J. P., Hasan, A., Goldemberg, D., Geven, K., & Iqbal, S. A. (2021). Simulating the potential impacts of COVID-19 school closures on schooling and learning outcomes: A set of global estimates. *The World Bank Research Observer, 36*(1), 1–40. https://doi.org/10.1093/wbro/lkab003

Bagozzi, R. P., & Yi, Y. (1988). On the evaluation of structural equation models. *Journal of the Academy of Marketing Science, 16*, 74–94. https://doi.org/10.1007/BF02723327

Bakhshi, H., Downing, J. M., Osborne, M. A., & Schneider, P. (2017). *The future of skills: Employment in 2030*. Pearson.

Bessadok, A. (2022). Analyzing student aspirations factors affecting e-learning system success using a structural equation model. *Education and Information Technologies, 27*(7), 9205–9230. https://doi.org/10.1007/s10639-022-11015-6

Bessadok, A., & Bardesi, H. (2023). Exploring human values and students' aspiration in E-learning adoption: A structural equation modeling analysis. *Sustainability, 15*(19), 14041. https://doi.org/10.3390/su151914041

Carlsen, L., & Bruggemann, R. (2022). The 17 United Nations' sustainable development goals: A status by 2020. *International Journal of Sustainable Development & World Ecology, 29*(3), 219–229. https://doi.org/10.1080/13504509.2021.1948456

Chagnon-Lessard, N., Gosselin, L., Barnabé, S., Bello-Ochende, T., Fendt, S., Goers, S., & Zhang, P. (2021). Smart campuses: Extensive review of the last decade of research and current challenges. *IEEE Access, 9*, 124200–124234. https://doi.org/10.1109/ACCESS.2021.3109516

Chankseliani, M., & McCowan, T. (2021). Higher education and the sustainable development goals. *Higher Education, 81*(1), 1–8. https://doi.org/10.1007/s10734-020-00652-w

Davis, F. D. (1989). Perceived usefulness, perceived ease of use, and user acceptance of information technology. *MIS Quarterly, 13*, 319–340. https://doi.org/10.2307/249008

DeLone, W. H., & McLean, E. R. (2003). The DeLone and McLean model of information systems success: A ten-year update. *Journal of Management Information Systems, 19*(4), 9–30. https://doi.org/10.1080/07421222.2003.11045748

Du, L., & Lv, B. (2024). Factors influencing students' acceptance and use generative artificial intelligence in elementary education: An expansion of the UTAUT model. *Education and Information Technologies*. https://doi.org/10.1007/s10639-024-12835-4

Fathelrahman, E. M., & AlBreiki, G. A. A. (2024). Quality of higher education in the Gulf Cooperation Council countries: A review on the progress towards 2030 SDGs. *Journal of Adult and Continuing Education*. https://doi.org/10.1177/1477971424125

Fornell, C., Tellis, G. J., & Zinkhan, G. M. (1982). Validity assessment: A structural equations approach using partial least squares. In *Proceedings of the American Marketing Association educators' conference* (Vol. 48, pp. 405–409). American Marketing Association.

Gado, S., Kempen, R., Lingelbach, K., & Bipp, T. (2022). Artificial intelligence in psychology: How can we enable psychology students to accept and use artificial intelligence? *Psychology Learning & Teaching, 21*(1), 37–56. https://doi.org/10.1177/14757257211037149

Gerbing, D. W., & Anderson, J. C. (1992). Monte Carlo evaluations of goodness of fit indices for structural equation models. *Sociological Methods & Research, 21*(2), 132–160. https://doi.org/10.1177/0049124192021002002

Grassini, S., Aasen, M. L., & Møgelvang, A. (2024). Understanding university students' acceptance of ChatGPT: Insights from the UTAUT2 model. *Applied Artificial Intelligence, 38*(1), 2371168. https://doi.org/10.1080/08839514.2024.2371168

Griebeler, J. S., Brandli, L. L., Salvia, A. L., Leal Filho, W., & Reginatto, G. (2022). Sustainable development goals: A framework for deploying indicators for higher education institutions. *International Journal of Sustainability in Higher Education, 23*(4), 887–914. https://doi.org/10.1108/IJSHE-03-2021-0088

Hair, J. F., Black, W. C., Babin, B. J., & Anderson, R. E. (2014). *Multivariate data analysis: Pearson new international edition*. Pearson Education Limited.

Hair, J. F., Black, W. C., Babin, B. J., Anderson, R. E., & Tatham, R. (2006). *Multivariate data analysis* (6th ed.). Pearson Prentice Hall.

Heleta, S., & Bagus, T. (2021). Sustainable development goals and higher education: Leaving many behind. *Higher Education, 81*(1), 163–177. https://doi.org/10.1007/s10734-020-00573-8
Holmes, W., Bialik, M., & Fadel, C. (2019). *Artificial intelligence in education: Promises and implications for teaching and learning*. Center for Curriculum Redesign.
Hutcheson, G. D., & Sofroniou, N. (1999). *The multivariate social scientist: Introductory statistics using generalized linear models*. SAGE. http://digital.casalini.it/9780857021908
Jöreskog, K. G. (1993). Testing structural equation models. In K. A. Bollen & J. S. Long (Eds.), *Sage focus editions* (Vol. 154, p. 294). Sage.
Jöreskog, K. G., & van Thillo, M. (1972). Lisrel: A general computer program for estimating a linear structural equation system involving multiple indicators of unmeasured variables. *ETS Research Bulletin Series, 12*, 1–71. https://doi.org/10.1002/j.2333-8504.1972.tb00827.x
Kline, R. B. (2016). Principles and practice of structural equation modeling. In *Methodology in the social sciences* (4th ed.). The Guilford Press.
Leal Filho, W., Lange Salvia, A., Beynaghi, A., Fritzen, B., Ulisses, A., Avila, L. V., & Nikolaou, I. (2024). Digital transformation and sustainable development in higher education in a post-pandemic world. *International Journal of Sustainable Development & World Ecology, 31*(1), 108–123. https://doi.org/10.1080/13504509.2023.2237933
Leal Filho, W., Ribeiro, P. C. C., Mazutti, J., Lange Salvia, A., Bonato Marcolin, C., Borsatto, L. S., & J. M., & Viera Trevisan, L. (2024). Using artificial intelligence to implement the UN sustainable development goals at higher education institutions. *International Journal of Sustainable Development & World Ecology*. https://doi.org/10.1080/13504509.2024.2327584
Leal Filho, W., Salvia, A. L., & Eustachio, J. H. P. P. (2023). An overview of the engagement of higher education institutions in the implementation of the UN Sustainable Development Goals. *Journal of Cleaner Production, 386*, 135694. https://doi.org/10.1016/j.jclepro.2022.135694
Lohmöller, J.-B. (1989). Predictive vs. structural modeling: PLS vs. ML. In *Latent variable path modeling with partial least squares* (pp. 199–226). Physica-Verlag. https://doi.org/10.1007/978-3-642-52512-4_5
Luckin, R., Holmes, W., Griffiths, M., & Forcier, L. B. (2016). *Intelligence unleashed: An argument for AI in education*. Pearson.
Ma, S., & Lei, L. (2024). The factors influencing teacher education students' willingness to adopt artificial intelligence technology for information-based teaching. *Asia Pacific Journal of Education, 44*(1), 94–111. https://doi.org/10.1080/02188791.2024.2305155
Naidoo, D. T. (2023). Integrating TAM and IS success model: Exploring the role of Blockchain and AI in predicting learner engagement and performance in E-learning. *Frontiers in Computer Science, 5*, 1227749. https://doi.org/10.3389/fcomp.2023.1227749
Omotosho, A. O., Akintolu, M., Kimweli, K. M., & Modise, M. A. (2023). Assessing the enactus global sustainability initiative's alignment with United Nations sustainable development goals: Lessons for higher education institutions. *Education Sciences, 13*(9), 935. https://doi.org/10.3390/educsci13090935
Parasuraman, A. (2000). Technology readiness index (TRI) a multiple-item scale to measure readiness to embrace new technologies. *Journal of Service Research, 2*(4), 307–320. https://doi.org/10.1177/109467050024001
Qahman, A. A., Zakaria, H. A. D. M. S., Hussin, M., Samra, Y. K., Aldaya, R. F., Din, R., & Othman, N. (2023). Adopting AI-enabled adaptive e-learning environment in Palestinian schools: Integrating extended technology acceptance and system success models. *International Journal of Academic Research in Business and Social Sciences, 13*(12), 3976–3990.
Serafini, P. G., de Moura, J. M., de Almeida, M. R., & de Rezende, J. F. D. (2022). Sustainable development goals in higher education institutions: A systematic literature review. *Journal of Cleaner Production, 370*, 133473. https://doi.org/10.1016/j.jclepro.2022.133473
Sharma, L., & Garg, P. K. (2021). *Artificial intelligence: Technologies, applications, and challenges*. Routledge. https://doi.org/10.1080/02188791.2024.2305155

Shenkoya, T., & Kim, E. (2023). Sustainability in higher education: Digital transformation of the fourth industrial revolution and its impact on open knowledge. *Sustainability, 15*(3), 2473. https://doi.org/10.3390/su15032473

Shwedeh, F., Salloum, S. A., Aburayya, A., Fatin, B., Elbadawi, M. A., Al Ghurabli, Z., & Al Dabbagh, T. (2024). AI adoption and educational sustainability in higher education in the UAE. In *Artificial intelligence in education: The power and dangers of ChatGPT in the classroom* (pp. 201–229). Springer. https://doi.org/10.1007/978-3-031-52280-2_14

Stolpe, K., & Hallström, J. (2024). Artificial intelligence literacy for technology education. *Computers and Education Open, 6*, 100159. https://doi.org/10.1016/j.caeo.2024.100159

Tang, M., Ren, P., & Zhao, Z. (2024). Bridging the gap: The role of educational technology in promoting educational equity. *The Educational Review, 8*(8), 1077–1086. https://doi.org/10.26855/er.2024.08.012

Tyagi, S., Gupta, V., & Mehndiratta, V. (2024). Building smart campuses: Integrating AI in higher education. In R. Arya, S. C. Sharma, A. K. Verma, & B. Iyer (Eds.), *Recent trends in artificial intelligence towards a smart world. Frontiers of artificial intelligence, ethics and multidisciplinary applications*. Springer. https://doi.org/10.1007/978-981-97-6790-8_15

UNESCO. (2020). *Artificial intelligence and education: Guidance for policy-makers*. United Nations Educational, Scientific and Cultural Organization. https://www.unesco.org/en/digital-education/artificial-intelligence

UNDP. (2022). *United Nations Development Programme*. https://www.undp.org/publications/undp-annual-report-2022

Van den Bergh, J. C. (2013). Environmental and climate innovation: Limitations, policies and prices. *Technological Forecasting and Social Change, 80*(1), 11–23. https://doi.org/10.1016/j.techfore.2012.08.004

Venkatesh, V., & Bala, H. (2008). Technology acceptance model 3 and a research agenda on interventions. *Decision Sciences, 39*(2), 273–315. https://doi.org/10.1111/j.1540-5915.2008.00192.x

Venkatesh, V., Morris, M. G., Davis, G. B., & Davis, F. D. (2003). User acceptance of information technology: Toward a unified view. *MIS Quarterly, 27*, 425–478. https://doi.org/10.2307/30036540

Venkatesh, V., Thong, J. Y., & Xu, X. (2012). Consumer acceptance and use of information technology: Extending the unified theory of acceptance and use of technology. *MIS Quarterly, 36*, 157–178. https://doi.org/10.2307/41410412

Wold, S., Ruhe, A., Wold, H., & Dunn, W. J. (1984). The collinearity problem in linear regression. The partial least squares (PLS) approach to generalized inverses. *SIAM Journal on Scientific and Statistical Computing, 5*, 735–743. https://doi.org/10.1137/0905052

Zaidan, E., Al-Saidi, M., & Hammad, S. H. (2019). Sustainable development in the Arab world–is the Gulf Cooperation Council (GCC) region fit for the challenge? *Development in Practice, 29*(5), 670–681. https://doi.org/10.1080/09614524.2019.1628922

GPSR Compliance

The European Union's (EU) General Product Safety Regulation (GPSR) is a set of rules that requires consumer products to be safe and our obligations to ensure this.

If you have any concerns about our products, you can contact us on ProductSafety@springernature.com

In case Publisher is established outside the EU, the EU authorized representative is:

Springer Nature Customer Service Center GmbH
Europaplatz 3
69115 Heidelberg, Germany

Batch number: 10725825

Printed by Printforce, the Netherlands